Introduction to
Matrix Methods in Optics

WILEY SERIES IN PURE AND APPLIED OPTICS

Advisory Editor

Stanley S. Ballard, University of Florida

Introduction to
Matrix Methods in Optics

A. Gerrard
University of Bath

J. M. Burch
National Physical Laboratory, Teddington

A Wiley-Interscience Publication

JOHN WILEY & SONS

London • New York • Sydney • Toronto

Copyright © 1975, by John Wiley & Sons, Ltd.

Library of Congress Cataloging in Publication Data:

Gerrard, Anthony.
Introduction to matrix methods in optics.

"A Wiley-Interscience publication"
1. Optics. 2. Matrices. I. Burch, James M., joint
author. II. Title.

QC355.2.G47   535   72-21192
ISBN 0 471 29685 6

Printed in Great Britain by Unwin Brothers Limited,
The Gresham Press, Old Woking, Surrey.

Preface

Our purpose in writing this book has been not to
present new results but to encourage the adoption of
simple matrix methods in the teaching of optics at
the undergraduate and technical college level. Many
of these methods have been known for some time but
have not found general acceptance; we believe that
the time has now come for lecturers to reconsider their
value. We believe this partly because the use of matrices is now being taught quite widely in schools, and
many students will already have glimpsed something of
the economy and elegance with which, for a linear system, a whole wealth of input-output relations can be
expressed by a single matrix.

A second reason is that, for more than a decade, the
field of optics has been enriched enormously by contributions from other disciplines such as microwave
physics and electrical engineering. Although an engineering student may be a newcomer to optics, he may
well have encountered matrix methods during his lec-
tures on electrical filters or transmission lines; we
think he will welcome an optics course which, instead
of barricading itself behind its own time-honoured
concepts, links itself recognizably to other disciplines.

Another barrier which we believe matrix methods may
help to blur is the classical separation of optics
into compartments labelled 'geometrical' and 'physical'.
The optics of today transcends all boundaries, and the
student may share our delight when a ray-transfer

matrix, based on purely geometrical considerations,
predicts with almost perfect accuracy the diffraction
behaviour of a Gaussian beam as it is generated in a
laser resonator or propagated through an external system. Surely the spirit of Gauss particularly must
rejoice at this versatility!

Hoping as we do that matrix methods may help to forge
links between the various branches of optics and other
subjects, we have sought to avoid any inconsistencies
of nomenclature. Out of the several types of ray-
transfer matrix that have been proposed, we have followed Sinclair (of The Institute of Optics, Rochester,
N.Y.) in choosing a form which is always unimodular,
and which is compatible with the (ynv) method of calculation and with modern work on laser resonators. In
contrast with Halbach's nomenclature, these matrices
are defined so that they tell us what output we shall
obtain for a given input; this is the choice favoured
by most workers, and it ensures consistency with the
rest of the book in which we describe the established
Jones and Mueller calculus for polarization problems.
The student will, of course, encounter situations, for
example in laser beam propagation, where he has to work
backwards and find what input is needed to produce a
given desired output. In nearly all such cases, however, he will be dealing with 2×2 unimodular matrices,
inversion of which he will learn to tackle with relish.

We shall now discuss some of the limitations and
omissions in what we have written, and then describe
briefly the arrangement of the chapters.

Because this is an introductory text which assumes
very little prior knowledge, we have confined our
attention to just two topics - namely paraxial imaging
and polarization. The first topic has the advantage
that the concepts required initially are almost in-
tuitive; the second serves to emphasize the transverse
nature of light waves but does not demand a knowledge
of electromagnetic theory. We should have liked to include a chapter on reflection and transmission of light
by thin films and stratified media, but to do this
properly we should have had to proceed via a derivation
from Maxwell's equations of the coupled behaviour of
the transverse electric and magnetic field components.

vi

It would have been possible to give a more superficial treatment, in which the illumination is restricted to be at normal incidence and a coefficient of reflection is assumed for each individual surface, but we feel it is better to omit the subject altogether and refer the student who is interested to existing coverage in the literature. Other topics which we rejected, but which might have been suitable for a more advanced book, are Wolf's coherency matrix and the use of 3×3 or 4×4 matrices to describe reflection from a series of variously oriented mirror surfaces, for example in a reflecting prism.

Our first chapter is intended for those who have no previous acquaintance with matrix algebra. Using numerous worked examples, it introduces the basic ideas of rectangular matrix arrays and gives the rules for adding them and for forming matrix products. The section on square matrices concentrates for simplicity on the 2×2 matrix. After the transpose matrix and the determinant have been introduced, the problem of matrix inversion is discussed. This leads into a brief treatment of diagonalization, and we conclude by showing how the Nth power of a matrix can be determined (without memorizing Sylvester's theorem).

Chapter II is devoted to the paraxial imaging properties of a centred optical system. Defining a ray in terms of its height and its optical direction-cosine, we show how a ray-transfer matrix can be used to describe the change that occurs in these two quantities as the ray traverses a system. The two basic types of matrix that represent the effect of a simple gap or of a single refracting surface are combined to form the equivalent matrix of a thin lens, a thick lens or a complete optical system. It is shown how the properties of a system can be inferred from knowledge of its matrix, and conversely how the matrix elements can be determined experimentally. At the end of the chapter we extend the ray-transfer matrix to include reflecting as well as refracting elements. The text is again supported by worked examples, and an appendix shows how the aperture properties of a system can be determined.

In the first part of chapter III we review and tabul-

ate the results so far obtained and use them to describe the radius of curvature of a wavefront, the
optical length of a ray path and the étendue of a beam.
We then consider optical resonators and show how a
round trip between the mirrors of a resonator can be
represented by a single equivalent matrix. In order
to consider the effect of repeated traversal of the
resonator, we now diagonalize its matrix and find that,
for the so-called 'unstable' case, both the eigenvalues and the eigenvectors are real; the former represent the loss per transit due to 'beam walk-off' and
the latter represent solutions for the radius of curvature of a self-perpetuating wavefront.

For the case of a 'stable' laser resonator, both
eigenvalues and eigenvectors are complex; the former
represent the phase shift per transit and the latter
can be interpreted in terms of Kogelnik's complex
curvature parameter to predict not only the divergence
but also the spot width of the Gaussian beam that the
laser will generate. Furthermore, if we have a mode-
matching problem in which we must calculate the diffraction of a laser beam as it is propagated externally,
this too can be solved very easily by using the ray-
transfer matrix. We conclude by indicating the extension of these methods to distributed lens-like media.
The use of an augmented matrix to handle residual misalignments is discussed in an appendix.

In chapter IV we consider two alternative matrix
methods for handling problems in polarization. After
reviewing the different kinds of polarized light, we
introduce first the Stokes parameters and the 4×4
Mueller matrices by which problems involving both polarized and unpolarized light can be tackled. The
discussion includes numerous worked examples and an
account of how both the Stokes parameters and the
elements of a Mueller matrix can be determined experimentally. The Mueller matrices that are likely to be
needed are tabulated and their derivation is given in
an appendix.

A similar discussion is then given for the Jones
calculus, which uses 2×2 complex matrices and is more
suitable for dealing with fully polarized light. The
material is so arranged that the student can, if he

wishes, concentrate exclusively either on the Jones or
on the Mueller method.

Other appendixes to this chapter contain a statistical
treatment of the Stokes parameters and a full analysis
of the connection between the elements of a Jones
matrix and those of the corresponding Mueller matrix.

Chapter V is concerned with the application of matrix
methods to the propagation of light in uniaxial crystals.
Although it is more advanced and demands some knowledge
of electromagnetic theory, it does not depend on the
contents of chapter IV and can be read separately. We
hope that the student who reads it will go on to some
of the topics that we have omitted. A bibliography is
provided.

Finally, since the chapters of this book have been
designed primarily as educational stepping-stones and
as illustrations of the matrix approach, only a limited
range of optics has been covered. How far is a student
likely to find this material of value during his sub-
sequent career?

For the small fraction of graduates who will spend
their life in optical design or manufacture, it has to
be admitted that problems involving polarization do not
often arise, and in most cases the contribution of
first-order optics is trivial; the real problems arise
either in using a computer to control third-order and
higher-order aberrations or in more practical aspects
of fabrication and assembly.

But for every professional optician there will be
many others whose practice it will be to buy their
optical equipment off the shelf and then incorporate
it into larger systems. Some of these workers will be
found in scientific research and others in new indus-
tries based on optoelectronics and laser engineering,
but many will be engaged in more traditional fields
such as mechanical engineering. We refer here not
only to photoelasticity and to established techniques
for optical inspection and alignment but also to more
recent developments in laser holography and speckle
interferometry. The era has already arrived where
optical methods of measurement can be applied to a wide
variety of unconventional tasks, and the engineer concerned usually has to work on a 'do it yourself' basis.

In holographic non-destructive testing, in the study of vibrations or in strain analysis, the component being viewed may be so irregular in shape that there is no point in trying to achieve well-corrected imaging over a wide flat field. It follows that money (as well as internal reflections) can often be saved by using the simplest of lenses; but before those cheap lenses are thrown together, with or without the help of a ray-transfer matrix, we hope that the reader of this book will at least remember to test each of them in a strain-viewer!

One of us, J. M. Burch, wishes to acknowledge the support of the Science and Technology Foundation of New York State which enabled him to spend a year as visiting professor at The Institute of Optics, University of Rochester. He is grateful to several of his Rochester colleagues, and notably to Douglas C. Sinclair, for discussions on some aspects of ray-transfer matrices.

Contents

I

Introduction to
Matrix Calculations

I.1 INTRODUCTORY DISCUSSION

In this book we consider how some simple ideas of matrix algebra can be applied with advantage to problems involving optical imaging and polarization. The discussion in this chapter is designed mainly for those readers who have not so far encountered matrices or determinants; the treatment is elementary and covers only what will be needed to understand the rest of the book.

Matrices were introduced in 1857 by the mathematician Cayley as a convenient shorthand notation for writing down a whole array of linear simultaneous equations. The rules for operating with matrix arrays are slightly different from those for ordinary numbers, but they were soon discovered and developed. Matrix methods became of great interest to the physicist in the 1920's when Heisenberg introduced the matrix form of quantum mechanics. They are used in many kinds of engineering calculation but their application to optics is more recent.

Determinants, with which we shall be concerned to a lesser extent, were introduced by Vandermonde as early as 1771. They were at first called 'eliminants' because they arose in solving equations by the method of successive elimination. In most of the optical problems with which we shall deal the determinants all have a value of unity, and this fact provides a convenient check at the end of a calculation.

Let us now consider how the notion of a matrix arises. Suppose we have a pair of linear equations

$$U = Ax + By$$

$$V = Cx + Dy$$

where A, B, C and D are known constants, and x and y are variables. These equations enable us to calculate U and V if x and y are known. It proves convenient, for many purposes, to separate the constants from the variables. We write the pair of equations thus:

$$\begin{bmatrix} U \\ V \end{bmatrix} = \begin{bmatrix} A & B \\ C & D \end{bmatrix} \begin{bmatrix} x \\ y \end{bmatrix}$$

a *single* equation which is defined as meaning exactly the same as the pair. We regard each of the groups of symbols enclosed between a pair of vertical brackets as a single entity, called a MATRIX. $\begin{bmatrix} U \\ V \end{bmatrix}$ and $\begin{bmatrix} x \\ y \end{bmatrix}$ are called 'column matrices' or, alternatively, 'column vectors', since each contains only a single column.

The general matrix is a rectangular array with the symbols arranged in rows and columns. The matrix $\begin{bmatrix} A & B \\ C & D \end{bmatrix}$, which has two rows and columns, is called a 'square matrix of order two'. Later we shall meet 'row matrices' (sometimes called 'row vectors') like $\begin{bmatrix} P & Q \end{bmatrix}$, in which the separate symbols, called 'matrix elements', are written horizontally in a single row. A matrix with only one element is just an ordinary number, or scalar quantity.

If we use a single symbol for each matrix, we can write the pair of equations even more briefly, thus:

$$C_2 = SC_1$$

where C_1 denotes the column matrix $\begin{bmatrix} x \\ y \end{bmatrix}$, C_2 denotes the column matrix $\begin{bmatrix} U \\ V \end{bmatrix}$ and S denotes the square matrix $\begin{bmatrix} A & B \\ C & D \end{bmatrix}$

Now let us suppose that U and V are linked in turn with another pair of variables, L and M, say, by another pair of linear equations, thus:

$$L = PU + QV$$

$$M = RU + TV$$

which we write in the form

$$\begin{bmatrix} L \\ M \end{bmatrix} = \begin{bmatrix} P & Q \\ R & T \end{bmatrix} \begin{bmatrix} U \\ V \end{bmatrix}$$

that is

$$C_3 = KC_2$$

where C_3 denotes $\begin{bmatrix} L \\ M \end{bmatrix}$ and K denotes $\begin{bmatrix} P & Q \\ R & T \end{bmatrix}$. We can, of course, find L and M in terms of x and y by substituting for U and V in the equations defining L and M. Thus:

$$L = P(Ax + By) + Q(Cx + Dy)$$

$$M = R(Ax + By) + T(Cx + Dy)$$

that is

$$L = (PA + QC)x + (PB + QD)y$$

$$M = (RA + TC)x + (RB + TD)y$$

which we write as

$$\begin{bmatrix} L \\ M \end{bmatrix} = \begin{bmatrix} PA + QC & PB + QD \\ RA + TC & RB + TD \end{bmatrix} \begin{bmatrix} x \\ y \end{bmatrix}$$

that is

$$C_3 = FC_1$$

where F denotes $\begin{bmatrix} PA + QC & PB + QD \\ RA + TC & RB + TD \end{bmatrix}$. But, on the other hand, we can write

$$C_3 = KC_2 = K(SC_1)$$

Now, if this were an equation in ordinary algebra, we could rewrite it as

$$C_3 = KSC_1 = (KS)C_1$$

merely changing the positions of the brackets. KS would be called the product of K and S.

Again, comparing the equations linking C_1 and C_3, we could write

$$C_3 = KSC_1 \quad \text{and} \quad C_3 = FC_1$$

Therefore

$$F = KS$$

and we would say that F was the *product* of K and S.

In matrices we wish to follow a similar method but we now need to *define* the product of two matrices, since only products of single numbers are defined in ordinary algebra.

I.2 MATRIX MULTIPLICATION

We *define* matrix multiplication so that the above formalism can be carried over from ordinary algebra to matrix algebra. Thus, we *define* the product of the matrices by stating that K multiplied by S gives the product matrix F; that is

$$\begin{bmatrix} P & Q \\ R & T \end{bmatrix} \begin{bmatrix} A & B \\ C & D \end{bmatrix} = \begin{bmatrix} PA + QC & PB + QD \\ RA + TC & RB + TD \end{bmatrix}$$

Examining the structure of the right-hand matrix (the product), it is easy to see how it is formed.

The top left-hand element is in the first row and the first column. It is produced by taking the first *row* of K, which is $\begin{bmatrix} P & Q \end{bmatrix}$, and the first *column* of S, which is $\begin{bmatrix} A \\ C \end{bmatrix}$, multiplying corresponding elements together (the first element of the row by the first element of the column, the second element of the row by the second element of the column), forming the products PA and QC, and then adding to get $PA + QC$.

The element in the first row and the second column of F is formed in the same way from the first row of K and the second column of S. The element in the second row and first column of F is formed from the second row of K and the first column of S. Finally, the element in the second row and second column of F is formed from the second row of K and the second column of S.

It proves useful, in some applications, to use a suffix notation for the elements of the matrices. We write a column matrix A, for instance, as

$$A = \begin{bmatrix} A_1 \\ A_2 \end{bmatrix}$$

the subscript indicating the position of the element in the column. A square matrix S we write as

$$S = \begin{bmatrix} S_{11} & S_{12} \\ S_{21} & S_{22} \end{bmatrix}$$

where the first subscript indicates which row an element is in and the second subscript indicates which column. If we re-express our two square matrices K and S in this suffix notation

$$K = \begin{bmatrix} K_{11} & K_{12} \\ K_{21} & K_{22} \end{bmatrix} \quad \text{and} \quad S = \begin{bmatrix} S_{11} & S_{12} \\ S_{21} & S_{22} \end{bmatrix}$$

then the product $F = KS$ becomes

$$\begin{bmatrix} F_{11} & F_{12} \\ F_{21} & F_{22} \end{bmatrix} = \begin{bmatrix} K_{11}S_{11} + K_{12}S_{21} & K_{11}S_{12} + K_{12}S_{22} \\ K_{21}S_{11} + K_{22}S_{21} & K_{21}S_{12} + K_{22}S_{22} \end{bmatrix}$$

that is

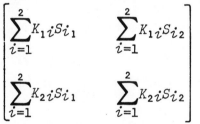

$$\begin{bmatrix} \sum_{i=1}^{2} K_{1i}S_{i1} & \sum_{i=1}^{2} K_{1i}S_{i2} \\ \sum_{i=1}^{2} K_{2i}S_{i1} & \sum_{i=1}^{2} K_{2i}S_{i2} \end{bmatrix}$$

This suggests a general formula for any element of the matrix:

$$F_{RT} = \sum_{i=1}^{i_{max}} K_{Ri} S_{iT}$$

where F_{RT} denotes the element in the Rth row and the Tth column of F, and similarly for K and S. (The summation sign used here indicates that the repeated suffix i takes on all possible values in succession; it is sometimes omitted.)

So far, we have confined our attention to two-by-two matrices and two-by-one columns; but the matrix idea is much more general than this. In this book we shall need two-by-two, three-by-three and four-by-four square matrices, two-by-one, three-by-one and four-by-one columns, and one-by-two, one-by-three and one-by four rows. The meaning of all these is defined in the same way. If, for example, we have four equations in four unknowns:

$$B_1 = K_{11}A_1 + K_{12}A_2 + K_{13}A_3 + K_{14}A_4$$

$$B_2 = K_{21}A_1 + K_{22}A_2 + K_{23}A_3 + K_{24}A_4$$

$$B_3 = K_{31}A_1 + K_{32}A_2 + K_{33}A_3 + K_{34}A_4$$

$$B_4 = K_{41}A_1 + K_{42}A_2 + K_{43}A_3 + K_{44}A_4$$

we express this in matrix form either as

$$
\begin{bmatrix} B_1 \\ B_2 \\ B_3 \\ B_4 \end{bmatrix}
=
\begin{bmatrix}
K_{11} & K_{12} & K_{13} & K_{14} \\
K_{21} & K_{22} & K_{23} & K_{24} \\
K_{31} & K_{32} & K_{33} & K_{34} \\
K_{41} & K_{42} & K_{43} & K_{44}
\end{bmatrix}
\begin{bmatrix} A_1 \\ A_2 \\ A_3 \\ A_4 \end{bmatrix}
$$

or in more abbreviated form as

$$B = KA$$

In this case we can calculate the Rth element of the column matrix B by the equation:

$$B_{R1} = \sum_{i=1}^{4} K_{Ri} A_{i1}$$

Note that the above rule for forming the elements of the product still applies with A in the place of S and B in the place of F; but because A and B are now four-by-one columns, the second subscript (T) on F and S takes only the value 1. We could therefore write more simply

$$B_R = \sum_{i=1}^{4} K_{Ri} A_{i}$$

(where B and A are evidently vectors).

This rule for matrix multiplication is very important and will be used frequently. It is vital that the student becomes thoroughly familiar with it by working many examples. The following are provided as illustrations.

I.2.1

If $A = \begin{bmatrix} 1 & 3 \\ 5 & 7 \end{bmatrix}$ and $B = \begin{bmatrix} 2 & 6 \\ 1 & -4 \end{bmatrix}$

then $AB = \begin{bmatrix} (1\times2)+(3\times1) & (1\times6)+(3\times-4) \\ (5\times2)+(7\times1) & (5\times6)+(7\times-4) \end{bmatrix}$

$= \begin{bmatrix} 2+3 & 6+(-12) \\ 10+7 & 30+(-28) \end{bmatrix} = \begin{bmatrix} 5 & -6 \\ 17 & 2 \end{bmatrix}$

but $BA = \begin{bmatrix} (2\times1)+(6\times5) & (2\times3)+(6\times7) \\ (1\times1)+(-4\times5) & (1\times3)+(-4\times7) \end{bmatrix}$

$= \begin{bmatrix} 2+30 & 6+42 \\ 1-20 & 3-28 \end{bmatrix} = \begin{bmatrix} 32 & 48 \\ -19 & -25 \end{bmatrix}$

The reader will note that AB and BA are quite different. In matrix multiplication the order of factors *must* be preserved: the equation $AB = BA$ does *not* hold. In mathematical language, matrix multiplication is non-commutative.

I.2.2

If $C = \begin{bmatrix} 3 & 1 & 4 \\ 2 & 1 & 6 \\ 1 & 3 & 4 \end{bmatrix}$ and $D = \begin{bmatrix} -5 \\ 3 \\ 1 \end{bmatrix}$

$CD = \begin{bmatrix} (3\times-5)+(1\times3)+(4\times1) \\ (2\times-5)+(1\times3)+(6\times1) \\ (1\times-5)+(3\times3)+(4\times1) \end{bmatrix}$

$= \begin{bmatrix} -15+3+4 \\ -10+3+6 \\ -5+9+4 \end{bmatrix} = \begin{bmatrix} -8 \\ -1 \\ 8 \end{bmatrix}$

If we try to form the product DC, we have first to multiply elements from the first row of D by those from the first column of C. There is only one element $\begin{bmatrix} -5 \end{bmatrix}$ in the first row of D, but *three* $\begin{bmatrix} 3 \\ 2 \\ 1 \end{bmatrix}$ in the first column of C. This product therefore *cannot be formed*. We can multiply two matrices C and D if (and only if) the number of columns in the premultiplier matrix C is the same as the number of rows in the postmultiplier matrix D; C and D are then called *compatible* for this multiplication, and the product CD can be formed.

I.2.3

Let $E = \begin{bmatrix} 3 & 1 & 4 \end{bmatrix}$ and $F = \begin{bmatrix} 1 & 5 & 9 \\ 2 & 4 & -3 \\ 6 & 1 & 3 \end{bmatrix}$

Then $EF = \begin{bmatrix} (3 \times 1) + (1 \times 2) + (4 \times 6) \\ (3 \times 5) + (1 \times 4) + (4 \times 1) \\ (3 \times 9) - (1 \times 3) + (4 \times 3) \end{bmatrix}$

$= \begin{bmatrix} (3 + 2 + 24) & (15 + 4 + 4) & (27 - 3 + 12) \end{bmatrix}$

$= \begin{bmatrix} 29 & 23 & 36 \end{bmatrix}$

If we reverse the order of multiplication, FE cannot be formed - there are three columns in F but only one row in E.

I.2.4

Let $H = \begin{bmatrix} 3 & 1 & 6 \end{bmatrix}$ and $K = \begin{bmatrix} 2 \\ 4 \\ 7 \end{bmatrix}$

If H is used as the premultiplier and K as the postmultiplier, then they are compatible since there are three columns in H and three rows in K. Thus

$$HK = \left[(3 \times 2) + (1 \times 4) + (6 \times 7)\right]$$
$$= (6 + 4 + 42) = 52$$

a single ordinary number.

Consider now what happens if we premultiply H by K. The two matrices are still compatible, since there is one column in K and one row in H. Thus

$$KH = \begin{bmatrix} 2 \\ 4 \\ 7 \end{bmatrix} \begin{bmatrix} 3 & 1 & 6 \end{bmatrix}$$

$$= \begin{bmatrix} 2 \times 3 & 2 \times 1 & 2 \times 6 \\ 4 \times 3 & 4 \times 1 & 4 \times 6 \\ 7 \times 3 & 7 \times 1 & 7 \times 6 \end{bmatrix} = \begin{bmatrix} 6 & 2 & 12 \\ 12 & 4 & 24 \\ 21 & 7 & 42 \end{bmatrix}$$

For this pair of matrices, HK is a single number but KH is a three-by-three square matrix.

I.3 NULL MATRICES

If $L = \begin{bmatrix} 0 & 0 \\ 0 & 0 \end{bmatrix}$ and $M = \begin{bmatrix} 7 & 1 \\ 4 & 2 \end{bmatrix}$

(or, indeed, any square matrix of order two), then LM and ML are both $\begin{bmatrix} 0 & 0 \\ 0 & 0 \end{bmatrix}$

The matrix $\begin{bmatrix} 0 & 0 \\ 0 & 0 \end{bmatrix}$ is called the 'null matrix of order two'. Null matrices are usually denoted by the symbol 0 and take the place of zero in ordinary algebra.

Any square or rectangular matrix possesses a null form 0 of which all the elements are zero. Either pre-multiplying or postmultiplying any matrix by a null matrix (of compatible form) produces a null matrix as a result.

I.4 UNIT MATRICES

I.4.1

If $P = \begin{bmatrix} 1 & 0 \\ 0 & 1 \end{bmatrix}$ and $Q = \begin{bmatrix} 3 & 5 \\ 2 & 6 \end{bmatrix}$

then both PQ and $QP = \begin{bmatrix} 3 & 5 \\ 2 & 6 \end{bmatrix}$, that is Q.

The matrix $\begin{bmatrix} 1 & 0 \\ 0 & 1 \end{bmatrix}$ has the property that, if we use it to premultiply any two-row matrix or postmultiply any two-column matrix, it leaves that matrix unchanged. We call it the 'unit matrix of order two'.

Other examples of the unit matrix, usually denoted by I, are the 3×3 matrix $\begin{bmatrix} 1 & 0 & 0 \\ 0 & 1 & 0 \\ 0 & 0 & 1 \end{bmatrix}$, the 4×4 matrix

$\begin{bmatrix} 1 & 0 & 0 & 0 \\ 0 & 1 & 0 & 0 \\ 0 & 0 & 1 & 0 \\ 0 & 0 & 0 & 1 \end{bmatrix}$, and so on.

The unit matrix of order n has n rows and columns. All its elements are zero except those in the 'principal diagonal', that is that running from the top left to bottom right; all of these diagonal elements have a value of unity.

I.5 DIAGONAL MATRICES

The unit matrix is a special case of a 'diagonal matrix'; the latter is defined as a square matrix for which all the non-diagonal elements are zero. The elements in the principal diagonal may take any value.

For example, $\begin{bmatrix} 1 & 0 \\ 0 & -1 \end{bmatrix}$ and $\begin{bmatrix} 3 & 0 & 0 \\ 0 & 1 & 0 \\ 0 & 0 & -\frac{1}{2} \end{bmatrix}$ are diagonal matrices.

If two diagonal matrices are multiplied together, the order of multiplication does not matter; the multiplication is extremely simple and the resulting matrix is also diagonal.

I.5.1

$$\text{If } A = \begin{bmatrix} a_1 & 0 & 0 \\ 0 & a_2 & 0 \\ 0 & 0 & a_3 \end{bmatrix} \text{ and } B = \begin{bmatrix} b_1 & 0 & 0 \\ 0 & b_2 & 0 \\ 0 & 0 & b_3 \end{bmatrix}$$

$$\text{then } AB = BA = \begin{bmatrix} a_1 b_1 & 0 & 0 \\ 0 & a_2 b_2 & 0 \\ 0 & 0 & a_3 b_3 \end{bmatrix}$$

(The reader will note that in this example we have used algebraic symbols instead of arithmetic quantities for the matrix elements. As in ordinary algebra, we can use one to denote an unknown value for the other.)

I.6 MULTIPLE PRODUCTS

If we wish to form the product of three matrices, L, M and N we can proceed in two ways:

(a) We can form the product (MN) and then premultiply it by L.
(b) We can form the product (LM) and then postmultiply it by N.

Provided that we preserve the order of the matrices, these two methods both give the same result. Thus $L(MN) = (LM)N$. We call either of these LMN, as in ordinary algebra.

I.6.1

$$\text{Let } L = \begin{bmatrix} 1 & 3 \\ 4 & 2 \end{bmatrix}, \ M = \begin{bmatrix} 2 & 1 \\ 3 & 1 \end{bmatrix} \text{ and } N = \begin{bmatrix} 4 & 2 \\ 1 & 3 \end{bmatrix}$$

$$\text{Then } L(MN) = \begin{bmatrix} 1 & 3 \\ 4 & 2 \end{bmatrix} \begin{bmatrix} 9 & 7 \\ 13 & 9 \end{bmatrix} = \begin{bmatrix} 48 & 34 \\ 62 & 46 \end{bmatrix} = LMN$$

and $(LM)N = \begin{bmatrix} 11 & 4 \\ 14 & 6 \end{bmatrix} \begin{bmatrix} 4 & 2 \\ 1 & 3 \end{bmatrix} = \begin{bmatrix} 48 & 34 \\ 62 & 46 \end{bmatrix} = LMN,$

as stated above.

Thus, although matrix multiplication is not 'commutative', it is nevertheless 'associative'. By extension of this principle, it is easy to show that for products of four or more matrices $PQRS = P(QR)S = (PQR)S$, etc.

I.7 MATRIX ADDITION AND SUBTRACTION

Provided that two matrices M and N have the same number of rows and columns, their sum or difference is obtained merely by adding or subtracting each corresponding pair of matrix elements. If $P = M + N$, then $P_{jk} = M_{jk} + N_{jk}$. Since all elements of the null matrix 0 are zero, we then have $M \pm 0 = M$ and $M - M = 0$. Matrices also obey the 'distributive law' $A(B + C) = AB + AC$.

I.7.1 Numerical Example

Let $A = \begin{bmatrix} 2 & 0 \\ 1 & 1 \end{bmatrix}$, $B = \begin{bmatrix} 3 & 0 \\ 1 & 2 \end{bmatrix}$ and $C = \begin{bmatrix} -1 & 1 \\ 2 & 0 \end{bmatrix}$

Then $A(B + C) = \begin{bmatrix} 2 & 0 \\ 1 & 1 \end{bmatrix} \begin{bmatrix} 2 & 1 \\ 3 & 2 \end{bmatrix} = \begin{bmatrix} 4 & 2 \\ 5 & 3 \end{bmatrix}$

and $AB + AC = \begin{bmatrix} 6 & 0 \\ 4 & 2 \end{bmatrix} + \begin{bmatrix} -2 & 2 \\ 1 & 1 \end{bmatrix} = \begin{bmatrix} 4 & 2 \\ 5 & 3 \end{bmatrix}$

It is clear from what we have stated that, if we add the same matrix to itself λ times, each of the matrix elements will be multiplied by the ordinary number (or 'scalar quantity') λ. This operation is sometimes referred to as 'scalar multiplication' of the matrix. The same result can also be obtained by matrix multiplication if we either premultiply or postmultiply by the diagonal matrix λI, the diagonal elements of which all have the value λ.

We have now considered the rules by which matrices can be multiplied, added and subtracted. The operation of dividing one matrix by another can also be useful, and we shall have frequent occasion in this book to use pairs of two-by-two square matrices, either of which is the 'reciprocal' of the other. Before this topic is considered, however, we shall need to discuss briefly the transpose of a matrix and the idea of a determinant.

I.8 TRANSPOSE MATRICES

The matrix obtained from another (say A) by interchanging its rows and columns is called its 'transpose', and will be denoted by A^T. If the matrix A has m rows and n columns, then its transpose A^T will have n rows and m columns.

I.8.1

$$\text{If } A = \begin{bmatrix} 3 & 4 \\ 2 & 1 \end{bmatrix} \quad \text{then } A^T = \begin{bmatrix} 3 & 2 \\ 4 & 1 \end{bmatrix}$$

I.8.2

$$\text{If } B = \begin{bmatrix} 3 \\ 1 \\ 4 \end{bmatrix} \quad \text{then } B^T = \begin{bmatrix} 3 & 1 & 4 \end{bmatrix}$$

I.8.3

$$\text{If } C = \begin{bmatrix} 5 & 7 & 1 \end{bmatrix} \quad \text{then } C^T = \begin{bmatrix} 5 \\ 7 \\ 1 \end{bmatrix}$$

There is an important theorem about transposes and their multiplication: *The transpose of the product of two matrices is the product of their transposes* IN REVERSED ORDER. That is

$$(AB)^T = B^T A^T$$

I.8.4

$$\text{If } A = \begin{bmatrix} 1 & 3 \\ 5 & 7 \end{bmatrix} \quad \text{and } B = \begin{bmatrix} 4 & 1 \\ 3 & 6 \end{bmatrix}$$

then $(AB) = \begin{bmatrix} 13 & 19 \\ 41 & 47 \end{bmatrix}$ so that $(AB)^T = \begin{bmatrix} 13 & 41 \\ 19 & 47 \end{bmatrix}$

Also $A^T = \begin{bmatrix} 1 & 5 \\ 3 & 7 \end{bmatrix}$ and $B^T = \begin{bmatrix} 4 & 3 \\ 1 & 6 \end{bmatrix}$

whence $B^T A^T = \begin{bmatrix} 13 & 41 \\ 19 & 47 \end{bmatrix}$

(the student should check these multiplications).

It will be noticed that, if A and B are rectangular matrices that are compatible only when A premultiplies B, then likewise B^T and A^T will be compatible only when B^T premultiplies A^T, as required by the above theorem.

By using the associative property of matrix multiplication it is easily shown that

$$(ABC)^T = \left((AB)C\right)^T = C^T(AB)^T = C^T B^T A^T$$

and similarly for products of any number of matrices

$$(ABCDEF)^T = F^T E^T D^T C^T B^T A^T, \text{ etc.}$$

I.9 DETERMINANTS

Associated with any square matrix there is a single number or quantity that is called its 'determinant'. For a matrix M the determinant is denoted either by enclosing the complete array between straight lines instead of brackets or by simply writing $\det(M)$. Thus

the determinant of $M = \begin{vmatrix} M_{11} & M_{12} & M_{13} \\ M_{21} & M_{22} & M_{23} \\ M_{31} & M_{32} & M_{33} \end{vmatrix} = \det(M)$.

For a square matrix of large order n the rule for calculating the determinant is quite complicated; it involves taking the sum of a very large number of alternatively positive and negative n-fold products of the matrix elements, and even for $n = 4$ there are $4! = 24$ products to be included.

In this book, however, we shall be concerned only with the determinants of 2×2 matrices, and to calculate these the rule is extremely simple: *Form the product of the two main diagonal elements (top left and bottom right) and subtract from this the product of the two other elements.*

Thus if the matrix $P = \begin{bmatrix} A & B \\ C & D \end{bmatrix}$ then det$(P) = \begin{vmatrix} A & B \\ C & D \end{vmatrix}$ $= (AD - BC)$ where A, B, C and D are ordinary numbers or scalar quantities.

I.9.1

If $P = \begin{bmatrix} 1 & 3 \\ 2 & 9 \end{bmatrix}$ then det$(P) = (1 \times 9) - (2 \times 3)$

$= 9 - 6 = 3$

We shall need the following theorem about determinants: *The determinant of the product of two square matrices is the product of their determinants.*

I.9.2

If $P = \begin{bmatrix} 6 & 3 \\ 5 & 7 \end{bmatrix}$ then det$(P) = (6 \times 7) - (3 \times 5)$

$= 42 - 15 = 27$

and if $Q = \begin{bmatrix} 1 & 2 \\ 3 & 4 \end{bmatrix}$ then det$(Q) = (1 \times 4) - (2 \times 3)$

$= 4 - 6 = -2$

Then (det P) × (det Q) = (27) × (−2) = − 54

If, on the other hand, we form the product matrix $PQ = \begin{bmatrix} 15 & 24 \\ 26 & 38 \end{bmatrix}$, we again obtain

det$(PQ) = (15 \times 38) - (24 \times 26)$

$= 570 - 624 = -54$

This theorem can be extended to the product of any number of matrices:

$$\det(PQRST, \text{etc.}) = \det(P)\det(Q)\det(R)\det(S)\det(T), \text{ etc.}$$

This fact often provides a convenient check on whether the multiplication of a matrix has been performed correctly.

It will be noticed that, although for square matrices the matrix product PQ is not in general the same as QP, the two matrices must have the same determinant. *As far as the determinant is concerned*, the order of multiplication makes no difference.

I.9.3

We found in section I.2.1 that if $A = \begin{bmatrix} 1 & 3 \\ 5 & 7 \end{bmatrix}$ and

$B = \begin{bmatrix} 2 & 6 \\ 1 & -4 \end{bmatrix}$ then $AB = \begin{bmatrix} 5 & -6 \\ 17 & 2 \end{bmatrix}$ but $BA = \begin{bmatrix} 32 & 48 \\ -19 & -25 \end{bmatrix}$.

In this case,

$\det(AB) = 10 + 102 = \underline{112}$

$\det(BA) = -800 + 912 = \underline{112}$

and, of course,

$(\det A) \times (\det B) = (-8) \times (-14) = \underline{112}$

For certain square matrices the determinant is zero and these matrices are said to be 'singular'. Nearly all of the matrices with which we shall deal in this book will be non-singular, that is they possess a non-vanishing determinant.

I.10 DIVISION OF MATRICES AND MATRIX INVERSION

In ordinary arithmetic, if we wish to divide a whole series of numbers by the same quantity k, it is often more convenient to form the reciprocal $(k)^{-1}$ and use this repeatedly as a multiplier. In the same way, if we wish to divide other matrices by a square matrix M, a useful way to proceed is to find a matrix R which is the 'reciprocal' of M. It turns out that, provided that M is non-singular, there exists one and only one reciprocal matrix R, the property of which is that both (MR) and (RM) are equal to the unit matrix I of the same order. Conveniently we now call this reciprocal matrix M^{-1}; we can use it either as a premultiplier to form $M^{-1}B$, in which case we have predivided B by M, or alternatively as a postmultiplier to form BM^{-1}, in which case we have postdivided - a different operation.

The process of finding a reciprocal matrix is known as 'inversion', and for a matrix of large order it is quite complicated. The rule is to find the 'adjugate matrix', take its transpose and then divide by the determinant.

In this book, however, we shall be concerned only with the inversion of 2×2 matrices, and the process is extremely simple. In this case the 'adjugate' matrix is formed merely by replacing each element by its diagonally opposite element, at the same time changing the sign of the top right and bottom left elements.

Thus for $M = \begin{bmatrix} A & B \\ C & D \end{bmatrix}$ we can write immediately

$\text{adj}(M) = \begin{bmatrix} D & -C \\ -B & A \end{bmatrix}$. We must now take the transpose

$(\text{adj } M)^T = \begin{bmatrix} D & -B \\ -C & A \end{bmatrix}$ and finally divide each element

by the determinant $\det(M) = (AD - BC)$.

Note that in taking the determinant any one of the above three matrices gives the same result.

I.10.1

To verify correctness of the above procedure, we can check as follows:

$$M(\text{adj } M)^T = \begin{bmatrix} A & B \\ C & D \end{bmatrix}\begin{bmatrix} D & -B \\ -C & A \end{bmatrix} = \begin{bmatrix} (AD - BC) & (-AB + BA) \\ (CD - DC) & (-BC + AD) \end{bmatrix}$$

$$= (AD - BC)\begin{bmatrix} 1 & 0 \\ 0 & 1 \end{bmatrix} = \det(M)\,I$$

(where I is the unit matrix).

As it happens, most of the 2×2 matrices we shall need to invert will also have a determinant of unity. For such matrices, sometimes termed 'unimodular', the rule for inversion can be stated even more simply: *To invert a 2×2 unimodular matrix, interchange the two main diagonal elements and change the sign of the other two, leaving them where they are.*

Thus $M = \begin{bmatrix} A & B \\ C & D \end{bmatrix}$ gives $M^{-1} = \begin{bmatrix} D & -B \\ -C & A \end{bmatrix}$

No calculation is needed!

Finally, just as for transpose matrices, to obtain the reciprocal of the product of several matrices we must multiply the individual reciprocals in reverse order.

$$(PQRS)^{-1} = S^{-1}R^{-1}Q^{-1}P^{-1}, \text{ etc.}$$

To confirm this, we form $(PQRS)(S^{-1}R^{-1}Q^{-1}P^{-1})$ and find that, starting with $SS^{-1} = I$, the central region of the chain yields a succession of unit matrices, collapsing as a final result to a single unit matrix.

I.11 MATRIX DIAGONALIZATION

Particularly in iterated systems where we wish to multiply repeatedly by the same non-singular matrix M, it proves helpful to find a 'diagonalizing matrix' F such that

$$M = F \Lambda F^{-1}$$

where Λ is a diagonal matrix and F^{-1} is the reciprocal of F. If we suppose that such matrices F and Λ have been found, then because $F^{-1}F = I$, the square of the matrix M becomes

$$M^2 = MM = (F\Lambda F^{-1})(F\Lambda F^{-1}) = F\Lambda I\Lambda F^{-1} = F\Lambda^2 F^{-1}$$

Likewise

$$M^3 = MM^2 = (F\Lambda F^{-1})(F\Lambda^2 F^{-1}) = F\Lambda^3 F^{-1}$$

and, in general,

$$M^N = F\Lambda^N F^{-1}.$$

Thus, once the diagonalizing transformation has been found the Nth power of the original matrix is obtained merely by taking the diagonal matrix to its Nth power;

to do the latter, all we have to do is to replace each rth diagonal element λ_r by λ_r^N.

The various diagonal elements λ_1 λ_2 . . . λ_r . . . of the matrix Λ are often termed the 'latent roots', or 'eigenvalues', of the original matrix M, and the individual columns of the diagonalizing matrix F are correspondingly called its 'latent vectors', or 'eigenvectors'.

To understand this a little further, let us pick out from the square matrix F its rth column F_r, an operation which can be achieved by postmultiplying F by a column vector C_r for which the rth element is unity while all the other elements are zero.

What happens if we operate with the matrix M on the column vector $F_r = FC_r$? We find

$$MF_r = (F\Lambda F^{-1})(FC_r) = F\Lambda C_r = F_r\lambda_r = \lambda_r F_r$$

(since λ_r is a scalar). In other words, *when M multiplies the vector F_r the result is the same vector multiplied by the scalar λ_r*. The column vector F_r is termed the rth eigenvector of the matrix M, and λ_r is the corresponding rth eigenvalue. For 2×2 matrices, which are all that we shall be diagonalizing in this book, there will be only two column eigenvectors F_1 and F_2 with eigenvalues λ_1 and λ_2.

(It is also possible, if M is used as a postmultiplier instead of a premultiplier, to obtain eigenvectors which will be row vectors taken from the reciprocal matrix F^{-1}. In this book we shall use column eigenvectors.)

I.12 EIGENVALUES AND EIGENVECTORS OF A 2×2 UNIMODULAR MATRIX

If we wish to diagonalize a square matrix M, we begin by determining the eigenvalues λ_1 λ_2 . . ., etc.; the usual prescription for obtaining these is to solve what is called the 'characteristic equation' of the matrix:

$$\det(\lambda I - M) = 0.$$

In the case of a 2×2 unimodular matrix, however, it may be more instructive to proceed from first prin-

ciples. Given the matrix $M = \begin{bmatrix} A & B \\ C & D \end{bmatrix}$, we require to

find a diagonalizing matrix F and a diagonal matrix Λ such that $M = F\Lambda F^{-1}$, whence $MF = F\Lambda$. Taking this latter equation as our starting point, we rewrite it in full and obtain the requirement

$$MF = \begin{bmatrix} A & B \\ C & D \end{bmatrix} \begin{bmatrix} F_{11} & F_{12} \\ F_{21} & F_{22} \end{bmatrix} = \begin{bmatrix} F_{11} & F_{12} \\ F_{21} & F_{22} \end{bmatrix} \begin{bmatrix} \lambda_1 & 0 \\ 0 & \lambda_2 \end{bmatrix}$$

where $(AD - BC) = 1$ since $\det(M) = 1$. On multiplying, we obtain

$$\begin{bmatrix} AF_{11} + BF_{21} & AF_{12} + BF_{22} \\ CF_{11} + DF_{21} & CF_{12} + DF_{22} \end{bmatrix} = \begin{bmatrix} F_{11}\lambda_1 & F_{12}\lambda_2 \\ F_{21}\lambda_1 & F_{22}\lambda_2 \end{bmatrix}$$

Since these matrices are required to be the same, we can now write down four separate equations which link elements on the left side to corresponding elements on the right:

$$AF_{11} + BF_{21} = F_{11}\lambda_1 \tag{I.1}$$

$$CF_{11} + DF_{21} = F_{21}\lambda_1 \tag{I.2}$$

$$AF_{12} + BF_{22} = F_{12}\lambda_2 \tag{I.3}$$

$$CF_{12} + DF_{22} = F_{22}\lambda_2 \tag{I.4}$$

If we divide equations (I.1) and (I.2) by F_{21}, we can eliminate the ratio (F_{11}/F_{21}) and obtain

$$(F_{11}/F_{21}) = B/(\lambda_1 - A) = (\lambda_1 - D)/C$$

Therefore,

$$(\lambda_1 - A)(\lambda_1 - D) = BC$$

Similarly, if we divide equations (I.3) and (I.4) by F_{22}, we eliminate the ratio (F_{12}/F_{22}) and obtain

$$(F_{12}/F_{22}) = B/(\lambda_2 - A) = (\lambda_2 - D)/C$$

Therefore,

$$(\lambda_2 - A)(\lambda_2 - D) = BC$$

In other words, λ_1 and λ_2 both obey the same equation

$$(\lambda - A)(\lambda - D) - BC = 0 \tag{I.5}$$

Although this determination is not fully rigorous, it is easily seen that the equation obtained is equivalent to the characteristic equation quoted above. We have

$$\det(\lambda I - M) = \begin{vmatrix} (\lambda - A) & -B \\ -C & (\lambda - D) \end{vmatrix}$$

$$= (\lambda - A) \times (\lambda - D) - BC = 0$$

Since we know that $\det(M) = (AD - BC) = 1$, this characteristic equation for the two values of λ can be simplified to become

$$\lambda^2 - (A + D)\lambda + 1 = 0 \qquad (\text{I.6})$$

We see immediately that the two solutions λ_1, λ_2, must be such that $(\lambda_1 + \lambda_2) = (A + D)$ and $\lambda_1\lambda_2 = 1$. Solving the quadratic, we find

$$\lambda = \tfrac{1}{2}\left[(A + D) \pm \sqrt{(A + D)^2 - 4}\right] \qquad (\text{I.7})$$

(λ_1 will be taken as the value with the positive sign in the above expression and λ_2 as the value with the negative sign.)

The quantity $(A + D)$, the sum of the main diagonal elements, is sometimes referred to as the 'trace', or 'spur', of the matrix. If the trace $(A + D)$ lies between 2 and -2 in value, then the two eigenvalues can be rewritten conveniently in terms of an angle θ, chosen so that it lies between 0 and π and such that $(A + D) = 2\cos\theta$. We then find that

$$\lambda_1 = \cos\theta + i\sin\theta = \exp(i\theta)$$

$$\lambda_2 = \cos\theta - i\sin\theta = \exp(-i\theta)$$

where exp denotes the exponential function and i is used to represent $\sqrt{-1}$

Conversely, if $(A + D)$ is greater than 2 or less than -2, we can choose a positive quantity t such that $(A + D) = 2\cosh t$ (or $-2\cosh(-t)$ if $(A + D)$ is negative). The eigenvalues then take the form

$$\lambda_1 = \exp(t) \quad (\text{or} - \exp(t) \text{ if } (A + D) \text{ is negative})$$

$$\lambda_2 = \exp(-t) \quad (\text{or} - \exp(-t) \text{ if } (A + D) \text{ is negative})$$

I.12.1

To complete the process of diagonalization, we must now determine the diagonalizing matrix F and its reciprocal F^{-1}. But it will be noticed that we have already determined the ratios

$$(F_{11}/F_{21}) = (\lambda_1 - D)/C = B/(\lambda_1 - A) \tag{I.8}$$

and

$$(F_{12}/F_{22}) = (\lambda_2 - D)/C = B/(\lambda_2 - A) \tag{I.9}$$

These are, in fact, the two important ratios that determine, except for a scalar multiplying factor, the nature of the two eigenvectors

$$F_1 = \begin{bmatrix} F_{11} \\ F_{21} \end{bmatrix} \quad \text{and } F_2 = \begin{bmatrix} F_{12} \\ F_{22} \end{bmatrix}$$

If we now choose arbitrarily that both F_{21} and F_{22} shall take the value C, we obtain as one possible F-matrix the form

$$F = \begin{bmatrix} F_{11} & F_{12} \\ F_{21} & F_{22} \end{bmatrix} = \begin{bmatrix} (\lambda_1 - D) & (\lambda_2 - D) \\ C & C \end{bmatrix}$$

for which the determinant is

$$\det(F) = (F_{11}F_{22}) - (F_{12}F_{21}) = C(\lambda_1 - \lambda_2)$$

If we wished we could convert F into a unimodular matrix by dividing each element by $\sqrt{C(\lambda_1 - \lambda_2)}$, but this is unnecessary. The reciprocal matrix will be $F^{-1} = (\text{adj } F)^T/\det(F)$ and is evidently

$$\begin{bmatrix} C & (D - \lambda_2) \\ -C & (\lambda_1 - D) \end{bmatrix} \quad \text{divided by the scalar quantity } C(\lambda_1 - \lambda_2)$$

The complete diagonalizing transformation that we have obtained is thus:

$$\begin{bmatrix} A & B \\ C & D \end{bmatrix} = \frac{\begin{bmatrix} (\lambda_1 - D) & (\lambda_2 - D) \\ C & C \end{bmatrix} \begin{bmatrix} \lambda_1 & 0 \\ 0 & \lambda_2 \end{bmatrix} \begin{bmatrix} C & (D - \lambda_2) \\ -C & (\lambda_1 - D) \end{bmatrix}}{C(\lambda_1 - \lambda_2)} \tag{I.}$$

where $(AD - BC) = 1$, and λ_1 and λ_2 are such that $\lambda_1\lambda_2 = 1$ and $(\lambda_1 + \lambda_2) = (A + D)$.

I.12.2

The student should check for himself that the triple product given above reduces so as to satisfy this result. We should note that, if the element C in the original matrix vanishes, then it is necessary to find the eigenvector ratios by using the alternative expressions $B/(\lambda_1 - A)$ and $B/(\lambda_2 - A)$. We then obtain

$$\begin{bmatrix} A & B \\ C & D \end{bmatrix} = \frac{\begin{bmatrix} B & B \\ (\lambda_1 - A) & (\lambda_2 - A) \end{bmatrix}\begin{bmatrix} \lambda_1 & 0 \\ 0 & \lambda_2 \end{bmatrix}\begin{bmatrix} (\lambda_2 - A) & -B \\ (A - \lambda_1) & B \end{bmatrix}}{- B(\lambda_1 - \lambda_2)} \quad (I.11)$$

(Obviously, if both C and B vanish, the matrix M is already diagonalized!)

II

Matrix Methods
in Paraxial Optics

II.1 INTRODUCTORY DISCUSSION

In this chapter we shall be considering how matrices can be used to describe the geometric formation of images by a centred lens system - a succession of spherical refracting surfaces all centred on the same optical axis. The results that we shall obtain will be valid only within the limits of two main approximations.

The first is the basic assumption of all geometric optics - that the wavelength of light is negligibly small and that propagation of light can be described not in terms of wavefronts but in terms of individual rays. (We shall return to this point in chapter III when we consider the propagation of a Gaussian beam, which is 'the closest approximation to a single ray that nature permits'.) As can be shown by means of Huygens' construction, if light waves are allowed to travel without encountering any obstacles they are propagated along a direction which is normal to the wavefronts. The concept of a geometric ray is an idealization of this wavenormal; in vector terms it can be thought of as the Poynting vector of the electromagnetic field, or as the gradient of the scalar function (eikonal) which describes the phase of the wave disturbance. A consequence of this is that each ray obeys Fermat's principle of least time; if we consider the neighbourhood of *any short section* of the raypath, the path which the ray chooses between a given entry point and a given exit point is that which *minimizes the time taken.*

Our second approximation is that we shall consider only paraxial rays - those that remain close to the

axis and almost parallel to it so that we can use
first-order approximations for the sines or tangents
of any angles that are involved. Our treatment will
therefore give no information about third-order effects
such as spherical aberration or the oblique aberrations
coma, astigmatism, field curvature and distortion. We
shall, however, make some mention of longitudinal and
transverse chromatic aberration.

The optics of paraxial imaging is often referred to
as Gaussian optics, since it was Carl Frederick Gauss
who in 1840 laid the foundations of the subject. In his
classical memoir, *Dioptrische Untersuchungen*, Gauss
showed that the behaviour of any lens system can be
determined from a knowledge of its six cardinal points
– namely two focal points, two nodal points of unit
angular magnification and two principal points of unit
linear magnification. Included in the paper were
recipes for experimentally determining the positions of
these points and iterative methods for calculating them
in terms of the surface curvatures, separations and
refractive indices of the lens system. In formulating
the latter, Gauss wrote down explicitly the two linear
simultaneous equations whereby the ray height and ray
angle of an output ray are linked to the corresponding
quantities for an input ray. But matrix formalism was
at that time unknown, and Gauss contented himself by
using an algorithm which he had learnt from Euler to
express the four coefficients of these equations in
convenient computational form. (The expressions that
he used, a shorthand form of continued fraction, are
now known as 'Gaussian brackets'; they are by no means
obsolete since very nearly the same economical course
of calculation is used in the modern 'ynv' method.)

As this chapter will show, matrices provide an alternative method for performing this type of calculation.
It would seem that they were first used in optics by
Sampson about sixty years ago, but it is only recently
that they have been widely adopted. The first books on
matrix methods were E.L. O'Neill's, *Introduction to
Statistical Optics*, Addison-Wesley, Reading, Mass.,
1963 and W. Brouwer's, *Matrix Methods in Optical
Instrument Design*, W.A. Benjamin Inc., New York, 1964.
During the next two years, discussions of the methods
appeared in papers by Halbach and by Richards in the
American Journal of Physics, but there was some dis-

agreement over the order in which the calculations should be arranged.

During 1965 Kogelnik published an important extension of the method whereby a ray-transfer matrix could be used to describe not only the geometric optics of paraxial rays but also the propagation of a diffraction-limited laser beam. We shall defer discussion of beams and resonators until the next chapter, but Kogelnik's work has had such an impact on the literature that we shall adopt his arrangement, and that followed more recently by Sinclair, in all our matrix calculations.

II.2 RAY-TRANSFER MATRICES

Let us now consider the propagation of a paraxial ray through a system of centred lenses. In conformity with modern practice we shall use a Cartesian coordinate system drawn so that Oz, which points from left to right, represents the optical axis of the system and also the general direction in which the rays travel. Of the transverse axes, Oy will be taken to point upwards in the plane of the diagram and Ox will be assumed to point normal to the diagram and away from the reader. In this chapter we shall not be considering skew rays, so our discussion can be confined to those rays that lie in the yz-plane and are close to the axis Oz.

The trajectory of a ray as it passes through the various refracting surfaces of the system will consist of a series of straight lines, each of which can be specified by stating (a) the coordinates of one point of it, and (b) the angle which it makes with the z-axis. If we choose in advance any plane perpendicular to the z-axis, a plane of constant z, we can call it a Reference Plane (RP). In terms of any particular reference plane we can then specify a ray by the height y at which it intersects the reference plane, and the angle v which it makes with the z-direction; v is measured in radians and is considered positive if an anticlockwise rotation turns a line from the positive direction of the z-axis to the direction in which light travels along the ray (see Figure II.1).

Although we could attempt to describe all the rays encountered in a calculation by referring them all to a *single* reference plane (for example the RP, $z = 0$), it is in practice much more convenient to choose a new

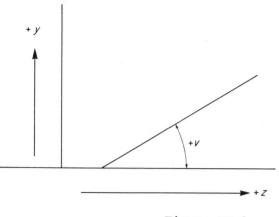

Figure II.1

RP for each stage of the initial calculation. This
means that ray data are continually transferred from
one RP to the next as we consider the various elements
of the system. However, once this initial calculation
has been done right through the system, we emerge with
an overall ray-transfer matrix which will convert all
the ray data we wish to consider from our chosen input
RP directly to our chosen output RP.

We have already pointed out that at any RP the specification of a ray can be made in terms of its ray
height y and its ray angle v. It will, however, make
the calculations more convenient if we replace the ray
angle v by the corresponding 'optical direction-cosine'
nv (or, strictly speaking, $n\sin v$) where n is the
refractive index of the medium in which the ray is
travelling. This optical direction cosine, which we
shall denote by V, has the property that, by Snell's
law, it will remain unchanged as it crosses a plane
boundary between two different media. (Although use
of the optical direction cosine $V = nv$ may seem a com-
plication, in most calculations where the initial and
final RPs are in air it will make little difference;
it greatly simplifies calculations involving plane-
parallel plates, and it ensures that *all* the matrices
involved are unimodular.)

As a ray passes through a refracting lens system,
there are only two basic types of process that we need
to consider in order to determine its progress:

(a) A *translation*, or gap, across which the ray simply launches itself in a straight line on its journey to the next refracting surface. In specifying a gap we need to know its thickness t and also the refractive index n of the medium through which the ray is passing.

(b) *Refraction* at the boundary surface between two regions of different refractive index. To determine how much bending the ray undergoes, we need to know the radius of curvature of the refracting surface and the two values of refractive index.

In the next two sections we shall investigate the effect which each of these two basic elements has on the y-value and V-value of a ray passing between the two reference planes, one on either side of the element. The ray first passes through RP_1, with values y_1 and V_1, then through the element and then through RP_2, with values y_2 and V_2. We seek equations giving y_2 and V_2 in terms of y_1 and V_1, and the properties of the element between the reference planes.

We shall find that for both types of element the equations are linear and can therefore be written in the matrix form

$$\begin{bmatrix} y_2 \\ V_2 \end{bmatrix} = \begin{bmatrix} A & B \\ C & D \end{bmatrix} \begin{bmatrix} y_1 \\ V_1 \end{bmatrix}$$

the matrix elements being such that the determinant $(AD - BC)$ equals unity.

Alternatively, if we wish to trace a ray backwards, the matrix equation can be inverted to yield

$$\begin{bmatrix} y_1 \\ V_1 \end{bmatrix} = \begin{bmatrix} D & -B \\ -C & A \end{bmatrix} \begin{bmatrix} y_2 \\ V_2 \end{bmatrix}$$

Finally, having verified that *each* element in a system can be represented by a ray-transfer matrix of this unimodular form, we shall multiply together in correct sequence all the elementary translation and refraction matrices to obtain the single ray-transfer matrix that represents the complete assembly; this may be anything from a single thin lens to a complicated optical system.

II.3 THE TRANSLATION MATRIX $\mathcal{T}$

Figures II.2a and II.2b show two examples of a ray which travels a distance t to the right between the two reference planes. Clearly the angle at which the ray is travelling will remain the same throughout the translation, but not so its distance from the axis. Figure II.2a has been drawn to illustrate the case when both y- and v-values remain positive, and Figure II.2b illustrates the case where the v-value is negative. In both figures the angles v are shown exaggerated for clarity – in practice the maximum v-value for a paraxial ray will be less than 0·1 (one-tenth of a radian or about 6°). Errors incurred in the paraxial approximation will then remain well below 1 per cent.

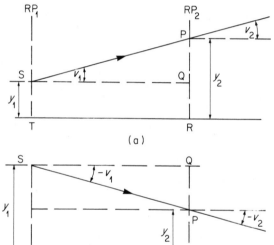

(a)

(b)

Figure II.2

Referring to Figure II.2a   *Referring to Figure II.2b*

$$y_2 = RP = RQ + QP$$
$$= TS + SQ \tan P\hat{S}Q$$
$$= y_1 + t \tan(v_1)$$
$$= y_1 + tv_1$$

$$y_2 = RP = RQ - PQ$$
$$= TS - SQ \tan P\hat{S}Q$$
$$= y_1 - t \tan(-v_1)$$
$$= y_1 + tv_1$$

(We have here made use of the paraxial approximation that (v) can be used to replace either $\tan(v)$ or $\sin(v)$ with negligible error.) It has already been pointed out that the ray quantities upon which our translation matrix is going to operate are the height of a ray and its optical direction cosine or V-value, rather than just its angle v. So if n is the refractive index of the medium between RP_1 and RP_2, the above equation needs to be rewritten

$$y_2 = y_1 + (t/n)(nv_1) = 1y_1 + TV_1$$

where $T = (t/n)$ is the 'reduced thickness' of the gap. It is immediately apparent from both diagrams that v_1 and v_2 are equal, so the equation for the new optical direction cosine V_2 can be written

$$V_2 = nv_2 = nv_1 = 0y_1 + 1V_1$$

The pair of equations that we have obtained can now be rewritten in matrix form

$$\begin{bmatrix} y_2 \\ V_2 \end{bmatrix} = \begin{bmatrix} 1 & T \\ 0 & 1 \end{bmatrix} \begin{bmatrix} y_1 \\ V_1 \end{bmatrix}$$

So the matrix representing a translation to the right through a reduced distance T is

$$\mathcal{T} = \begin{bmatrix} 1 & T \\ 0 & 1 \end{bmatrix}$$

Its determinant $\det(\mathcal{T})$ is obviously unity.

II.3.1 *Compound layers and plane-parallel plates*

It is worth noting that, if we arbitrarily imagine that we have divided a gap t into adjoining gaps t_1 and t_2, still with the same index of refraction n, we shall obtain two successive translation matrices

$$\mathcal{T}_1 = \begin{bmatrix} 1 & T_1 \\ 0 & 1 \end{bmatrix} \quad \text{and} \quad \mathcal{T}_2 = \begin{bmatrix} 1 & T_2 \\ 0 & 1 \end{bmatrix}$$

where $T_1 = (t_1/n)$, $T_2 = (t_2/n)$ and $t_1 + t_2 = t$.

If we then premultiply or postmultiply $\mathcal{T}_1$ by $\mathcal{T}_2$, we obtain

$$\mathcal{T}_2\mathcal{T}_1 = \mathcal{T}_1\mathcal{T}_2 = \begin{bmatrix} 1 & (T_1 + T_2) \\ 0 & 1 \end{bmatrix} = \mathcal{T}$$

as before (since $T_1 + T_2 = T$).

A similar situation applies where a given gap of total thickness t consists of several separate layers such that each ith layer has its own thickness t_i and its own refractive index n_i. Provided that all the boundary surfaces are flat and perpendicular to the axis, the ray height and the optical direction cosine will remain unchanged at each boundary. It follows that we do not have to worry about any of the refraction matrices, since each of them is merely a unit matrix $\begin{bmatrix} 1 & 0 \\ 0 & 1 \end{bmatrix}$ which can be neglected. (This will be verified in the next section.)

We can now see that, provided that each ith layer is represented by its reduced thickness $T_i = (t_i/n_i)$, all the individual translation matrices can be multiplied together to produce a single matrix for the effect of the whole gap. As we have already seen, these translation matrices have the useful property that the order in which they are multiplied together does not matter. The T-value appearing in the product matrix is just the sum of the T-values in the individual matrices:

$$\begin{bmatrix} 1 & T_1 \\ 0 & 1 \end{bmatrix}\begin{bmatrix} 1 & T_2 \\ 0 & 1 \end{bmatrix}\begin{bmatrix} 1 & T_3 \\ 0 & 1 \end{bmatrix} \cdots \begin{bmatrix} 1 & T_n \\ 0 & 1 \end{bmatrix} = \begin{bmatrix} 1 & \sum_i T_i \\ 0 & 1 \end{bmatrix}$$

Having seen that translation matrices produce the same product in whatever order they are taken, let us consider the optical corollary to this situation. If we are looking perpendicularly through a whole series of plane-parallel plates, then moving the plates or even interchanging them may affect the amount of light that is reflected, but the geometry of the transmitted images will remain exactly the same.

(When we say that a plane-parallel glass plate of refractive index n and thickness t has a reduced thickness (t/n) we are using quite appropriate language. If

we are looking at an object on the far side of the
plate the light actually takes longer to reach us than
if the plate were absent, but the object certainly
looks closer. Replacing a layer of air by the same
thickness of glass makes the world seem closer to the
observer by a distance $(t/1) - (t/n) = t \cdot (n - 1)/n$,
which is about one-third the thickness of the plate.
For an object submerged in water the factor $(n - 1)/n$
is only about one-quarter, but the depth t may be considerable - even the bear fishing with his paw needs
to know about reduced thickness!)

II.4 THE REFRACTION MATRIX $\mathscr{R}$

We now investigate the action of a curved surface
separating two regions of refractive index n_1 and n_2.
The radius of curvature of the surface will be taken
as positive if the centre of curvature lies to the
right of the surface. The situation illustrated in
Figure II.3 shows a surface of positive curvature with

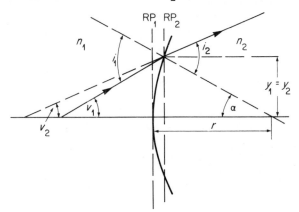

Figure II.3

the refractive index n_2 on the right of the surface
greater than that on the left (n_1). The ray illustrated
is also one that has positive y- and V-values on both
sides of the surface.

As in Figure II.2, the angles are shown greatly exaggerated; a consequence of this is that RP_1, drawn
through the pole of the surface, appears well separated
from RP_2, drawn through the entry point of the ray to
the surface. But for paraxial rays the separation
between these two planes will be $r(1 - \cos\alpha)$ and
therefore negligible since α, like v_1 and v_2, is

assumed to be a small angle. We therefore have that $y_2 = y_1$.

Applying Snell's law in the diagram, we have

$$n_1 \sin i_1 = n_2 \sin i_2 \quad \text{or} \quad n_1 i_1 = n_2 i_2$$

to paraxial approximation. But, by the exterior angle theorem,

$$i_1 = v_1 + \alpha = v_1 + y_1/r \text{ and } i_2 = v_2 + \alpha = v_2 + y_1/r$$

Hence

$$n_1(v_1 + y_1/r) = n_2(v_2 + y_1/r)$$

or $\quad V_1 + n_1 y_1/r = V_2 + n_2 y_1/r$

So, rearranging the equations in matrix form, we obtain finally

$$\begin{bmatrix} y_2 \\ V_2 \end{bmatrix} = \begin{bmatrix} 1 & 0 \\ -(n_2 - n_1)/r & 1 \end{bmatrix} \begin{bmatrix} y_1 \\ V_1 \end{bmatrix}$$

The quantity $(n_2 - n_1)/r$ is usually termed the refracting power of the surface.

We have obtained the refraction matrix

$$\mathscr{R} = \begin{bmatrix} 1 & 0 \\ -(n_2 - n_1)/r & 1 \end{bmatrix}$$

by considering one special case in which all the quantities are positive. But if a thorough analysis is made, investigating the other cases where the change of refractive index or the curvature is reversed, or where the y- or V-values are negative, the same refraction matrix $\mathscr{R}$ will be found to give consistent and correct results for the change in ray direction that is produced.

We shall defer until the end of this chapter the important case of a reflecting surface. As far as refracting surfaces are concerned, if a ray encounters a plane boundary where r is infinite, the refracting power $(n_2 - n_1)/r$ contributed by the surface also vanishes, and the refraction matrix degenerates into the trivial case of a unit matrix. We have already used this fact in the previous section, when telescoping together the translation matrices of several plane-parallel plates.

II.4.1 Thin lens approximation

A similar telescoping of refraction matrices is possible when several refracting surfaces are so close together that the intervening gaps are negligible (in this case it is the translation matrix that degenerates into a unit matrix). If each ith refracting surface possesses a curvature r_i and refractive indices n_i and n_{i+1}, we shall be able to represent its refracting power by $P_i = (n_{i+1} - n_i)/r_i$ and the telescoped matrix for the thin lens combination will be

$$\begin{bmatrix} 1 & 0 \\ -\sum_i P_i & 1 \end{bmatrix}$$

independent of the order in which the individual powers are added.

The student should verify for himself that

$$\mathscr{R}_2\mathscr{R}_1 = \begin{bmatrix} 1 & 0 \\ -P_2 & 1 \end{bmatrix}\begin{bmatrix} 1 & 0 \\ -P_1 & 1 \end{bmatrix} = \begin{bmatrix} 1 & 0 \\ -(P_1+P_2) & 1 \end{bmatrix} = \mathscr{R}_1\mathscr{R}_2$$

The refraction matrix of a single thin lens, for example, is the same whichever way round it is mounted. Its refracting power will be

$$P = P_1 + P_2 = (n - 1)/r_1 + (1 - n)/r_2$$

$$= (n - 1)(1/r_1 - 1/r_2) = 1/f$$

The refracting power P is usually measured in dioptres, the focal length f and radii of curvature r_1 and r_2 being given in metres.

In general, if calculations are to be made with a series of thin lenses whose focal lengths are given, it is more convenient to replace each focal length f_i by its corresponding refracting power or focal power $P_i = 1/f_i$. The refraction matrix for the ith lens is then

$$\mathscr{R}_i = \begin{bmatrix} 1 & 0 \\ -P_i & 1 \end{bmatrix} = \begin{bmatrix} 1 & 0 \\ -1/f_i & 1 \end{bmatrix}$$

As we have seen, it is a simple matter to telescope together either a succession of $\mathscr{R}$-matrices or a succession of $\mathscr{T}$-matrices into a single matrix. In the general case, however, we shall encounter an alternating sequence of $\mathscr{R}$- and $\mathscr{T}$-matrices and must consider carefully the order in which they arise. For any refraction-translation product, matrix multiplication is non-commutative. We have, for example,

$$\mathscr{R}\mathscr{T} = \begin{bmatrix} 1 & 0 \\ -P & 1 \end{bmatrix}\begin{bmatrix} 1 & T \\ 0 & 1 \end{bmatrix} = \begin{bmatrix} 1 & T \\ -P & (1 - PT) \end{bmatrix}$$

whereas
$$\mathscr{T}\mathscr{R} = \begin{bmatrix} 1 & T \\ 0 & 1 \end{bmatrix}\begin{bmatrix} 1 & 0 \\ -P & 1 \end{bmatrix} = \begin{bmatrix} (1 - PT) & T \\ -P & 1 \end{bmatrix}$$

The latter matrix is different, since the two main diagonal elements have been interchanged.

II.5 THE RAY-TRANSFER MATRIX FOR A SYSTEM

II.5.1 Numbering of reference planes

Let us consider propagation of a paraxial ray through an optical system containing n refracting surfaces separated by $(n - 1)$ gaps. If only because each lens surface will tend to be enclosed in a cell or tube, it may be convenient to choose our first 'input' reference plane RP_1 at a distance d_a to the left of the first refracting surface. RP_2 and RP_3 will then lie immediately to the left and to the right of the first refracting surface; RP_4 and RP_5 will lie on either side of the second surface, and so on until we reach RP_{2n} and RP_{2n+1} on either side of the nth surface. Our final 'output' reference plane RP_{2n+2} will then be taken to lie a distance d_b to the right of this last refracting surface. We seek now to obtain an overall ray-transfer matrix M that will enable us directly to convert an input ray vector $\begin{bmatrix} y_1 \\ V_1 \end{bmatrix}$ into an output ray vector $\begin{bmatrix} y_{2n+2} \\ V_{2n+2} \end{bmatrix}$ (see Figure II.4 overleaf).

II.5.2 Numbering of matrices

The next step is to write down translation or refraction matrices that represent each of the elements between the various reference planes. Again working from left to right across the diagram we number these

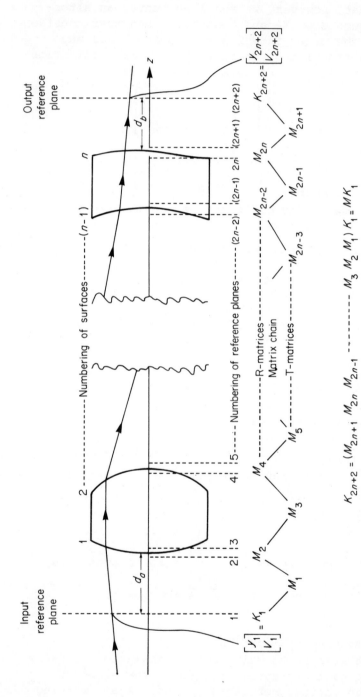

$$K_{2n+2} = (M_{2n+1} \ M_{2n} \ M_{2n-1} \ \text{------} \ M_3 \ M_2 \ M_1) \ K_1 = MK_1$$

where $M = (M_{2n+1} \ M_{2n} \ M_{2n-1} \ \text{------} \ M_3 \ M_2 \ M_1)$

the product of the matrix chain taken in descending order

Figure II.4

matrices $M_1 M_2 M_3 M_4 \ldots M_{2n+1}$, the rule being that the number assigned to each matrix is the same as for the reference plane on its left. If we use the symbol K_r to denote the ray vector $\begin{bmatrix} y_r \\ V_r \end{bmatrix}$ for a ray traversing the rth reference plane, then for transfer of ray data from RP_r to RP_{r+1} we have the basic recurrence relation $K_{r+1} = M_r K_r$, and similarly $K_r = M_{r-1} K_{r-1}$, etc.

II.5.3 *Calculation of output ray which will be produced by a given input ray*

By repeatedly using this recurrence relation, and also the associative property of matrix multiplication, we now find

$$K_{2n+2} = M_{2n+1} K_{2n+1} = M_{2n+1} (M_{2n} K_{2n})$$

$$= (M_{2n+1} M_{2n})(M_{2n-1} K_{2n-1})$$

$$= (M_{2n+1} M_{2n} M_{2n-1} M_{2n-2} \cdot \cdot \cdot M_3 M_2 M_1) K_1$$

Hence $K_{2n+2} = MK_1$ where M represents the complete matrix product *taken in descending order.*

$$M = (M_{2n+1} M_{2n} M_{2n-1} M_{2n-2} M_{2n-3} \cdot \cdot \cdot M_3 M_2 M_1)$$

It is important to note here that, as shown by the suffices, the individual matrices are written in these equations in the reverse order to that in which we numbered them. It may be helpful to visualize this order as that which is seen by an observer looking back from the output reference plane towards the light source; it is also perhaps the natural order for writing the equation - once we have decided to start with the output ray vector, then the matrices which follow can be regarded as the links in a mathematical chain that brings us back to the input.

II.5.4 *Calculation of input ray required to produce a given output ray*

If we wish to proceed from first principles, then we can invert each of the individual matrices and use the inverted recurrence relation $K_r = M_r^{-1} K_{r+1}$. We then obtain

$$K_1 = M_1^{-1} K_2 = M_1^{-1} M_2^{-1} K_3 = (M_1^{-1} M_2^{-1} M_3^{-1} M_4^{-1} \ldots M_{2n+1}^{-1}) K_{2n+2}$$

(In this case the links of the chain appear in *ascending* order, but each of them has been inverted.)

Usually, however, we shall already have calculated the overall matrix M and can proceed directly by inversion of the equation $K_{2n+2} = M K_1$ to obtain $K_1 = M^{-1} K_{2n+2}$. This represents exactly the same result, since, as was shown in section I.10, the reciprocal of a matrix product is the same as the product of the individual reciprocals taken in reversed order.

II.5.5 *Formation of product matrix M*

By virtue of the associative property, there are several ways in which we can multiply out the complete matrix product. But if more than three matrices are involved it may save some writing to take them in pairs initially; the matrix for each $\mathcal{R}\mathcal{T}$ product can be written down almost at sight, since only the fourth element $(1 - PT)$ needs any calculation. As the product matrices in a long chain are built up it is advisable occasionally to check that their determinants are still unity - if they are not then a mistake must have been made since all of the $\mathcal{R}$- and $\mathcal{T}$-matrices are themselves unimodular.

Once the overall matrix M for a complete system has been found, it is usually convenient to forget about all the intervening elements and to renumber the output reference plane as RP_2. (It is, after all, now possible to transfer data from RP_1 to the output in a single step.)

In some cases, however, one half of a system may be separated from the other by a variable gap, and it will be better to calculate separate matrices for the two known portions of the system. When this has been done the two calculated matrices can be linked by a $\begin{bmatrix} 1 & t \\ 0 & 1 \end{bmatrix}$ matrix representing the variable value of the gap. The four elements of the overall matrix will then be simple linear functions of t.

It frequently happens also that a complicated lens assembly is used with variable object and image distances. Here again it will pay to calculate first a single matrix for the known central portion of the system.

II.6 DERIVATION OF PROPERTIES OF A SYSTEM FROM ITS MATRIX

Let us suppose that we have numerically evaluated the matrix M for a complete optical system, and that the equation $K_2 = MK_1$ when written out fully reads

$$\begin{bmatrix} y_2 \\ V_2 \end{bmatrix} = \begin{bmatrix} A & B \\ C & D \end{bmatrix} \begin{bmatrix} y_1 \\ V_1 \end{bmatrix}$$

where $(AD - BC) = 1$. In order to understand better the significance of these four quantities A, B, C and D, let us consider what happens if one of them vanishes.

(a) If $D = 0$, the equation for V_2 reads $V_2 = Cy_1 + 0V_1$ $= Cy_1$. This means that all rays entering the input plane from the same point (characterized by y_1) emerge at the output plane making the same angle $V_2 = Cy_1$ with the axis, no matter at what angle V_1 they entered the system. It follows that the input plane RP$_1$ *must be the first focal plane* of the system (see Figure II.5).

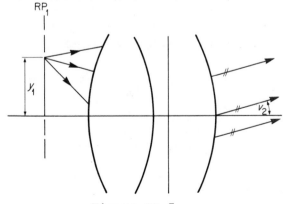

Figure II.5

(b) If $B = 0$, the equation for y_2 reads $y_2 = Ay_1 + OV_1$ $= Ay_1$. This means that all rays leaving the point O (characterized by y_1) in RP$_1$ will pass through the same point I (characterized by y_2) in RP$_2$. Thus O and I are object and image points, so that *RP$_1$ and RP$_2$ are now conjugate planes.* In addition, $A = y_2/y_1$ gives the *magnification* produced by the system in these circumstances (see Figure II.6).

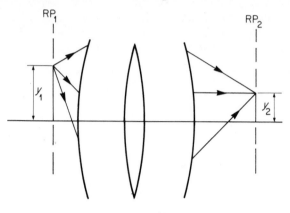

Figure II.6

(c) If $C = 0$, then $V_2 = DV_1$. This means that all rays which enter the system parallel to one another (V_1 gives their direction on entry) will emerge parallel to one another in a new direction V_2. A lens system like this, which transforms a parallel beam into another parallel beam in a different direction, is called an *afocal* or *telescopic* system. In this case $(n_1 D/n_2) = (v_2/v_1)$ is the *angular magnification* produced by the system (see Figure II.7).

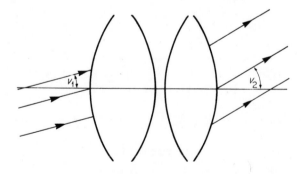

Figure II.7

(d) If $A = 0$, the equation for y_2 reads $y_2 = BV_1$. This means that all rays entering the system at the same angle (characterized by V_1) will pass through the same point (characterized by y_2) in the output plane RP_2. Thus, the system brings bundles of parallel rays to focus at points in RP_2, that is RP_2 *is the second focal plane* of the system (see Figure II.8).

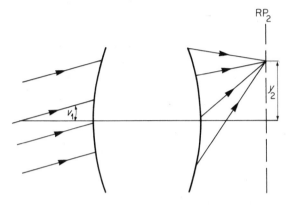

Figure II.8

(e) It is worth remembering that, if either A or D in a ray-transfer matrix vanishes, the equation $(AD - BC) = 1$ then requires that $BC = -1$. Likewise, if either B or C vanishes, A must be the reciprocal of D.

The fact that B vanishes when the object-image relationship holds between the first and second reference planes, and that A or $(1/D)$ is then the transverse magnification, leads to an experimental method for finding the matrix elements of an optical system without dismantling it or measuring its individual components. We will describe this method after using the matrix method to solve some problems.

In general, unless the V-value of a ray is zero it will cut the axis somewhere. With respect to a given reference plane in which the ray height is y and the ray angle v is (V/n), the z-position where it cuts the axis will be $(-y/v) = (-ny/V)$. If y and V are both positive or both negative, the point will be to the left of the reference plane (see Figure II.9).

42

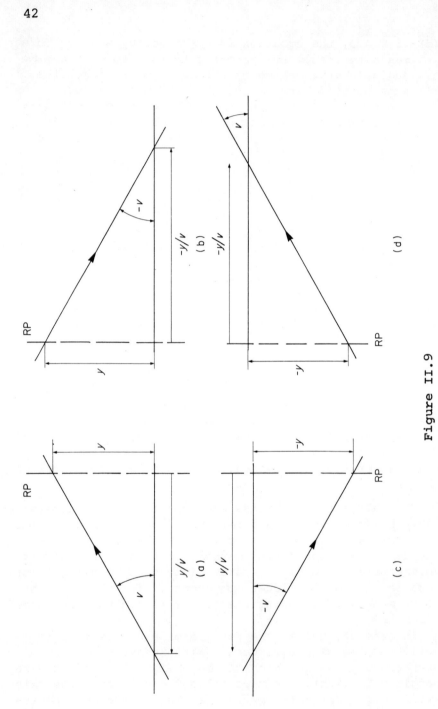

Figure II.9

II.7 ILLUSTRATIVE PROBLEMS

Problem 1

The left end of a long plastic rod of refractive index 1·56 is ground and polished to a convex (outward) spherical surface of radius 2·8 cm. An object 2 cm tall is located in the air and on the axis at a distance of 15 cm from the vertex. (See Figure II.10) Find the position and size of the image inside the rod.

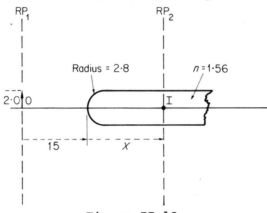

Figure II.10

Solution

Using cm as the unit of length for both $\mathcal{R}$- and $\mathcal{T}$-matrices, we proceed as follows. The $\mathcal{R}$-matrix of the surface is

$$\begin{bmatrix} 1 & 0 \\ -(n_2 - n_1)/r & 1 \end{bmatrix} = \begin{bmatrix} 1 & 0 \\ (-1\cdot56 + 1)/2\cdot8 & 1 \end{bmatrix} = \begin{bmatrix} 1 & 0 \\ -0\cdot2 & 1 \end{bmatrix}$$

If the image is at X cm to the right of the curved surface, the matrix chain leading from this image back to the object is

$$M = \begin{bmatrix} 1 & X/n \\ 0 & 1 \end{bmatrix} \begin{bmatrix} 1 & 0 \\ -0\cdot2 & 1 \end{bmatrix} \begin{bmatrix} 1 & 15 \\ 0 & 1 \end{bmatrix}$$

(from image to surface)	(surface refraction)	(from surface to object)

$$= \begin{bmatrix} 1 & X/1\cdot56 \\ 0 & 1 \end{bmatrix} \begin{bmatrix} 1 & 15 \\ -0\cdot2 & -2 \end{bmatrix} = \begin{bmatrix} (1 - X/7\cdot8) & (15 - X/0\cdot78) \\ -0\cdot2 & -2 \end{bmatrix}$$

Check

det(M) = $(AD - BC)$ = $- 2 + (X/3 \cdot 9) + 3 - (X/3 \cdot 9) = 1$

For the object-image relationship to hold, the top right hand element, B, must vanish:

$$15 - \frac{X}{0 \cdot 78} = 0 \quad \therefore X = 11 \cdot 7$$

The image is thus 11·7 cm inside the rod. The lateral magnification is given by either A or $1/D$ and is - 0·5. Therefore the image is (0·5 × 2) cm = 1 cm tall, and is inverted.

Problem 2

A glass rod 2·8 cm long and of index 1·6 has both ends ground to spherical surfaces of radius 2·4 cm, and con- vex to the air. An object 2 cm tall is located on the axis, in the air, 8 cm to the left of the left-hand vertex. (See Figure II.11). Find the position and size of the final image.

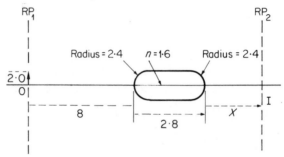

Figure II.11

Solution

If the final image is X cm to the right of the rod, the matrix chain is

$$M = \begin{bmatrix} 1 & X \\ 0 & 1 \end{bmatrix} \begin{bmatrix} 1 & 0 \\ \dfrac{-(1-1 \cdot 6)}{-2 \cdot 4} & 1 \end{bmatrix} \begin{bmatrix} 1 & \dfrac{2 \cdot 8}{1 \cdot 6} \\ 0 & 1 \end{bmatrix} \begin{bmatrix} 1 & 0 \\ \dfrac{-(1 \cdot 6-1)}{2 \cdot 4} & 1 \end{bmatrix} \begin{bmatrix} 1 \\ 0 \end{bmatrix}$$

| (image to rod) | (right end of rod) | (length of rod) | (left end of rod) | (rod obje |

$$= \begin{bmatrix} 1 & X \\ 0 & 1 \end{bmatrix} \begin{bmatrix} 1 & 0 \\ -0\cdot25 & 1 \end{bmatrix} \begin{bmatrix} 1 & 1\cdot75 \\ 0 & 1 \end{bmatrix} \begin{bmatrix} 1 & 0 \\ -0\cdot25 & 1 \end{bmatrix} \begin{bmatrix} 1 & 8 \\ 0 & 1 \end{bmatrix}$$

$$= \begin{bmatrix} 1 & X \\ 0 & 1 \end{bmatrix} \times \begin{bmatrix} 1 & 1\cdot75 \\ -0\cdot25 & 0\cdot5625 \end{bmatrix} \times \begin{bmatrix} 1 & 8 \\ -0\cdot25 & -1 \end{bmatrix}$$

$$= \begin{bmatrix} 1 & X \\ 0 & 1 \end{bmatrix} \times \begin{bmatrix} 0\cdot5625 & 6\cdot25 \\ -0\cdot391 & -2\cdot56 \end{bmatrix}$$

Therefore,

$$\begin{bmatrix} A & B \\ C & D \end{bmatrix} = \begin{bmatrix} 0\cdot5625 - 0\cdot391X & 6\cdot25 - 2\cdot56X \\ -0\cdot391 & -2\cdot56 \end{bmatrix}$$

For the object-image relationship, we must have $2\cdot56X = 6\cdot25$. Therefore $X = 2\cdot44$ (cm to right of rod). The magnification is $1/D = -1/2\cdot56 = -0\cdot39$. The image is therefore $(2 \times 0\cdot39)$ cm $= 0\cdot78$ cm tall, and inverted.

Problem 3
A parallel beam of light enters a clear plastic spherical bead whose diameter is 2 cm and refractive index 1·4. (See Figure II.12). At what point beyond the bead is the light brought to a focus?

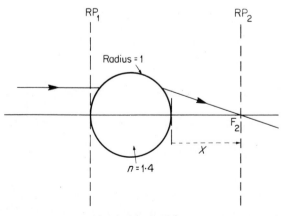

Figure II.12

Solution

If we choose an output reference plane RP_2 X cm to the right of the bead, the matrix chain is

$$M = \begin{bmatrix} 1 & X \\ 0 & 1 \end{bmatrix} \begin{bmatrix} 1 & 0 \\ \dfrac{-(1-1\cdot4)}{-1} & 1 \end{bmatrix} \begin{bmatrix} 1 & \dfrac{2}{1\cdot4} \\ 0 & 1 \end{bmatrix} \begin{bmatrix} 1 & 0 \\ \dfrac{-(1\cdot4-1)}{1} & 1 \end{bmatrix}$$

(air   (right face)   (between   (left face)
gap)   faces)

$$= \begin{bmatrix} 1 & X \\ 0 & 1 \end{bmatrix} \times \begin{bmatrix} 0\cdot429 & 1\cdot429 \\ -0\cdot571 & 0\cdot429 \end{bmatrix}$$

$$= \begin{bmatrix} 0\cdot429 - 0\cdot571X & 1\cdot429 + 0\cdot429X \\ -0\cdot571 & 0\cdot429 \end{bmatrix} = \begin{bmatrix} A & B \\ C & D \end{bmatrix}$$

The condition for parallel light to come to focus in RP_2 is that the element A shall vanish:

$$0\cdot571X = 0\cdot429 \quad \therefore \ X = 0\cdot75$$

Alternatively, this result could be derived directly from the matrix for the bead,

$$\begin{bmatrix} 0\cdot429 & 1\cdot429 \\ -0\cdot571 & 0\cdot429 \end{bmatrix}$$

For, if $V_1 = 0$,

$$\begin{bmatrix} y_2 \\ V_2 \end{bmatrix} = \begin{bmatrix} A & B \\ C & D \end{bmatrix} \begin{bmatrix} y_1 \\ 0 \end{bmatrix} = \begin{bmatrix} Ay_1 \\ Cy_1 \end{bmatrix}$$

Therefore, the distance at which the $\begin{bmatrix} y_2 \\ V_2 \end{bmatrix}$ ray cuts the axis is $- ny_2/V_2 = - A/C$ to the right of the bead

$$= 0\cdot429/0\cdot571 = 0\cdot75.$$

Caution

This calculation refers only to the central paraxial portion of the beam - the outer region will show spherical aberration.

Problem 4

A lantern slide 2 inches tall is located 10·5 feet from a screen (see Figure II.13). What is the focal length of the lens which will project an image measuring 40 inches on the screen, and where must it be placed?

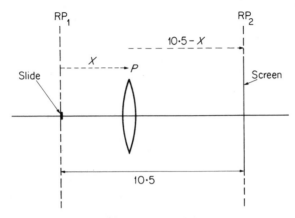

Figure II.13

Solution

Although the object and image sizes are given in inches, it may be more convenient here to use feet for calculating the matrix elements. The lens will be taken to be a thin positive lens whose power P is measured in reciprocal foot units. Let it be placed X feet from the slide and $(10·5 - X)$ feet from the screen. The matrix chain from screen to slide is then

$$M = \begin{bmatrix} 1 & 10·5 - X \\ 0 & 1 \end{bmatrix} \begin{bmatrix} 1 & 0 \\ -P & 1 \end{bmatrix} \begin{bmatrix} 1 & X \\ 0 & 1 \end{bmatrix}$$

(screen to   (lens) (lens to
  lens)   slide)

$$= \begin{bmatrix} 1 & 10·5 - X \\ 0 & 1 \end{bmatrix} \begin{bmatrix} 1 & X \\ -P & 1 - PX \end{bmatrix}$$

Therefore,

$$\begin{bmatrix} A & B \\ C & D \end{bmatrix} = \begin{bmatrix} 1 - 10·5P + PX & X + (10·5 - X)(1 - PX) \\ -P & 1 - PX \end{bmatrix}$$

The real image produced from a real object by a single lens is always inverted, so the magnification specified must be $- 40/2 = - 20$. In the above matrix, therefore, $A = 1/D = - 20$ and $B = 0$, to secure the object-image relationship. Since $D = 1 - PX = - 0 \cdot 05$, the equation for B becomes

$$B = X - 0 \cdot 05 (10 \cdot 5 - X) = 0$$

Therefore,

$1 \cdot 05X = 0 \cdot 05 (10 \cdot 5)$, so that $X = 0 \cdot 5$.

The equation for D now becomes

$1 - 0 \cdot 5P = - 0 \cdot 05$, so that $P = + 2 \cdot 1$.

The focal length of the lens is therefore $(1/2 \cdot 1)$ feet $= 5 \cdot 7$ inches.

Answer
The lens needs to have a positive focal length of $5 \cdot 7$ inches. It should be placed 6 inches from the slide.

Problem 5
A positive (converging) lens of focal length + 8 cm is mounted 6 cm to the left of a negative (diverging) lens of focal length - 12 cm. If an object 3 cm tall is located on the axis 24 cm to the left of the positive lens, find the position and size of the image. (See Figure II.14).

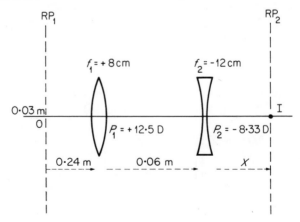

Figure II.14

Solution

We shall work in metres and dioptres. The powers of
the lenses are $+ 100/8 = + 12 \cdot 5$ dioptres for the
positive lens and $- 100/12 = - 8 \cdot 33$ dioptres for the
negative. If the image is located at a distance X
metres to the right of the negative lens, the matrix
chain from image back to object is

$$M = \begin{bmatrix} 1 & X \\ 0 & 1 \end{bmatrix}\begin{bmatrix} 1 & 0 \\ 8 \cdot 33 & 1 \end{bmatrix}\begin{bmatrix} 1 & 0 \cdot 06 \\ 0 & 1 \end{bmatrix}\begin{bmatrix} 1 & 0 \\ -12 \cdot 5 & 1 \end{bmatrix}\begin{bmatrix} 1 & 0 \cdot 24 \\ 0 & 1 \end{bmatrix}$$

   (image to   (negative   (lens   (positive (positive
   negative   lens)   separation)   lens)   lens to
   lens)   object)

$$= \begin{bmatrix} 1 & X \\ 0 & 1 \end{bmatrix} \times \begin{bmatrix} 1 & 0 \cdot 06 \\ 8 \cdot 33 & 1 \cdot 5 \end{bmatrix} \times \begin{bmatrix} 1 & 0 \cdot 24 \\ -12 \cdot 5 & -2 \end{bmatrix}$$

Therefore,

$$\begin{bmatrix} A & B \\ C & D \end{bmatrix} = \begin{bmatrix} 1 & X \\ 0 & 1 \end{bmatrix} \times \begin{bmatrix} 0 \cdot 25 & 0 \cdot 12 \\ -10 \cdot 42 & -1 \end{bmatrix}$$

$$= \begin{bmatrix} 0 \cdot 25 - 10 \cdot 42X & 0 \cdot 12 - X \\ -10 \cdot 42 & -1 \end{bmatrix}$$

For the object-image relationship to hold, we must have
$B = 0 \cdot 12 - X = 0$, so $X = 0 \cdot 12$ metres. The magnification is $1/D = - 1$, so the final image is 3 cm tall, and
inverted. It is located 12 cm to the right of the
negative lens.

Problem 6

An object is located at a distance U to the left of a
thin lens and its image is formed a distance V to the
left of the same lens (see Figure II.15). If now the
object is displaced axially a small distance dU to the
left, find an expression for the corresponding displacement dV of the image; dV/dU is called the longitudinal magnification. Show that it is the square of
the lateral magnification.

50

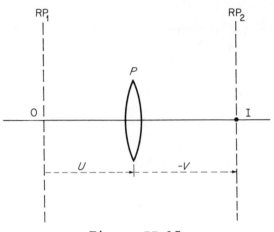

Figure II.15

Solution

(Note that if the image formed is a real one, V will
have a negative value; but, whether the thin lens is
positive or negative, there will be some object distances for which the image is virtual, in which case V
is positive. Note also that, in this and the next
problem, V is used to denote an image distance, and
not an optical direction-cosine.)

The matrix chain from image back to object is

$$M = \begin{bmatrix} 1 & -V \\ 0 & 1 \end{bmatrix}\begin{bmatrix} 1 & 0 \\ -P & 1 \end{bmatrix}\begin{bmatrix} 1 & U \\ 0 & 1 \end{bmatrix} = \begin{bmatrix} 1 & -V \\ 0 & 1 \end{bmatrix}\begin{bmatrix} 1 & U \\ -P & 1-PU \end{bmatrix}$$

   (image to   (lens)   (lens to
   lens with V object)
   negative as
   measured in
  - z-direction)

$$= \begin{bmatrix} 1+PV & U-V+PUV \\ -P & 1-PU \end{bmatrix}$$

For the object-image relationship, the top right-hand
element ($U - V + PUV$) must vanish. Hence

$$V = \frac{U}{1 - PU}$$

and, differentiating,

$$\frac{dV}{dU} = \frac{(1 - PU)(1) + UP}{(1 - PU)^2}$$

$$= \frac{1}{(1 - PU)^2}$$

As the bottom right-hand element indicates, one expression for the lateral magnification is $1/(1 - PU)$, whose square agrees with our expression for dV/dU. Other expressions that can be found for the lateral magnification are $(1 + PV)$ and V/U. (It will be noted that, in obtaining this solution, we have determined a basic relationship between P, U and V for a thin lens. Since $(U - V + PUV) = 0$, we can divide by UV and rearrange to obtain the familiar lens formula

$$\frac{1}{U} - \frac{1}{V} = P = \frac{1}{f}$$

where f is the focal length of the lens.)

Problem 7
Prove that the distance between the real object and real image formed by a thin positive lens cannot be less than four times the focal length. (See Figure II.16).

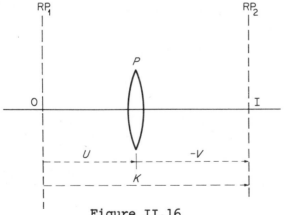

Figure II.16

Solution
As in Problem 6, the top right-hand element in the image-to-object matrix is $(U - V + PUV)$, and this must vanish. Furthermore, since U and V are both measured to the left of the lens, U must be positive, but V

must be negative for a real image to be formed from a real object. If the object-image distance is K, then $K = U - V$ and $V = U - K$. Substituting for V, therefore, we have

$$U - (U - K) + PU(U - K) = 0$$

whence

$$K = \frac{U}{PU - 1} + U$$

Therefore,

$$\frac{dK}{dU} = \frac{(PU - 1).1 - UP}{(PU - 1)^2} + 1 = 1 - \frac{1}{(PU - 1)^2}$$

For dK/dU to vanish, we must have $PU - 1 = 1$, whence $PU = 2$ or $PU - 1 = -1$, whence $PU = 0$ (a trivial case). Since both P and U are assumed to be positive, the case $PU = 2$ needs to be considered further by considering the second derivative.

$$\frac{d^2K}{dU^2} = \frac{d}{dU}\left[\frac{-1}{(PU - 1)^2}\right] = \frac{-(-2)P}{(PU - 1)^3} = \frac{2P}{(PU - 1)^3}$$

Now, for the case $PU = 2$, the denominator is unity and the numerator is positive; so d^2K/dU^2 is positive and the value of K must be passing through a minimum. If we now substitute $U = 2/P$ in the expression for K, we find

$$K_{min} = \frac{2/P}{1} + 2/P = 4/P = 4f$$

where $f = 1/P$ again represents the focal length of the lens.

II.8 EXPERIMENTAL DETERMINATION OF THE MATRIX ELEMENTS OF AN OPTICAL SYSTEM

We shall assume initially that the system with which we are dealing is a positive lens system located in air. The first step is to choose two conveniently accessible planes near the first and last surfaces of the system: we shall regard these as our input reference plane RP_1 and our output reference plane RP_2. We require to determine elements A, B, C and D such that

$$\begin{bmatrix} y_2 \\ V_2 \end{bmatrix} = \begin{bmatrix} A & B \\ C & D \end{bmatrix} \begin{bmatrix} y_1 \\ V_1 \end{bmatrix}$$

The procedure that we shall follow is to locate an object of known size at various distances to the left of RP_1 and then to measure the size and position of each of the resultant real images.

In the general case, let R represent the displacement in the $+z$-direction from the object to RP_1, and let S represent the displacement, again in the $+z$-direction, from RP_2 to the real image. We shall operate so that R and S remain positive.

The ray transfer matrix chain from the plane containing the image back to the plane containing the object is then

$$M = \begin{bmatrix} 1 & S \\ 0 & 1 \end{bmatrix} \begin{bmatrix} A & B \\ C & D \end{bmatrix} \begin{bmatrix} 1 & R \\ 0 & 1 \end{bmatrix} = \begin{bmatrix} 1 & S \\ 0 & 1 \end{bmatrix} \begin{bmatrix} A & AR+B \\ C & CR+D \end{bmatrix}$$

$$= \begin{bmatrix} A+SC & AR+B+S(CR+D) \\ C & CR+D \end{bmatrix}$$

For this matrix, we know that its determinant is unity and that its upper right-hand element vanishes because of the object-image relationship. The upper left-hand element $(A + SC)$ represents the transverse magnification and the lower right-hand element $(CR + D)$ represents its reciprocal, which we shall denote by α.

The quantities that we can measure experimentally are the distances R and S, and the ratio (height of object)/(height of image) $= \alpha = CR + D$. (Note that if the image is inverted, as it often is in practice, then α is negative.)

If α is measured for several values of the object distance R and the results plotted as a graph of α against R, this graph will be a straight line whose slope-tangent is C and whose intercept on the α-axis is D. Thus, C and D are determined.

From the vanishing of the upper right-hand element, we have $AR + B = -S(CR + D) = -S\alpha = \beta$, say, where β is known. Once again, if β is plotted against R, the slope of the graph will be A and the intercept on the β-axis will be B. Once all four elements have been determined, a check should be made that the determinant $(AD - BC)$ is approximately unity.

II.8.1 Negative lens systems

If the system to be measured is a divergent, negative
one, it will be necessary to use an auxiliary lens to
project real images of known size and location to the
right of RP_1. If a suitable range of negative R-values
is chosen, it may still be possible to produce real
images that can be located and measured to the right
of RP_2. The procedure will be considerably less convenient, however.

II.8.2 Immersed systems

If the optical system is immersed, so that n_1 represents the refractive index of the medium to the left of
RP_1 and n_2 the index of the medium to the right of RP_2,
then the experimental procedure required will undoubt-
edly be difficult. As far as the analysis is concerned,
however, all that is necessary is to replace the distances R and S wherever they appear (for example in the
translation matrices) by their reduced values R/n_1 and
S/n_2. When the new image-object matrix is calculated,
with this substitution, its lower right-hand element
will still correspond to the measured value of α.

II.9 LOCATING THE CARDINAL POINTS OF A SYSTEM

Let us assume that, either by calculation or by the
experimental method just described, we have determined,
for a given system, the matrix $M = \begin{bmatrix} A & B \\ C & D \end{bmatrix}$ which links
a chosen output plane RP_2 to a chosen input plane RP_1.
We now seek to locate, with respect to these reference
planes, the two focal points, the principal planes of
unit transverse magnification and the nodal planes of
unit angular magnification. For the sake of generality,
we shall assume that n_1 and n_2 respectively are the
refractive indices to the left and to the right of the
system.

(a) Consider first, as in Figure II.17a, a ray which
enters the system parallel to the axis at the height
y_1. According to the usual ideas of Gaussian optics,
it arrives with the same height at the second principal
plane H_2, and is there effectively bent in such a way
that it emerges through the second focal point F_2. For
this ray, v_1 and therefore V_1 both vanish, so that at
RP_2 our ray-transfer matrix indicates that $y_2 = Ay_1$ and

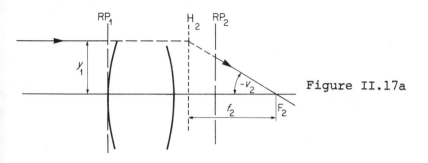

Figure II.17a

$v_2 = (V_2/n_2) = (Cy_1/n_2)$.

If t_2 is the *displacement* measured in the + z-direction from RP$_2$ to F$_2$, then we must have $t_2 = - y_2/v_2 = - n_2A/C$. This locates the second focus.

When the ray that we have been considering emerged from the second principal plane H$_2$ its y-coordinate was y_1. Therefore, if we define the second focal length f_2 as the *displacement* from H$_2$ to F$_2$, its value must be $f_2 = - y_1/v_2$. Therefore $f_2 = - n_2y_1/Cy_1 = - n_2/C$, and the second focal length is determined.

The displacement from RP$_2$ to the second principal plane is therefore $s_2 = t_2 - f_2 = n_2(1 - A)/C$.

(b) Consider now the ray which enters the system at the angle v_1 after passing through the first focus F$_1$ (see Figure II.17b). It is effectively bent parallel to the axis at the first principal plane H$_1$ and continues through RP$_2$ with v_2, and therefore V_2, equal to zero.

We can therefore write $V_2 = Cy_1 + Dn_1v_1 = O$ and $y_1 = - Dn_1v_1/C$. From the diagram, the *displacement* t_1 of F$_1$ from RP$_1$ is given by $t_1 = - y_1/v_1 = n_1D/C$. This locates the first focus.

Still considering the same ray, we recall that at

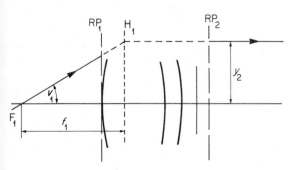

Figure II.17b

56

the first principal plane its y-coordinate must be $y_2 = Ay_1 + Bn_1v_1$. Therefore, if the first focal length f_1 is defined as the *displacement* from F_1 to H_1, we must have $f_1 = y_2/v_1 = - DAn_1/C + Bn_1 = - n_1(AD - BC)/C$. Remembering that $(AD - BC) = 1$, we obtain finally $f_1 = - n_1/C$ for the first focal length.

The displacement s_1 of the first principal plane H_1 from RP_1 is then $s_1 = t_1 + f_1 = n_1(D - 1)/C$.

(c) Finally, we wish to determine two nodal points L_1 and L_2 such that any ray entering the system directed towards L_1 appears on emergence as a ray coming from L_2 and making the same angle $v_2 = v_1$ with the axis

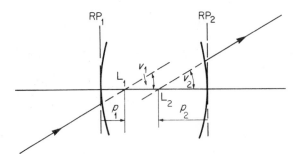

Figure II.17c

(see Figure II.17c). Let the *displacements* of L_1 and L_2 from RP_1 and RP_2 respectively be denoted by p_1 and p_2 respectively. The matrix chain linking the second nodal point L_2 back to the first nodal point L_1 will then be:

$$\begin{bmatrix} 1 & p_2/n_2 \\ 0 & 1 \end{bmatrix} \begin{bmatrix} A & B \\ C & D \end{bmatrix} \begin{bmatrix} 1 & -p_1/n_1 \\ 0 & 1 \end{bmatrix}$$

$$(\text{L}_2 \text{ to} \qquad (\text{RP}_2 \text{ to} \qquad (\text{RP}_1 \text{ to}$$
$$\text{RP}_2) \qquad \text{RP}_1) \qquad \text{L}_1)$$

$$= \begin{bmatrix} 1 & p_2/n_2 \\ 0 & 1 \end{bmatrix} \begin{bmatrix} A & -(Ap_1/n_1) + B \\ C & -(Cp_1/n_1) + D \end{bmatrix}$$

$$= \begin{bmatrix} A + (p_2C/n_2) & -(Ap_1/n_1) + B + p_2\left[-(Cp_1/n_1) + D\right]/n_2 \\ C & -(Cp_1/n_1) + D \end{bmatrix}$$

(Note that reduced displacements have been used and that $-p_1$ rather than p_1 is used for the $\mathcal{T}$-matrix from RP_1 to L_1.)

Let us denote the matrix just obtained by ϕ and use y_0, v_0 to denote the coordinates of a ray entering the L_1 plane and y_3, v_3 to denote the coordinates of a ray leaving the L_2 plane. We then have

$$y_3 = \phi_{11}y_0 + \phi_{12}V_0 = \phi_{11}y_0 + \phi_{12}(n_1v_0)$$
and
$$n_2v_3 = V_3 = \phi_{21}y_0 + \phi_{22}V_0 = \phi_{21}y_0 + \phi_{22}(n_1v_0)$$

But if L_1 and L_2 are nodal points and $y_0 = 0$, then whatever the value of v_0 we must have $y_3 = 0$ and $v_3 = v_0$. This will be true only if $\phi_{12} = 0$ and $\phi_{22}(n_1/n_2) = 1$, that is if the matrix ϕ is an object-image matrix with a transverse (linear) magnification $1/\phi_{22} = (n_1/n_2)$.

From the equation $\phi_{22} = (n_2/n_1)$ we obtain $p_1 = (Dn_1 - n_2)/C$. Substituting for p_1 in the equation $\phi_{12} = 0$, we obtain eventually $p_2 = (n_1 - An_2)/C$.

For convenience we now recapitulate these results in tabular form.

System parameter described	Measured From	To	Function of matrix elements	Special case $n_1 = n_2 = 1$
First focal point	RP_1	F_1	$n_1 D/C$	D/C
First focal length	F_1	H_1	$-n_1/C$	$-1/C$
First principal point	RP_1	H_1	$n_1(D-1)/C$	$(D-1)/C$
First nodal point	RP_1	L_1	$(Dn_1 - n_2)/C$	$(D-1)/C$
Second focal point	RP_2	F_2	$-n_2A/C$	$-A/C$
Second focal length	H_2	F_2	$-n_2/C$	$-1/C$
Second principal point	RP_2	H_2	$n_2(1-A)/C$	$(1-A)/C$
Second nodal point	RP_2	L_2	$(n_1 - An_2)/C$	$(1-A)/C$

The right-hand column of this table shows that, for the most frequently encountered case of an optical system located in the air, the nodal points coincide with the principal points; this is because the conditions for unit angular and unit linear magnification are then the same. Secondly, both focal lengths are given by $-1/C$. (Where the refractive index differs from unity, it is the 'reduced focal length' that is given by $-1/C$.)

To complete our elementary treatment of paraxial refracting systems, let us see what happens to an $\begin{bmatrix} A & B \\ C & D \end{bmatrix}$ matrix if we transform it so that, instead of converting ray data from RP_1 to RP_2, it operates (a) between the two principal planes or (b) between the two focal planes.

(a) For operation between the two principal planes, the new matrix will be

$$\begin{bmatrix} 1 & \dfrac{1-A}{C} \\ 0 & 1 \end{bmatrix} \begin{bmatrix} A & B \\ C & D \end{bmatrix} \begin{bmatrix} 1 & \dfrac{1-D}{C} \\ 0 & 1 \end{bmatrix}$$

$(H_2$ to $RP_2)$ $(RP_2$ to $RP_1)$ $(RP_1$ to H_1 – note sign reversal)

$$= \begin{bmatrix} 1 & \dfrac{1-A}{C} \\ 0 & 1 \end{bmatrix} \begin{bmatrix} A & \dfrac{A(1-D)}{C} + B \\ C & \dfrac{C(1-D)}{C} + D \end{bmatrix}$$

$$= \begin{bmatrix} 1 & \dfrac{(1-A)}{C} \\ 0 & 1 \end{bmatrix} \begin{bmatrix} A & \dfrac{(A-1)}{C} \\ C & 1 \end{bmatrix} \qquad \text{(since } AD - BC = 1\text{)}$$

$$= \begin{bmatrix} 1 & 0 \\ C & 1 \end{bmatrix}$$

Between these two planes, we have a refraction matrix which is the same as for a thin lens of power $P = -C = 1/f$. As we should expect, there is an object-image relationship between RP_1 and RP_2, with transverse magnification unity.

(b) For operation between the two focal planes, the new matrix will be

$$\begin{bmatrix} 1 & -A/C \\ 0 & 1 \end{bmatrix} \begin{bmatrix} A & B \\ C & D \end{bmatrix} \begin{bmatrix} 1 & -D/C \\ 0 & 1 \end{bmatrix}$$

(F_2 to RP_2) (RP_2 to RP_1) (RP_1 to F_1 — note sign reversal)

$$= \begin{bmatrix} 1 & -A/C \\ 0 & 1 \end{bmatrix} \begin{bmatrix} A & -\dfrac{AD}{C} + B \\ C & -\dfrac{CD}{C} + D \end{bmatrix}$$

$$= \begin{bmatrix} 1 & -A/C \\ 0 & 1 \end{bmatrix} \begin{bmatrix} A & -1/C \\ C & 0 \end{bmatrix}$$

$$= \begin{bmatrix} 0 & -1/C \\ C & 0 \end{bmatrix}$$

$$= \begin{bmatrix} 0 & f \\ -1/f & 0 \end{bmatrix}$$

This matrix expresses the well-known result that the ray height in the second focal plane depends only on the ray angle in the first, while the ray angle in the second focal plane depends only on the ray height in the first. Furthermore, if we use y_1/V_1 to represent the focal distance z_1 of an object point (measured to the left from F_1) and y_2/V_2 to represent the focal distance z_2 of a corresponding image point, we find immediately that

$$z_1 z_2 = (y_1/V_1)(y_2/V_2) = -f^2 \qquad \text{(Newton's equation)}$$

Again as we should expect, the vanishing of the principal elements of the matrix indicates that RP_1 and RP_2 are located at the focal planes. Given that the equivalent focal length is unchanged by this transformation, the element C remains the same and the upper right-hand element has to equal $-1/C$ for the matrix to remain unimodular.

II.10 FURTHER PROBLEMS

Problem 8

A glass hemisphere of radius r and refractive index n
is used as a lens, rays through it being limited to
those nearly coinciding with the axis. Show that, if
the plane surface faces to the left, the second principal point coincides with the intersection of the
convex surface with the axis (its vertex). Show also
that the first principal point is inside the lens at a
distance r/n from the plane surface and that the focal
length of the lens is equal to $r/(n - 1)$. (See Figure
II.18.)

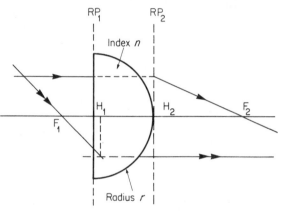

Figure II.18

Solution

If we locate our reference planes at the two surfaces
of the hemisphere, the matrix chain is

$$\begin{bmatrix} 1 & 0 \\ \dfrac{-(1-n)}{-r} & 1 \end{bmatrix} \begin{bmatrix} 1 & r/n \\ 0 & 1 \end{bmatrix} \begin{bmatrix} 1 & 0 \\ 0 & 1 \end{bmatrix} = \begin{bmatrix} 1 & r/n \\ \dfrac{-(n-1)}{r} & \dfrac{1}{n} \end{bmatrix}$$

(refraction at curved surface RP₂)	(reduced thickness of lens)	(a trivial case of refraction by plane surface at RP₁)

(Although this is a very simple calculation, it is
still wise to check the determinant.) The required
results can now be found in the right-hand column of
the results tabulated in the last section. The dis-

tance from RP_2 to the second principal point is $(1 - A)/C$, which vanishes. The distance from RP_1 to the first principal point is

$$(D - 1)/C = (1/n - 1) \times r/-(n - 1) = r/n.$$

Since $n > 1$, this distance is less than the thickness r of the hemispherical lens.

Finally, the focal length measured in air in either direction is $- 1/C = r/(n - 1)$.

Problem 9
A thin lens of positive focal length 10 cm is separated by 5 cm from a thin negative lens of focal length - 10 cm (see Figure II.19). Find the equivalent focal length of the combination and the position of the foci and principal planes.

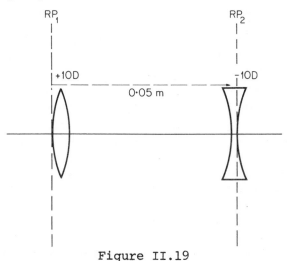

Figure II.19

Solution
We will assume that the positive lens is on the left of the system and our reference planes will be located at the lens positions. Converting distances to metres and focal lengths to dioptres, we obtain the system matrix

$$M = \begin{bmatrix} 1 & 0 \\ 10 & 1 \end{bmatrix} \begin{bmatrix} 1 & 0{\cdot}05 \\ 0 & 1 \end{bmatrix} \begin{bmatrix} 1 & 0 \\ -10 & 1 \end{bmatrix}$$

$\quad\quad$ (negative $\quad\quad\quad$ (lens $\quad\quad\quad$ (positive
$\quad\quad\quad$ lens at $\quad\quad$ separation) $\quad\quad$ lens at
$\quad\quad\quad\quad$ RP_2) $\quad\quad\quad\quad\quad\quad\quad\quad\quad$ RP_1)

$$= \begin{bmatrix} 1 & 0 \cdot 05 \\ 10 & 1 \cdot 5 \end{bmatrix} \begin{bmatrix} 1 & 0 \\ -10 & 1 \end{bmatrix}$$

Therefore

$$\begin{bmatrix} A & B \\ C & D \end{bmatrix} = \begin{bmatrix} 0 \cdot 5 & 0 \cdot 05 \\ -5 & 1 \cdot 5 \end{bmatrix}$$

Checking the determinant, we find $(AD - BC) =$ $(0 \cdot 5)(1 \cdot 5) + (5)(0 \cdot 05) = 1$. Referring again to the tabulated results of section II.9, we find that the focal length of the combination must be $- 1/C = + 0 \cdot 2$ metres $= 20$ cm (focal power $= + 5$ dioptres).

The first focal point is located a distance D/C $= - 0 \cdot 3$ metres to the right of RP_1, that is 30 cm to the left of the positive lens. Since the focal length is 20 cm, we can see immediately that the first principal plane will be 10 cm to the left of the positive lens. (According to the formula, we obtain $(D - 1)/C = 0 \cdot 5/-5$ for the distance to the right of RP_1 in metres.)

For the second focal point, its distance to the right of RP_2 - the negative lens - will be $- A/C$ $= - 0 \cdot 5/-5 = 0 \cdot 1$ metres $= 10$ cm. Either by direct reasoning, or by using the formula $(1 - A)/C$, the second principal plane is now found to be 10 cm to the left of the negative lens, that is 5 cm to the left of the positive lens.

The system used this way round is, in effect, a crude form of telephoto lens, for which the focal length is longer than the tube required to house the lens and its focused image plane.

Problem 10

Two positive thin lenses of powers P_1 and P_2 are used with a lens separation t to form an eyepiece for a reflecting telescope (see Figure II.20). Show that the equivalent focal length of the eyepiece is $(P_1 + P_2 - P_1P_2t)^{-1}$. If both lenses are made of the same glass, how can transverse chromatic aberration be avoided? Will such an eyepiece be free from longitudinal colour error?

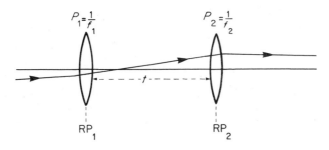

Figure II.20

Solution

As in the previous problem, we choose the two lenses as our reference planes, and obtain the system matrix

$$\begin{bmatrix} 1 & 0 \\ -P_2 & 1 \end{bmatrix} \begin{bmatrix} 1 & t \\ 0 & 1 \end{bmatrix} \begin{bmatrix} 1 & 0 \\ -P_1 & 1 \end{bmatrix}$$

$$= \begin{bmatrix} 1 & t \\ -P_2 & 1 - P_2 t \end{bmatrix} \begin{bmatrix} 1 & 0 \\ -P_1 & 1 \end{bmatrix}$$

$$= \begin{bmatrix} 1 - P_1 t & t \\ -P_2 - P_1 + P_1 P_2 t & 1 - P_2 t \end{bmatrix}$$

From the lower left-hand element, the equivalent focal length is evidently given by

$$1/f = (P_1 + P_2 - P_1 P_2 t)$$

Since the eyepiece is to be used with an achromatic reflecting objective, the angular magnification of an astronomical object observed with the resultant telescope will remain independent of wavelength only if the equivalent focal length of the eyepiece also remains unaffected by the small changes in refractive index that any glass exhibits as the wavelength varies.

Now, as was shown in section II.4, the power P of any thin lens is the product of $(n - 1)$ and a purely geometrical factor G representing the difference between the two surface curvatures. We can therefore write for the combination:

$$1/f = (n - 1)(G_1 + G_2) - (n - 1)^2 G_1 G_2 t$$

Differentiating, we find

$$\frac{d}{dn}(1/f) = G_1 + G_2 - 2(n - 1)G_1G_2t$$

For this derivative to vanish, we must have

$$t = \frac{G_1 + G_2}{2(n - 1)G_1G_2} = \frac{1}{2}\left[\frac{1}{(n - 1)G_2} + \frac{1}{(n - 1)G_1}\right]$$

$$= \frac{1}{2}\left[\frac{1}{P_2} + \frac{1}{P_1}\right] = (f_1 + f_2)/2$$

This condition is, of course, the basis of the classical Ramsden and Huygens eyepieces.

Unfortunately, although the condition just obtained avoids transverse colour error, or chromatic difference of magnification, it does not ensure that the eyepiece is free from longitudinal error. As indicated in the table of section II.9, the first focal point of this system will be located a distance D/C to the left of RP_1. Although we have ensured that the denominator C in this expression remains independent of small changes in n, this is clearly not true for the numerator $D = 1 - P_2t = 1 - (n - 1)G_2t$. In observing a star image, therefore, the best focus setting for this eyepiece will vary slightly with the wavelength.

(Apart from these questions of colour error that can be treated by paraxial methods, it should be remembered that, even for small angular fields, the spherical aberration of such a simple eyepiece may be objectionable for use with a Newtonian reflecting telescope of short focal ratio.)

Problem 11
A glass lens 3 cm thick along the axis has one face of radius 5 cm and convex outward, while the other face is of radius 2 cm and concave outward. The former face is on the left and in contact with the air and the latter is in conect with a liquid of refractive index 1·4. If the refractive index of the glass is 1·5, find the positions of the foci, the principal points and the nodal points, and calculate the focal lengths of the system. (See Figure II.21.)

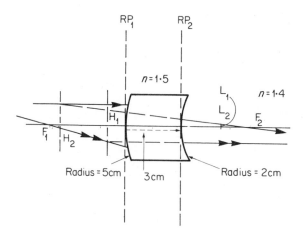

Figure II.21

Solution

Once again we obtain the matrix for the system and
then use the results tabulated in section II.9, remembering this time to use the central column. If we
choose reference planes located at the two surfaces of
the lens and work in cm, the matrix is

$$M = \begin{bmatrix} 1 & 0 \\ \dfrac{-(1\cdot 4 - 1\cdot 5)}{+\,2} & 1 \end{bmatrix} \begin{bmatrix} 1 & \dfrac{3}{1\cdot 5} \\ 0 & 1 \end{bmatrix} \begin{bmatrix} 1 & 0 \\ \dfrac{-(1\cdot 5 - 1)}{+\,5} & 1 \end{bmatrix}$$

(immersed concave   (lens   (convex lens
lens surface)   thickness)   surface)

$$= \begin{bmatrix} 1 & 0 \\ 0\cdot 05 & 1 \end{bmatrix} \begin{bmatrix} 1 & 2 \\ 0 & 1 \end{bmatrix} \begin{bmatrix} 1 & 0 \\ -0\cdot 1 & 1 \end{bmatrix}$$

$$= \begin{bmatrix} 1 & 2 \\ 0\cdot 05 & 1\cdot 1 \end{bmatrix} \begin{bmatrix} 1 & 0 \\ -0\cdot 1 & 1 \end{bmatrix}$$

Therefore

$$M = \begin{bmatrix} A & B \\ C & D \end{bmatrix} = \begin{bmatrix} 0\cdot 8 & 2 \\ -0\cdot 06 & 1\cdot 1 \end{bmatrix}$$

Check

$$\det(M) = (0 \cdot 8 \times 1 \cdot 1) + (2 \times 0 \cdot 06) = 1$$

Remembering that in this problem $n_1 = 1$ but $n_2 = 1 \cdot 4$, we now obtain from the table in section II.9 the following solutions.

Input side of system : $n_1 = 1$

Cardinal point	Location (measured in cm to the right of RP_1)
First focus F_1	$D/C = 1 \cdot 1/-0 \cdot 06 = -18 \cdot 3$
First principal point H_1	$\dfrac{(D-1)}{C} = 0 \cdot 1/-0 \cdot 06 = -1 \cdot 67$
First nodal point L_1	$\dfrac{Dn_1 - n_2}{C} = \dfrac{1 \cdot 1 - 1 \cdot 4}{-0 \cdot 06} = +5 \cdot 0$

The first focal length $f_1 = -1/C = -1/(-0 \cdot 06)$

$$= +16 \cdot 7 \text{ cm } (+ 6 \text{ dioptres})$$

Output side of system : $n_2 = 1 \cdot 4$

Cardinal point	Location (measured in cm to the right of RP_2)
Second focus F_2	$-n_2A/C = \dfrac{-1 \cdot 4 \times 0 \cdot 8}{-0 \cdot 06} = +18 \cdot 7$
Second principal point H_2	$\dfrac{n_2(1-A)}{C} = \dfrac{1 \cdot 4 \times (1 - 0 \cdot 8)}{-0 \cdot 06}$ $= -4 \cdot 67$
Second nodal point L_2	$\dfrac{n_1 - An_2}{C} = \dfrac{1 - (1 \cdot 4 \times 0 \cdot 8)}{-0 \cdot 06}$ $= +2 \cdot 0$

The second focal length $f_2 = -n_2/C = -1 \cdot 4/-0 \cdot 06$

$$= +23 \cdot 3 \text{ cm}$$

For this monocentric optical system the nodal points L_1 and L_2 coincide, being located at the common centre of curvature of the two spherical surfaces.

II.11 EXTENSION OF RAY-TRANSFER METHOD TO REFLECTING SYSTEMS

Partly in order to simplify the presentation, our discussion of paraxial systems has so far confined itself to through-type optical systems that contain only refracting surfaces. Where performance must be achieved regardless of cost, however, there is increasing use of catadioptric systems, and we shall certainly need in the next chapter to consider the reflecting surfaces of optical resonators. At this stage, therefore, we extend our treatment to include reflecting surfaces.

There is a single basic rule that will enable us to achieve this while still using the ray vectors and the $\mathcal{T}$- and $\mathcal{R}$-matrices already described: *Whenever a light ray is travelling in the − z-direction, the refractive index of the medium through which it is passing must be taken as negative.* Let us now consider the effect of this rule on each of the three components of our ray-transfer matrix formalism.

II.11.1 Specification of ray vectors

When interpreting $\begin{bmatrix} y \\ V \end{bmatrix}$ ray data, we must remember that it is not V but V/n which specifies the geometric angle v between the ray direction and the optical axis. For example, when a ray is reflected from a plane surface normal to the axis (see Figure II.22), its ray direction v will be reversed in sign, but so will the value of n for the returning ray, so that its optical direction-cosine V will remain unchanged. This means that the law of reflection, like the paraxial form of Snell's law of refraction, is already contained in the statement that across any plane mirror or any plane refracting surface $\begin{bmatrix} y_2 \\ V_2 \end{bmatrix} = \begin{bmatrix} y_1 \\ V_1 \end{bmatrix}$. We shall return briefly to this point when considering $\mathcal{R}$-matrices.

II.11.2 𝒯-matrices

When light rays are travelling across a gap from the plane $z = z_1$ to the plane $z = z_2$, the gap value $(z_2 - z_1)$ will be positive for a ray travelling in the + z-direction but negative for a ray travelling in the

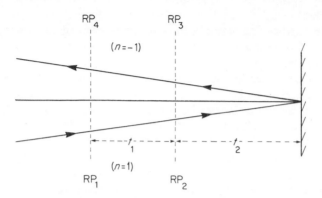

(For rays travelling in the $-z$-direction, the refractive index changes its sign.)
For transfer between planes,

$$M_{21} = \begin{bmatrix} 1 & t_1 \\ 0 & 1 \end{bmatrix}$$

$$M_{32} = \begin{bmatrix} 1 & 2t_2 \\ 0 & 1 \end{bmatrix}$$

$$M_{43} = \begin{bmatrix} 1 & (-t_1/-1) \\ 0 & 1 \end{bmatrix} = \begin{bmatrix} 1 & t_1 \\ 0 & 1 \end{bmatrix}$$

Figure II.22

$-z$-direction. On the other hand, in the latter case, the index of the medium must be taken as negative, so that the *reduced* value of the gap will remain positive, being the ratio of two negative quantities. Provided therefore that a matrix chain is being calculated in the usual way from output back to input, the T-values representing reduced gaps in the various $\mathscr{T}$-matrices will all appear positive.

II.11.3 $\mathscr{R}$-matrices

We shall now use the term $\mathscr{R}$-matrix to describe the effect of a reflecting as well as a refracting surface. To calculate the power P of a reflecting curved surface, we modify the formula $P = (n_2 - n_1)/r$, replacing the refractive index of the second medium n_2 by the negative of the refractive index n for the medium in which the reflector is immersed and in which the ray continues to travel after its reflection. The power P is thus $- 2n/r$ and the $\mathscr{R}$-matrix is $\begin{bmatrix} 1 & 0 \\ 2n/r & 1 \end{bmatrix}$.

It is important to remember here that the radius of curvature of a mirror, like that of a lens surface, is reckoned to be positive if the centre of curvature is located to the right of its surface. This will be the case, for example, if a convex mirror is arranged to face to the left so as to reflect rays back into the − z-direction. As in the case of a diverging lens, the power of such a convex (diverging) mirror will be reckoned to be negative, so the lower left-hand element of the $\mathscr{R}$-matrix will be positive.

Note that if the same mirror is turned round to face the other way, both the curvature radius r and the index n will need to be sign-changed, so the matrix itself remains unchanged. The power P of a diverging (convex) mirror always appears negative, while that of a concave (converging) mirror always appears as a positive quantity.

II.11.4 *Succession of plane surfaces*

It will now be clear that, as was the case for refraction, reflection by a plane surface for which r is infinite can be represented in trivial fashion by a unit matrix. Therefore, if light is reflected backwards and forwards by a succession of plane surfaces, all perpendicular, we are entitled as before to telescope together the various $\mathscr{T}$-matrices, summing together all the reduced thicknesses that are involved. If the matrix chain is being calculated in the usual way, from output back to input, all these reduced thicknesses make a contribution that is positive. If, for example, light emerges from a plane-parallel glass plate of thickness t and index n after suffering $2k$ internal reflections, we can represent the effect of the plate by combining the original transmission $\mathscr{T}$-matrix

$$\begin{bmatrix} 1 & t/n \\ 0 & 1 \end{bmatrix}$$ with a $2k$-fold 'reflection' $\mathscr{T}$-matrix

$$\begin{bmatrix} 1 & 2kt/n \\ 0 & 1 \end{bmatrix}.$$ (See Figure II.23.) Apart from the attenuation produced by each reflection, the geometry of the emerging rays will be the same as if the light had been transmitted once through a single plate $(2k + 1)$ times as thick.

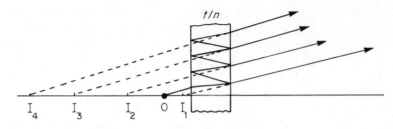

t/n

$I_4 \quad I_3 \quad I_2 \quad O \quad I_1$

Virtual images of O are at I_1, I_2, I_3 and I_4, because of internal reflection.

Figure II.23

To conclude this chapter, we shall now calculate an overall matrix for a simple catadioptric system.

Problem 12

(a) Light enters from the left a solid glass sphere of radius r and refractive index n (see Figure II.24). When it reaches the right-hand surface, some of it is reflected back and emerges again on the left. Taking reference planes at the left-hand surface, calculate a ray-transfer matrix for this sequence.

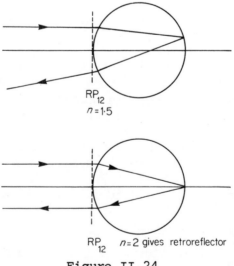

RP_{12}
$n = 1.5$

$RP_{12} \quad n = 2$ gives retroreflector

Figure II.24

(b) Transform this matrix to refer to reference planes passing through the centre of the sphere. Interpret your result and discuss particularly the case $n = 2$.

Solution (a)

With respect to reference planes taken on the left-hand surface, the matrix chain is

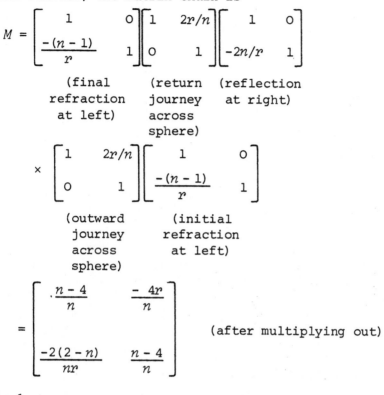

$$M = \begin{bmatrix} 1 & 0 \\ \dfrac{-(n-1)}{r} & 1 \end{bmatrix} \begin{bmatrix} 1 & 2r/n \\ 0 & 1 \end{bmatrix} \begin{bmatrix} 1 & 0 \\ -2n/r & 1 \end{bmatrix}$$

       (final   (return   (reflection
      refraction   journey   at right)
      at left)   across
                  sphere)

$$\times \begin{bmatrix} 1 & 2r/n \\ 0 & 1 \end{bmatrix} \begin{bmatrix} 1 & 0 \\ \dfrac{-(n-1)}{r} & 1 \end{bmatrix}$$

      (outward   (initial
      journey   refraction
      across   at left)
      sphere)

$$= \begin{bmatrix} \dfrac{n-4}{n} & \dfrac{-4r}{n} \\ \dfrac{-2(2-n)}{nr} & \dfrac{n-4}{n} \end{bmatrix} \qquad \text{(after multiplying out)}$$

Check

$$\det(M) = \frac{n^2 - 8n + 16}{n^2} - \frac{16 - 8n}{n^2} = 1$$

Solution (b)

If we wished to convert this matrix to refer to reference planes located at a distance t to the left of the left-hand surface of the sphere, the effect of the lengthened journey between reference planes would be represented by adding the same $\mathcal{T}$-matrix $\begin{bmatrix} 1 & t \\ 0 & 1 \end{bmatrix}$ at either end of the matrix chain. If our reference planes are to be moved to the right, however, the notional journey for the light rays is *shortened*. To bring the reference planes to the centre of the sphere we must both premultiply and postmultiply the matrix M

by $\begin{bmatrix} 1 & -r \\ 0 & 1 \end{bmatrix}$.

With respect to the centre of the sphere, therefore, our new matrix is

$$M_1 = \begin{bmatrix} 1 & -r \\ 0 & 1 \end{bmatrix} \begin{bmatrix} \dfrac{n-4}{n} & \dfrac{-4r}{n} \\ \dfrac{-2(2-n)}{nr} & \dfrac{n-4}{n} \end{bmatrix} \begin{bmatrix} 1 & -r \\ 0 & 1 \end{bmatrix}$$

$$= \begin{bmatrix} -1 & 0 \\ \dfrac{-2(2-n)}{nr} & -1 \end{bmatrix}$$

Obviously, $\det(M_1) = 1$.

Some care is needed in interpreting this result. The vanishing of the upper right-hand element indicates that the centre of the sphere is imaged onto itself (as must be the case for *any* series of reflections or refractions in a monocentric system). But because the upper left-hand element shows that the lateral magnification is not unity, but minus unity, the principal planes of the system are not, as one might think, at the centre of the sphere.

For values of refractive index between 1 and 2, the equivalent focal length of the system - $1/C$ will be positive, and since $(1 - D) = 2$, the principal planes will be located two focal lengths away from the sphere centre on the right.

Finally, for the case $n = 2$, the element C vanishes and the system becomes an afocal one for which the ray-transfer matrix is $\begin{bmatrix} -1 & 0 \\ 0 & -1 \end{bmatrix}$. This means that, although the lateral magnification $y_2/y_1 = A$ is still minus unity, nevertheless, for angular magnification, as was shown in section II.6, we have

$$v_2/v_1 = \frac{n_1}{n_2}\left(\frac{V_2}{V_1}\right) = \frac{n_1}{n_2}D = +1$$

(since $n_2 = -n_1$ for the returning ray). A given ray therefore emerges with its y-value reversed, but it returns to infinity in the same direction from whence it came. Another form of optical element exhibiting this retro-reflective property is the cube-corner reflector (each face of which reverses one of the three direction-cosines of a ray). (Unlike the normal cat's-eye type of reflector, a high-index spherical bead needs no alignment; it operates equally well as a retroreflector in almost any direction. Vast numbers of them are now used on retroreflective number plates. Because of the effects of spherical aberration, the choice $n = 2$ does not necessarily give the best photometric performance.)

II.11.5 *Case of coincident reference planes*

When calculations are to be made on a reflecting system it often proves convenient to choose input and output reference planes that are located at the same z-position. In such a situation there is an inherent symmetry and the ray-transfer matrix obtained will always be such that its main diagonal elements A and D are identical. The student may have noticed that this was so in the problem just considered.

To see that this must be so in the general case, we shall make use of the concept that the path of any light-ray is reversible.

Let us choose a common reference plane at some convenient distance to the left of a reflecting optical system, and let us imagine that an arbitrarily chosen trial ray $\begin{bmatrix} y_1 \\ V_1 \end{bmatrix}$ launched in the $+ z$-direction returns as a reflected ray $\begin{bmatrix} y_2 \\ V_2 \end{bmatrix}$ travelling in the $- z$-direction, where $\begin{bmatrix} y_2 \\ V_2 \end{bmatrix} = \begin{bmatrix} A & B \\ C & D \end{bmatrix} \begin{bmatrix} y_1 \\ V_1 \end{bmatrix}$ as usual.

We seek to prove that $A = D$ in the above matrix.

74

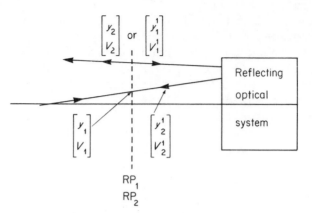

Figure II.25

Let us now take this reflected ray $\begin{bmatrix} y_2 \\ V_2 \end{bmatrix}$ and send it back along its tracks so that it forms a second trial ray $\begin{bmatrix} y_1^1 \\ V_1^1 \end{bmatrix}$. (See Figure II.25.) Because of the sign convention concerning n, the V-value for this ray will be reversed so that $\begin{bmatrix} y_1^1 \\ V_1^1 \end{bmatrix} = \begin{bmatrix} y_2 \\ -V_2 \end{bmatrix} = \begin{bmatrix} 1 & 0 \\ 0 & -1 \end{bmatrix} \begin{bmatrix} y_2 \\ V_2 \end{bmatrix}$. Operating on this second trial ray, the reflecting system will produce a second reflected ray

$$\begin{bmatrix} y_2^1 \\ V_2^1 \end{bmatrix} = \begin{bmatrix} A & B \\ C & D \end{bmatrix} \begin{bmatrix} y_1^1 \\ V_1^1 \end{bmatrix} = \begin{bmatrix} A & B \\ C & D \end{bmatrix} \begin{bmatrix} 1 & 0 \\ 0 & -1 \end{bmatrix} \begin{bmatrix} y_2 \\ V_2 \end{bmatrix}$$

$$= \begin{bmatrix} A & B \\ C & D \end{bmatrix} \begin{bmatrix} 1 & 0 \\ 0 & -1 \end{bmatrix} \begin{bmatrix} A & B \\ C & D \end{bmatrix} \begin{bmatrix} y_1 \\ V_1 \end{bmatrix}$$

But because ray tracks can be traversed in either direction, this second reflected ray must be emerging (in the $-z$-direction) along the path of our original trial ray $\begin{bmatrix} y_1 \\ V_1 \end{bmatrix}$ — in other words $\begin{bmatrix} y_1 \\ V_1 \end{bmatrix} = \begin{bmatrix} 1 & 0 \\ 0 & -1 \end{bmatrix} \begin{bmatrix} y_2^1 \\ V_2^1 \end{bmatrix}$.

Putting all this information together, we find

$$\begin{bmatrix} y_1 \\ V_1 \end{bmatrix} = \begin{bmatrix} 1 & 0 \\ 0 & -1 \end{bmatrix} \begin{bmatrix} A & B \\ C & D \end{bmatrix} \begin{bmatrix} 1 & 0 \\ 0 & -1 \end{bmatrix} \begin{bmatrix} A & B \\ C & D \end{bmatrix} \begin{bmatrix} y_1 \\ V_1 \end{bmatrix}$$

$$= \begin{bmatrix} A & B \\ -C & -D \end{bmatrix} \begin{bmatrix} A & B \\ -C & -D \end{bmatrix} \begin{bmatrix} y_1 \\ V_1 \end{bmatrix}$$

$$= \begin{bmatrix} (A^2 - BC) & B(A - D) \\ -C(A - D) & (D^2 - BC) \end{bmatrix} \begin{bmatrix} y_1 \\ V_1 \end{bmatrix}$$

Since this equation must be true for an arbitrarily chosen trial vector $\begin{bmatrix} y_1 \\ V_1 \end{bmatrix}$, the matrix calculated must reduce to a unit matrix. Since $(AD - BC) = 1$ this will be so if and only if the two main diagonal elements A and D of the original matrix are equal. This fact can provide a useful check on calculations to find the overall matrix for a catadioptric system.

The authors gratefully acknowledge permission to use the following copyright material:

To Messrs. McGraw-Hill Book Company for problem 3 on page 42, problems 11 and 13 on page 43, and problems 10 and 11 on page 60 from *Fundamentals of Optics*, 3rd Edition, 1957, by Jenkins and White. Copyright 1937 and 1950 by the McGraw-Hill Book Company, Inc. Used by permission of McGraw-Hill Company.

To R.A. Houston, and to Messrs. Longmans, Green and Co., for problem 2 on page 48, problem 12 on page 31, and problem 9 on page 49 of *Treatise on Light*, 5th Edition, 1928, by R.A. Houston.

III

Optical Resonators and Laser Beam Propagation

One of the pleasant surprises of modern optics has been the ease with which geometrical ray-transfer methods can be adapted to describe the generation and propagation of a laser beam. To give some account of this development we shall need to use some of the concepts of wave theory, but we shall try to explain these as they arise and we hope that the chapter will be read even by those who have not yet received an introduction to physical optics. Our starting point will be the matrices and ray vectors discussed in the last chapter.

III.1 REVIEW OF RESULTS OBTAINED FOR PARAXIAL IMAGING
    SYSTEMS

We summarize below in Table 1 the ray-transfer matrices corresponding to eight frequently occurring situations.

For optical systems that represent a combination of the above situations, we multiply together the appropriate matrices, remembering to work backwards from output to input. If $\begin{bmatrix} A & B \\ C & D \end{bmatrix}$ represents the resultant overall matrix, we check that its determinant is unity, and then use it in the ray-transfer equation

$$\begin{bmatrix} y_2 \\ V_2 \end{bmatrix} = \begin{bmatrix} y_2 \\ n_2 v_2 \end{bmatrix} = \begin{bmatrix} A & B \\ C & D \end{bmatrix} \begin{bmatrix} y_1 \\ V_1 \end{bmatrix} = \begin{bmatrix} A & B \\ C & D \end{bmatrix} \begin{bmatrix} y_1 \\ n_1 v_1 \end{bmatrix}$$

Table 1

Number	Description	Optical Diagram	Ray-transfer matrix
1	Translation ($\mathcal{T}$-matrix)		$\begin{bmatrix} 1 & t/n \\ 0 & 1 \end{bmatrix} = \begin{bmatrix} 1 & T \\ 0 & 1 \end{bmatrix}$
2	Refraction at single surface ($\mathcal{R}$-matrix)		$\begin{bmatrix} 1 & 0 \\ -\dfrac{(n_2 - n_1)}{r} & 1 \end{bmatrix} = \begin{bmatrix} 1 & 0 \\ -P & 1 \end{bmatrix}$
3	Reflection at single surface (for convention see section II.11)		$\begin{bmatrix} 1 & 0 \\ \dfrac{2n}{r} & 1 \end{bmatrix} = \begin{bmatrix} 1 & 0 \\ -P & 1 \end{bmatrix}$
4	Thin lens in air (focal length f, power P)		$\begin{bmatrix} 1 & 0 \\ -(n-1)\left(\dfrac{1}{r_1} - \dfrac{1}{r_2}\right) & 1 \end{bmatrix} = \begin{bmatrix} 1 & 0 \\ -1/f & 1 \end{bmatrix}$

Table 1 continued

Number	Description	Optical Diagram	Ray-transfer matrix
5	Between principal planes of lens system in air (focal length f)		$\begin{bmatrix} 1 & 0 \\ -1/f & 1 \end{bmatrix} = \begin{bmatrix} 1 & 0 \\ -P & 1 \end{bmatrix}$
6	Between focal planes of lens system in air (focal length f)		$\begin{bmatrix} 0 & f \\ -1/f & 0 \end{bmatrix}$
7	Imaging between two conjugate planes with lateral magnification m and focal length f		$\begin{bmatrix} m & 0 \\ -1/f & 1/m \end{bmatrix}$
8	Afocal system with lateral magnification m		$\begin{bmatrix} m & 0 \\ 0 & 1/m \end{bmatrix}$

III.2 DESCRIPTION OF WAVE PROPAGATION IN TERMS OF GEOMETRICAL OPTICS

It is perhaps useful to expand here on some of the points made at the beginning of the previous chapter. As was first shown by Huygens, the propagation of light is best understood in terms of wavefronts. If we consider the light waves coming from a point source, we can visualize the wavefronts expanding outwards, rather like the ripples on a pond, and gradually acquiring a larger radius of curvature. If we place a positive lens in the path of these waves, we can cause them to reconverge onto another point - the image. Alternatively, if we use a slightly weaker lens, we can produce collimated light, consisting of nominally plane wavefronts. If we introduce boundaries or obstacles into the path of the wavefronts, we break them up and obtain the effects of diffraction; in the absence of such obstacles, however, the light energy is propagated in a direction normal to the wavefronts.

The concept of a geometrical ray is an idealization of this wavenormal. Provided that we avoid focal or near-focal regions, we can visualize the rays and their associated wavefronts as orthogonally intersecting families of straight lines and curved surfaces.

Provided that a wavefront is known to be spherical, its shape is completely specified when we know the ray vectors for two of its intersecting normals. Furthermore, if it is known that the centre of curvature of the spherical wavefront lies somewhere on the optical axis, a *single* additional parameter will suffice to locate it.

We return now to our geometric ray model, but instead of considering individual rays, we extend our thinking to the idea of a ray pencil, a bundle or family of rays all emerging from the same point. In Figure III.1, for example, we show several paraxial rays diverging from an axial object point O_1 at a distance r_1 to the left of the reference plane RP_1. For all the members of this family, the ratio of ray height y_1 to ray angle v_1 will be equal to r_1, that is

$$r_1 = \left(\frac{y_1}{v_1}\right)$$

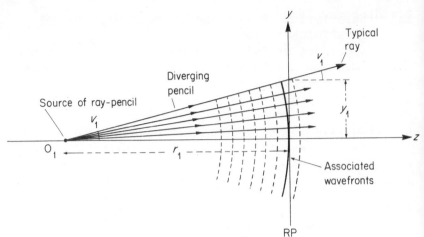

Figure III.1

If we now regard each of the rays in this pencil as the normal to a wavefront, the wavefront so obtained will be spherically curved, with its centre of curvature located at O_1, a distance r_1 to the left of RP_1; and, since the rays in our pencil are travelling from left to right, in the $+ z$-direction, the wavefront associated with this pencil must be expanding and diverging.

In specifying the curvature of a wavefront, we shall find it convenient to call this positive when the wave is diverging and to use a 'reduced value' $R = (r/n)$ $= (y/V)$, for the radius of curvature. By using reduced values, we gain the advantage that, whenever a ray pencil crosses a plane boundary between two media, there is no change in the R-value for the associated wavefront. (This must be so, because for every ray crossing such a boundary, both the ray height y and the optical direction-cosine V remain constant.) For a centred spherical wavefront and a given reference plane, the R-value is the single parameter required to specify its shape.

Let us now see what happens to the (y/V)-value or R-value associated with a ray pencil when it is propagated through an optical system. We have directly from the ray-transfer matrix the two equations

$$y_2 = Ay_1 + BV_1$$

$$V_2 = Cy_1 + DV_1$$

Dividing the first equation by the second, we obtain

$$\frac{y_2}{V_2} = \frac{Ay_1 + BV_1}{Cy_1 + DV_1} = \frac{A\left(\frac{y_1}{V_1}\right) + B}{C\left(\frac{y_1}{V_1}\right) + D}$$

On rewriting (y/V)-values as R-values, we obtain for R_2, the reduced value for the radius of curvature of the associated wavefront after passing through the system, the important equation

$$R_2 = \frac{AR_1 + B}{CR_1 + D}$$

This is a rule which is well worth remembering and it has been called 'the ABCD rule'. It enables us to calculate very simply how the curvature of a centred spherical wavefront changes from one reference plane to the next.

III.2.1 Examples of the ABCD rule

(a) If we consider propagation of a diverging wave across a gap, insertion of the $\mathscr{T}$-matrix in the ABCD formula gives immediately

$$R_2 = \frac{1.R_1 + T}{0.R_1 + 1} = R_1 + T$$

(an obvious result for a diverging wavefront).

(b) For refraction by a thin positive lens of power P, insertion of the $\mathscr{R}$-matrix in the ABCD formula gives

$$R_2 = \frac{1.R_1 + 0}{-PR_1 + 1} = \frac{R_1}{1 - PR_1}$$

or, equivalently, $1/R_2 = 1/R_1 - P$. (The divergence of the wavefront has been reduced by the power of the converging lens.) We shall return to the ABCD rule later, when we discuss laser beam propagation.

So far, we have considered only the shape, or curvature, of the light waves, but it may be important to consider also the optical distances that they are called upon to travel. This is especially so if two waves are separated and then recombined, so that the possibility of interference effects arises. The expression usually used for optical path calculations is called the 'eikonal'; it is a formula for the optical distance measured along a light ray between two reference planes.

In specifying such a ray, it is possible to use either the input or the output y- and V-values, but the choice usually preferred is to work in terms of the input ray height y_1 and the output ray height y_2. Let us now see how the ray-transfer matrix can be used to derive an eikonal expression $W(y_1, y_2)$ for the total path between the two reference planes.

If we are given only the y_1- and y_2-values associated with a ray, we can nevertheless obtain the corresponding V_1 and V_2 by solving the simultaneous equations

$$y_2 = Ay_1 + BV_1$$

$$V_2 = Cy_1 + DV_1$$

for V_1 and V_2. Evidently,

$$V_1 = \frac{y_2 - Ay_1}{B} \quad \text{and} \quad V_2 = Cy_1 + \frac{D(y_2 - Ay_1)}{B}$$

that is

$$V_2 = \frac{(Dy_2 - y_1)}{B}$$

since $(AD - BC) = 1$.

The axial intersection point O_1 of the input ray is therefore located at a reduced distance $R_1 = y_1/V_1$ $= By_1/(y_2 - Ay_1)$ to the left of RP_1. Similarly, the axial intersection point O_2 of the output ray is located at a reduced distance R_2 to the left of RP_2, where $R_2 = y_2/V_2 = By_2/(Dy_2 - y_1)$ (see Figure III.2).

At this stage, we imagine that our given input and output rays are representatives of ray pencils diverging from O_1 in the object space and from O_2 in the image space respectively. It will be clear that O_2 is

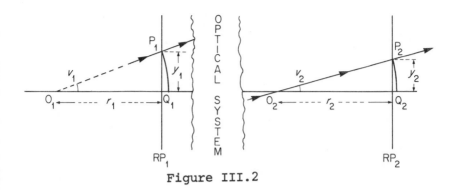

Figure III.2

the image of O_1, and we shall now use Malus' theorem
that, in any stigmatic imaging system, the optical
path measured along all the rays involved is the same –
in other words, a spherical wave leaving O_1 generates
another spherical wave centred on O_2.

But, if we consider a ray travelling straight down
the axis, from O_1 to O_2 in Figure III.2, the optical
distance involved is equal to $\left[n_1 (O_1 Q_1) - n_2 (O_2 Q_2) + K \right]$,
where K represents the total optical path measured
along the axis from RP_1 to RP_2. The total optical
path from O_1 to O_2 via P_1 and P_2 must, by Malus'
theorem, be the same as the expression bracketed above;
to obtain the portion between P_1 and P_2 we need to
subtract $n_1 \times (O_1 P_1)$ and to add $n_2 \times (O_2 P_2)$. We thus
obtain, for the total optical path between P_1 and P_2,
the eikonal

$$W(y_1, y_2) = \left[K - n_1 (O_1 P_1 - O_1 Q_1) + n_2 (O_2 P_2 - O_2 Q_2) \right]$$

The expressions in small brackets represent the
sagittae of the spherical waves reaching RP_1 and RP_2.
If we use the reduced value $R = r/n$ for the radius of
curvature of each wave, the expression $y^2/2R = ny^2/2r$
correctly represents both the geometrical sagitta and
the refractive index of the medium in which the wave
is travelling. We thus obtain

$$W(y_1, y_2) = \left[K - y_1{}^2/2R_1 + y_2{}^2/2R_2 \right]$$

$$= \left[K - \frac{y_1 (y_2 - Ay_1)}{2B} + \frac{y_2 (Dy_2 - y_1)}{2B} \right]$$

$$= \left[K + \frac{(Ay_1{}^2 + Dy_2{}^2 - 2y_1 y_2)}{2B} \right]$$

It will be remembered that K is the total optical path, $\int n(z)\,dz$, integrated along the axis between RP_1 and RP_2. Since K is independent of y_1 and y_2, its main importance will arise in interferometer calculations when two separate paths are being compared.

For many calculations involving optical diffraction, it is the transverse variation of W with y_1 and y_2 that is usually of interest. Let us suppose that we are given a complex function $A_1(y_1)$ which represents the amplitude and phase of the light waves passing through the input plane RP_1 and that we wish to calculate the corresponding complex amplitude $A_2(y_2)$ that will be generated in the output plane.

If we restrict ourselves to a one-dimensional calculation and neglect minor corrections due to slowly varying obliquity effects, then the Fresnel-Huygens theory gives us the following diffraction equation:

$$A_2(y_2) = \left(\frac{-i}{\lambda B}\right)^{\frac{1}{2}} \int A_1(y_1)\ \exp\left[\frac{2\pi i}{\lambda} \times W(y_1, y_2)\right] dy_1$$

In this expression, the constant term $(-i/\lambda B)^{\frac{1}{2}}$ is a normalizing factor which is needed to make the equation dimensionally correct and which takes account of the way in which the disturbance in the output plane will be weakened if B is very large. If calculations are made in the x-direction as well as in the y-direction, then the amplitude will fall off linearly with B, and for the irradiance we shall have the usual inverse square law. (Strictly speaking, a point source located in RP_1 generates an image (B/D) to the left of RP_2, but the lateral magnification is $1/D$, so it is only the element B that is needed for insertion in the normalizing factor.)

Turning now to the integrand, we find from the argument of the complex exponential term that the contribution which each element of RP_1 makes to the wave amplitude in RP_2 has a phase angle ϕ equal to 2π times the optical path length W divided by the wavelength λ. To visualize how this integral is built up, we can regard it as the summation of a large number of elementary vectors drawn at the corresponding angle ϕ on an Argand diagram. When all of these vectors are joined end-to-end, the line joining the beginning to the end of the resultant curve represents vectorially the resultant amplitude in the output plane, and the square of its length represents the corresponding resultant irradiance.

Let us now consider a few simple examples of the eikonal function between two reference planes.

III.2.2 Propagation across an airgap of thickness t

$$\begin{bmatrix} A & B \\ C & D \end{bmatrix} = \begin{bmatrix} 1 & t \\ 0 & 1 \end{bmatrix}$$

Substituting these matrix elements in the eikonal formula, we find

$$W(y_1, y_2) = \left[t + \frac{1y_1^2 + 1y_2^2 - 2y_1y_2}{2t} \right]$$

$$= \left[t + \frac{(y_1 - y_2)^2}{2t} \right]$$

(If we calculate by direct application of Pythagoras' theorem, we find that the exact distance is $\left[t^2 + (y_2 - y_1)^2 \right]^{\frac{1}{2}}$. But for a paraxial ray joining the two planes, the ratio $(y_2 - y_1)/t$ must be small, so the approximation is a good one.)

By inserting the above expression for the eikonal into the diffraction equation, we can now handle problems in Fresnel diffraction. For example, if RP_1 is illuminated by a plane wave of unit amplitude and we cover half of RP_1 with an absorbing knife-edge, we produce a situation where $A_1(y_1) = 1$ for $y_1 > 0$ and $A_1(y_1) = 0$ for $y_1 \lesssim 0$. The amplitude distribution in the shadow pattern generated in RP_2 will then be given by

$$A_2(y_2) = \left(\frac{-i}{\lambda t} \right)^{\frac{1}{2}} \int_{-\infty}^{+\infty} A_1(y_1) \exp\left[\frac{2\pi i}{\lambda} \left(t + \frac{(y_2 - y_1)^2}{2t} \right) \right] \cdot dy_1$$

$$= \left(\frac{-i}{\lambda t} \right)^{\frac{1}{2}} \exp\left[\frac{2\pi i t}{\lambda} \right] \int_0^{+\infty} \exp\left[\frac{i\pi}{\lambda t} (y_2 - y_1)^2 \right] dy_1$$

The integrals involved in this calculation are called Fresnel's integrals; when they are represented graphically in the Argand diagram, they generate the Cornu spiral.

Since the eikonal expression for a gap is a function only of the difference $(y_2 - y_1)$, the new amplitude distribution $A_2(y_2)$, derived by the diffraction formula, can be regarded as a convolution of the original amplitude distribution $A_1(y_1)$, with the quadratic phase variation produced by the eikonal term; $A_2(y_2)$ is sometimes called a 'shadow transform' of $A_1(y_1)$. Transforms of this kind were important in Gabor's early work on 'in-line' holography.

III.2.3 Transfer between the two focal planes of a lens

$$\begin{bmatrix} A & B \\ C & D \end{bmatrix} = \begin{bmatrix} 0 & f \\ -1/f & 0 \end{bmatrix}$$

Since A and D both vanish, we obtain

$$W(y_1, y_2) = K - (y_1 y_2/f)$$

where K is the optical distance between the two focal points.

For any given value of y_1, this eikonal shows a linear variation across the output plane - in other words, a point source y_1 off axis generates on the other side of the lens a plane wavefront whose tilt is $\partial W/\partial y_2 = - (y_1/f)$ radians.

If we insert this eikonal into the diffraction equation, we find that, because the quadratic phase terms have vanished, we are now handling the special case of Fraunhofer diffraction. The relationship between $A_1(y_1)$ and $A_2(y_2)$ is a Fourier transform relationship:

$$A_2(y_2) = \left(\frac{-i}{\lambda f}\right)^{\frac{1}{2}} \int_{-\infty}^{\infty} A_1(y_1) \exp\left[\frac{2\pi i}{\lambda}(K - y_1 y_2/f)\right] dy_1$$

$$= \left(\frac{-i}{\lambda f}\right)^{\frac{1}{2}} \exp\left[\frac{2\pi i K}{\lambda}\right] \int_{-\infty}^{\infty} A_1(y_1) \exp\left[- 2\pi i (y_1 y_2/f\lambda)\right] d$$

If, for example, RP$_1$ is illuminated by a plane wave of unit amplitude, and we introduce there a slit-shaped aperture which transmits only in the central region, $a > y_1 > -a$, the amplitude generated in RP$_2$, where the light would normally be focused to a point, will be given by

$$A_2(y_2) = \left(\frac{-i}{\lambda f}\right)^{\frac{1}{2}} \exp\left[\frac{2\pi i K}{\lambda}\right] \int_{-a}^{a} \exp\left[-2\pi i (y_1 y_2/f\lambda)\right] dy_1$$

$$= \left(\frac{-i}{\lambda f}\right)^{\frac{1}{2}} 2a \exp\left[\frac{2\pi i K}{\lambda}\right]\left[\frac{\sin(2\pi a y_2/f\lambda)}{(2\pi a y_2/f\lambda)}\right]$$

Because of the effects of diffraction, the light waves in the focal plane of the lens produce an image of finite width. The irradiance distribution $I_2(y_2)$ that we shall actually observe in the image is given by

$$I_2(y_2) = A_2(y_2)\, A_2^{*}(y_2) = \left(\frac{4a^2}{\lambda f}\right)\left[\frac{\sin(2\pi a y_2/f\lambda)}{(2\pi a y_2/f\lambda)}\right]^2$$

When y_2 is very small, the expression in the square brackets reaches its maximum value of unity, because

$$\lim_{z\to 0}\left(\frac{\sin z}{z}\right) = 1$$

We therefore find that, *for coherent illumination*, the brightness at the centre of the image increases as the *square* of the slit width. Two factors are at work here; opening up the width of the slit not only admits more energy to the lens but also reduces the diffraction spread of the image. We shall return to questions of resolving power in the next section, but it should be evident from the formula that the brightness of the Fraunhofer diffraction pattern must fall to zero when the argument of the sine function becomes equal to $\pm\,\pi$, that is, when $y_2 = \pm\,(f\lambda/2a)$. (See Figure III.3.)

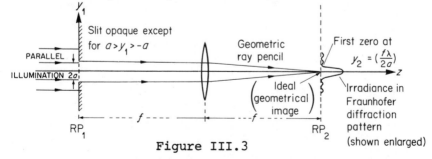

Figure III.3

For this value of y_2, the eikonal tells us that one edge of the slit in RP₁ is exactly one wavelength further away than the other edge. All possible phases

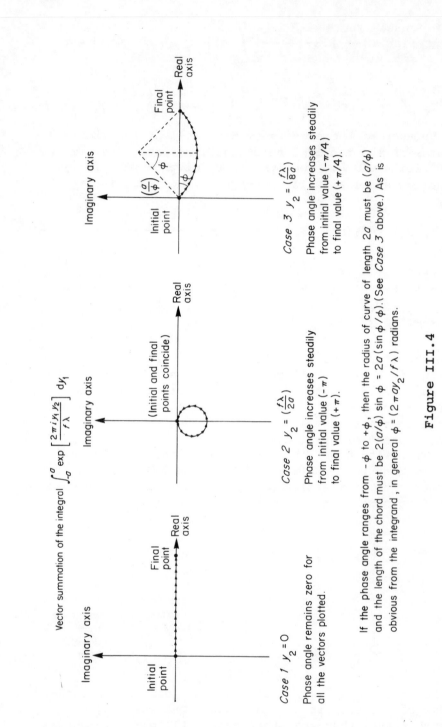

Vector summation of the integral $\int_{-a}^{a} \exp\left[\frac{2\pi i y_1 y_2}{f\lambda}\right] dy_1$

Case 1 $y_2 = 0$

Phase angle remains zero for all the vectors plotted.

Case 2 $y_2 = \left(\frac{f\lambda}{2a}\right)$

Phase angle increases steadily from initial value $(-\pi)$ to final value $(+\pi)$.

(Initial and final points coincide)

Case 3 $y_2 = \left(\frac{f\lambda}{8a}\right)$

Phase angle increases steadily from initial value $(-\pi/4)$ to final value $(+\pi/4)$.

If the phase angle ranges from $-\phi$ to $+\phi$, then the radius of curve of length $2a$ must be $\left(a/\phi\right)$ and the length of the chord must be $2\left(a/\phi\right) \sin\phi = 2a\left(\sin\phi/\phi\right)$. (See *Case 3* above.) As is obvious from the integrand, in general $\phi = \left(2\pi a y_2/f\lambda\right)$ radians.

Figure III.4

are equally represented, and there is complete local cancellation of the wave disturbance. If we visualize this situation on the Argand diagram, we find that, for $y_2 = 0$, the result produced by summing all the individual vectors is a straight line; as the magnitude of y_2 increases, however, we obtain a curve which bends round into a single closed circle for $y_2 = \pm (f\lambda/2a)$. For each of these curves, if $2a$ represents the length measured round the curve, the length of the chord representing the resultant amplitude will be $[2a\sin(2\pi ay_2/f\lambda)]/(2\pi ay_2/f\lambda)$ and the radius of curvature will be $(f\lambda/2\pi y_2)$. (See Figure III.4.)

III.2.4 *The case where B vanishes*

If $B = 0$ in the ray-transfer matrix, there must be an imaging relationship between RP_1 and RP_2 and there is only one y_2-value for which a valid ray trajectory can be drawn from a given y_1. Not surprisingly, therefore, the eikonal function that we have obtained 'blows up'; it cannot give a useful answer to an impossible question. In order to handle this situation, we must use an eikonal which is expressed, not as a function of y_1 and y_2, but as a function of y_1 and V_1 for example,

In recalculating the eikonal $W(y_1,V_1)$ for this special case, we shall use the simplified ray-transfer matrix $\begin{bmatrix} m & 0 \\ -1/f & 1/m \end{bmatrix}$, listed as entry 7 in Table 1 for an imaging system.

Considering a typical ray as before, we have the same equation

$$W = (K - y_1^2/2R_1 + y_2^2/2R_2) = (K - y_1V_1/2 + y_2V_2/2)$$

but we now require to express the product (y_2V_2) in terms of y_1 and V_1. Hence,

$$W(y_1,V_1) = [K - y_1V_1/2 + (my_1/2)(- y_1/f + V_1/m)]$$

$$= (K - my_1^2/2f)$$

Notice that in this case the dependence on V_1 disappears and the eikonal is a function of y_1 only; this follows directly from Malus' theorem - the time taken for light to travel from the object point at y_1 to its image point in RP$_2$ will be the same along *all* the various ray directions that could be specified by V_1.

For some purposes, such as optical image processing a very faithful relay system may be required. The designer may then need to avoid the quadratic second term in the eikonal by employing an afocal imaging system; but the effects of third-order aberrations will also need to be considered and the lenses will be expensive to fabricate.

The eikonal function that we have been considering above has been defined with respect to two separated reference planes. Wherever possible it is preferable to specify it symmetrically as a function of ray height in the two reference planes. When we include the other transverse direction (normal to the plane of the diagram), the eikonal $W(x_1,y_1,x_2,y_2)$ depends on four parameters. If we keep RP$_1$ fixed in the plane $z = 0$, and replace RP$_2$ by a variable reference plane which is shifted gradually outwards from RP$_1$ in the + z-direction, this enables us to develop a more general three-dimensional eikonal function $W(x_1,y_1,x,y,z)$. For a point source located at (x_1,y_1) in the reference plane RP$_1$, the wavefronts that will be generated in any other region of space are the surfaces over which the value of W remains constant: the local gradient of the scalar W-function is then a three-dimensional momentum vector which specifies not only the ray direction in which the waves are travelling but also the rate at which they change phase with distance.

At the beginning of this section, we showed how the *ABCD* rule enables us to follow the progress of the well-defined wavefronts associated with a ray pencil, or generated by an ideal point source. Nevertheless, as our discussion of the eikonal has shown, the blurring effects of Fraunhofer diffraction make it impossible for the light waves to concentrate themselves back into an ideal point image. Near a focus, therefore, the very small R-values predicted by the *ABCD* rule may be seriously misleading.

In the next section we consider in more detail the limitations on the degree to which the direction of travel of a light beam can be specified. These limitations are important, not only for ordinary inco-

herent light, but also for the spatially coherent
light which is generated inside a laser resonator.

III.3 RESOLVING POWER, ETENDUE AND THE SPACE-
BANDWIDTH PRODUCT

Because the wavelength of light is not infinitesimal, the image formed by a converging pencil of
rays is never a point; it is a spot whose finite
width is determined by Fraunhofer diffraction. In
the last section, we found that for parallel light
passing through a slit of width $2a$ and focused by a
lens of focal length f the bright central spot has
an irradiance that falls off to zero at a radius
$(0 \cdot 5 f\lambda/a)$. For the case of a circular aperture of
the same diameter $2a$, evaluation of the diffraction
integral yields an expression which is similar except
that the sine function is replaced by a first-order
Bessel function. The central spot of this distribution
is known as the Airy disc, and the radius at which its
brightness falls to zero is $(0 \cdot 61 f\lambda/a)$ - about 22 per
cent greater than would be obtained for a square aperture of side $2a$. The larger the aperture, the smaller
the image, provided of course that the lens is free
from all aberrations.

If an optical system is arranged to form diffraction-
limited images of two equally bright point sources, it
is usually said that the images are 'just resolved' if
the centre of one of them coincides with the first dark
ring of the other.

(a) For the case of a well-corrected telescope objective of diameter d, the image separation in the
focal plane will be $(1 \cdot 22 f\lambda/d)$ and the angular
separation of the two sources will be $(1 \cdot 22\lambda/d)$
radians.

(b) For a microscope objective the relevant diaphragm
which determines the resolving power is located in
the back focal plane, so that, when seen from the
specimen, the entrance pupil seems to be at infinity. If we use θ to denote its angular radius,
then, for small θ, the least resolved distance in
the plane of the specimen will be $(0 \cdot 61\lambda/\theta)$. In
practice, θ may be quite large, and it becomes
necessary to use the more accurate expression
$(0 \cdot 61\lambda/\sin\theta)$, or $(0 \cdot 61\lambda/n\sin\theta)$ if the specimen is

immersed in a medium of refractive index n. The quantity ($n\sin\theta$) is usually referred to as the Numerical Aperture (N.A.) of the microscope objective. (We can also regard it as the maximum V-value for a ray leaving an axial object point.)

(c) Finally, for a diffraction-limited photographic objective of diameter d and focal ratio $(f/d) = F$, the corresponding least-resolved image separation is given by $\Delta y = (1 \cdot 22 F \lambda)$. For example, an $F/6$ lens operating in green light will just resolve the images of points or lines 4 μm apart. (This formula assumes self-luminous objects, or incoherent illumination. To achieve the same resolution with coherent light, it may be necessary to use oblique illumination.) For 'well-resolved' images, however, a spacing closer to 8 μm will be needed and even at this spatial frequency, which corresponds to 125 lines per mm, calculation of the modulation transfer function shows that the contrast of a grid pattern will be reproduced with only 50 per cent of its original value.

Let us now consider the problem of combining a small pinhole source with a well-corrected collimator objective of focal ratio F, so as to produce an 'artificial star'. Even if the pinhole is less than a wavelength in diameter, its apparent angular size when measured by means of the light emerging from the collimator will be ($1 \cdot 22 \lambda / d$) radians, which corresponds to a linear radius of ($1 \cdot 22 \lambda F$) when referred back to the focal plane of the collimator. It is therefore sensible to use a physical pinhole of about this diameter - anything smaller wastes a lot of light without appreciably improving the directionality. The light illuminating the pinhole will normally be a focused image of the brightest source conveniently available - sun, zirconium arc, tungsten-halogen lamp, etc. We should not forget, however, that much higher brightness still can be achieved by using a laser source focused on to the same pinhole.

For some problems of illumination, for example in a searchlight, it may be desirable to use a beam whose angular spread is appreciably greater than that determined by diffraction. In that case, much of the advantage of the laser may be lost, and the conven-

tional incoherent light source may be the better
choice.

In order to describe the directionality of a light
beam, we may take a reference plane at any convenient
distance R to the right of the light source. If V_{max}
is the maximum angle accepted by the collimator, then
for any ray leaving the *centre* of the source, we have
a ray vector $\begin{bmatrix} RV \\ V \end{bmatrix}$, where V lies between V_{max} and $-V_{max}$.

We can regard the family of rays in this pencil as the
'core' of the light beam.

Let us now consider some other ray which traverses
the reference plane with vector components y' and V'.
By how many 'resolution widths' does this trial ray
depart from the core of the light beam?

In order to refer the light back to the source
plane, we multiply by the (negative) translation
matrix $\begin{bmatrix} 1 & -R \\ 0 & 1 \end{bmatrix}$ and find that the y-value representing
the radius on the source from which our ray has come
must be $(y' - RV')$. If we divide this radius by the
diffraction-limited radius $(0{\cdot}61\lambda/V_{max})$, we obtain the
dimensionless ratio $(y'V_{max} - RV'V_{max})/0{\cdot}61\lambda$. This
can be rewritten as

$$\frac{1}{0{\cdot}61\lambda} \det \begin{bmatrix} y' & RV_{max} \\ V' & V_{max} \end{bmatrix}$$

Here, we have taken the trial ray vector $\begin{bmatrix} y' \\ V' \end{bmatrix}$ together
with an extreme member of the central ray pencil to
form a new kind of 2×2 matrix, the determinant of
which provides a means of telling whether the trial
ray belongs to the same family. If it does, then,
since the two column vectors are characterized by the
same (y/V) ratio, the determinant vanishes. If it
'belongs to a different ray pencil', however, the determinant has a finite value, and, if that value appreciably exceeds $0{\cdot}61\lambda$, then the new ray must be contrib-
uting light that departs measurably from the core of
the beam.

It is important at this point to realize that the value of this determinant depends only on what pair of rays is chosen, and not on the choice of reference plane. We are really asking the question, 'by what fraction of a diffraction width do these two rays miss each other when referred back to the source plane?' We shall relate the determinant presently to the quantity known as the Lagrangian invariant, and show that its constancy can be understood in terms of energy conservation. But to prove that it remains constant from one reference plane to the next, we can rely upon the fact that whenever a *unimodular* ray-transfer matrix transforms $\begin{bmatrix} y_1 & y_1' \\ V_1 & V_1' \end{bmatrix}$ into $\begin{bmatrix} y_2 & y_2' \\ V_2 & V_2' \end{bmatrix}$ the matrix elements may be changed, but the determinant *must* remain the same - zero if our two chosen rays $\begin{bmatrix} y_1 \\ V_1 \end{bmatrix}$ and $\begin{bmatrix} y_1' \\ V_1' \end{bmatrix}$ belong to the same ray pencil, non-zero but constant if they do not.

If we choose two rays that differ from each other as much as possible, the value of this determinant gives us, for any given imaging system, a measure of the number of laterally distinguishable ways in which light can be propagated through that system. A convenient procedure is to work in terms of the entrance window, or field plane of the system, and its entrance pupil. (For definitions of these terms see appendix A.) The two rays that we shall choose are, firstly, a 'principal ray' which passes from a point on the edge of the field to the centre of the entrance pupil, and secondly a 'marginal ray' which passes from the centre of the field to a point on the edge of the entrance pupil.

If the field has a radius a_1 and the entrance pupil of radius a_2 is located at a distance b to the right, then the principal ray referred to the field plane will have a ray vector $\begin{bmatrix} a_1 \\ -a_1/b \end{bmatrix}$, and the marginal ray referred to the same plane will have a ray vector $\begin{bmatrix} 0 \\ a_2/b \end{bmatrix}$. From this pair of ray vectors, we obtain the

determinant $\det \begin{bmatrix} a_1 & 0 \\ -a_1/b & a_2/b \end{bmatrix} = a_1a_2/b$. Dividing
this as before by $0{\cdot}61\lambda$, we obtain the dimensionless
parameter $(a_1a_2/0{\cdot}61b\lambda)$, which represents half the total
number of lines that this particular imaging system
can just resolve, even if it is free from all aberrations.

Instead of imagining lines with separation $(0{\cdot}61\lambda b/a_2)$
stacked across a field diameter, let us consider the
situation in two dimensions. If the field radius is
a_1, and if we take the radius of an image dot to be
$(0{\cdot}3\lambda b/a_2)$, then the total number of just-resolvable
picture elements will be $\pi a_1{}^2/\pi(0{\cdot}3\lambda b/a_2)^2$, or
(approximately) $\pi^2 a_1{}^2 a_2{}^2/b^2\lambda^2$.

The exact value of this number depends on what crit-
erion for resolution or image contrast we choose to
adopt. But the last expression that we have given can
be rewritten as

$$\frac{1}{\lambda^2} \times (\pi a_1{}^2) \times (\pi a_2{}^2/b^2) = \frac{1}{\lambda^2} A\Omega$$

where A denotes the area occupied by the beam in the
image plane and Ω denotes the solid angle subtended by
the exit pupil.

The product $A\Omega$ is sometimes referred to as the
Lagrangian invariant. The fact that it remains constant from one image plane to the next follows immediately from the form of the imaging ray-transfer matrix
$\begin{bmatrix} m & 0 \\ -1/f & 1/m \end{bmatrix}$. The lateral magnification m is always
the reciprocal of the angular magnification, so that
when A is transformed into m^2A the new value of the
solid angle becomes Ω/m^2.

For imaging systems that are forced to work at low
levels of illumination and for many instruments such
as spectrometers and monochromators, the product $A\Omega$ is
an important and useful measure of how much energy can
be accepted and transmitted through the system. Since
the term 'Lagrangian invariant' is somewhat obscure, a
more descriptive name has been sought for this product;
in preference to terms like 'light grasp' and 'through-
put', the French word, 'étendue', introduced by Connes,
is now widely accepted.

Let us now suppose that an imaging system is illuminated by a uniformly bright extended source, whose 'radiant sterance' can be expressed as S watts radiated per square meter per steradian. The total radiative power in watts accepted by the system is then $SA\Omega$: if the system lenses are of ideal transparency, then the radiant sterance will be conserved from one image plane to the next and all of the power will be transmitted to the output.

We have already seen that, if the étendue is divided by λ^2, we obtain a dimensionless number which describes the maximum number of resolvable picture elements that can be handled by an imaging system, *operating under diffraction-limited conditions*. It must be emphasized that very few optical systems operating with incoherent illumination achieve this resolution, and the reason for using the largest possible lens is usually to improve the brightness rather than the resolution of the image. But in modern astronomical cameras, a remarkably perfect correction of oblique aberrations is sometimes achieved, and the field may contain as many as 10^8 resolvable images.

With the development of laser light sources, however, there are now several types of optical system that employ coherent illumination, and for all of these the value of $(A\Omega/\lambda^2)$ is of real significance. In the optical processing of information, for example, this number is usually referred to as the 'space-bandwidth product', and it represents the effective number of independent parallel channels over which the processor can handle information. Because all of these channels operate simultaneously, there is an enormous potential advantage in using an optical system, as opposed to a system in which all the data have to be scanned sequentially.

Another example arises when a fine-grained photographic plate is being used for holography, or for photography of a laser speckle pattern. If the plate is 100 mm square and receives incoming light over a solid angle of perhaps one steradian, then it is readily capable of recording over 10^{10} laterally distinguishable elements, even if all this information arrives in an exposure time lasting only a few nanoseconds. If the same information were transmitted over a television link with a video bandwidth of 10 Megahertz, the time needed to record it all would be about a quarter of an hour.

Paradoxically enough, the laser light which makes it easy to obtain these very large space-bandwidth products depends for its own generation on optical resonators for which the product $(A\Omega/\lambda^2)$ is severely restricted and not much larger than unity. This restriction is imposed deliberately, the object being to provide a spatial filter, so that, when laser action begins, the light waves build up into only one well-defined standing-wave pattern - a 'single transverse mode' of oscillation.

In calculations on optical resonators, the usual practice is to work in terms of the parameter $(a_1 a_2/b\lambda)$ already considered above and to refer to this as the Fresnel number N. The two-dimensional space-bandwidth product $(A\Omega/\lambda^2)$ is then numerically equal to the square of (πN). For a typical low-power helium-neon gas laser, the resonator length b may be 300 mm and the radius a of the discharge tube 0·5 mm. For a wavelength of 633 nanometers, the Fresnel number N is then $(0\cdot 5(0\cdot 5)10^6 / (300)(633)$, or approximately 1·3.

In principle, if we regard the discharge tube as a spatial filter, about a dozen just-distinguishable spots of light could be propagated from one end to the other; for an oscillating laser, such an output distribution would be undesirable, but fortunately a process of natural selection operates and only the central spot is sufficiently free from diffraction losses to build up into a self-sustaining wave-pattern. It is the geometry of this wave-pattern that we shall be considering in later sections of this chapter.

III.4 MATRIX REPRESENTATION OF AN OPTICAL RESONATOR

As was mentioned briefly in the last section, the function of an optical resonator is to ensure that, when a laser oscillates, the light waves that are generated have the required shape and spacing. For the quantum electronics engineer who designs such light sources, there are really three important aspects that need to be considered:

(a) A suitable laser material has to be found, such that the stimulated downward transitions occurring between two of its allowed energy states can provide radiation of the wavelength desired.

(b) The material has to be transformed from one that absorbs light of this particular wavelength into one that amplifies with an adequate gain coefficient. This is achieved usually by some kind of 'pump' which supplies energy selectively in such a way as to produce a 'population inversion' between the two energy states.

(c) To turn the light amplifier so obtained into a light oscillator, it is necessary to apply 'optical feedback' so that the same light travels repeatedly through the amplifier. Provided the gain produced by the amplifier exceeds the total losses produced in the resonator, regenerative build-up occurs, starting with the natural spontaneous emission that is always present. For most continuously operating lasers, at any rate, the resulting pattern of standing light waves stabilizes itself eventually at a power level determined by the rate at which the amplifier can be fed with the necessary pump energy.

For the purposes of this chapter, we shall take it for granted that the problems of finding a laser material and of pumping it have been solved. We shall assume a cylindrical laser amplifier - for example a glass rod or a gas discharge tube, which produces a uniform gain across its aperture and which introduces no aberrations or changes in the shape of the wavefronts that it is called upon to amplify. Although this is an ideal situation, it is approached quite closely in many low-power gas lasers and also in some neodymium glass and dye laser systems.

As far as the third process of applying optical feedback is concerned, we shall ignore the many important transient effects that can occur and consider only the propagation of light in a steadily operating laser, in which the losses of the resonator are balanced exactly by the gains of the amplifier. Another aspect that we shall take for granted is the process of axial mode selection, which chooses exactly the right spacing for the light waves so that, when they have completed a 'round trip' inside the resonator, they return with the same phase as before. All that we shall consider

will be the question, 'What happens to the transverse
shape of the light waves as they travel inside the
resonator,'

To answer this question, we need to know the geometry
of the optical resonator. In the vast majority of
laser systems, the light is shuttled backwards and
forwards between two end-mirrors. It is also possible
to operate ring lasers, in which the light circulates
around a cyclic path, in either a clockwise or a
counter-clockwise direction. But such systems are
rare, and we shall defer their consideration until
later.

Let us now consider Figure III.5 which illustrates
the arrangement of a typical laser resonator. A laser

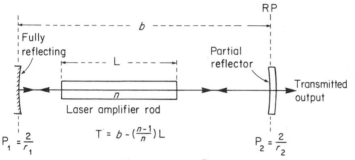

Figure III.5

amplifier rod of length L is placed between two end-
reflectors, which are spaced a distance b apart. Since
the laser rod is equivalent to a plane-parallel plate,
the translation matrix representing the gap between
the two mirrors will contain the reduced thickness

$$T = \frac{(b - L)}{1} + \frac{L}{n} = b - \frac{(n - 1)L}{n}$$

where n is the refractive index of the laser material.

Before we write down the ray-transfer matrices to
describe this system, we shall comment briefly on the
function of the partially transmitting mirror to the
right of the system. It is usually called the 'output
mirror', since it enables a suitable fraction of the
light energy stored in the resonator to escape as a
useful output for external use. The decision of what
reflectivity to use for an output mirror is rather
like that facing the cereal farmer who needs to retain
just enough seed for next year's harvest. With the

high gain provided by a neodymium laser, it sometimes happens that only 10 per cent of the energy is retained, while 90 per cent is coupled to the output: with helium-neon systems, on the other hand, the gain available from the laser amplifier is *very* small, and one cannot afford to couple out more than about 1 per cent. It is for this reason that such lasers refuse to oscillate unless they are built with high-grade multilayer reflectors and ultra-clean surfaces.

Since we shall be interested in the shape of the waves that will emerge from the output, it will be convenient to locate RP$_1$ at the partially reflecting mirror surface and to consider an initial ray $\begin{bmatrix} y_1 \\ V_1 \end{bmatrix}$ which arrives at RP$_1$ in the $+ z$-direction after emerging from the amplifier. That portion of the ray which is reflected by the output mirror then travels back through the amplifier to the left-hand mirror and then returns, again through the amplifier, to the output mirror.

If we now locate our final reference plane RP$_2$ so as to coincide with RP$_1$, we can write down the overall ray-transfer matrix M connecting this 'pair' of reference planes and representing a 'single round trip' through the resonator.

If we use P_1 and P_2 to denote the powers of the two end-mirrors, we have

$$\begin{bmatrix} y_2 \\ V_2 \end{bmatrix} = \begin{bmatrix} 1 & T \\ 0 & 1 \end{bmatrix} \begin{bmatrix} 1 & 0 \\ -P_1 & 1 \end{bmatrix} \begin{bmatrix} 1 & T \\ 0 & 1 \end{bmatrix} \begin{bmatrix} 1 & 0 \\ -P_2 & 1 \end{bmatrix} \begin{bmatrix} y_1 \\ V_1 \end{bmatrix}$$

$$= \begin{bmatrix} 1 - P_1 T - 2P_2 T + P_1 P_2 T^2 & T(2 - P_1 T) \\ -P_1 - P_2 + P_1 P_2 T & 1 - P_1 T \end{bmatrix} \begin{bmatrix} y_1 \\ V_1 \end{bmatrix}$$

$$= \begin{bmatrix} M \end{bmatrix} \begin{bmatrix} y_1 \\ V_1 \end{bmatrix} = \begin{bmatrix} A & B \\ C & D \end{bmatrix} \begin{bmatrix} y_1 \\ V_1 \end{bmatrix}$$

Having checked that the determinant of this matrix is unity, we have all that is needed to calculate the effect of single or, indeed, repeated traversals of the resonator.

In order to represent the effect of N successive round trips through the resonator, we need to raise the overall ray-transfer matrix M to the Nth power: to achieve this, we proceed according to the methods of matrix diagonalization described in sections 1.11 and 1.12. We seek a transformation such that

$$M = F\Lambda F^{-1}$$

where

$$\Lambda = \begin{bmatrix} \lambda_1 & 0 \\ 0 & \lambda_2 \end{bmatrix}$$

is a diagonal matrix and where

$$M^N = F\Lambda^N F^{-1}.$$

In order to identify the eigenvalues λ_1 and λ_2, we consider the trace $(A + D)$ of the matrix that we have calculated and obtain

$$(A + D) = (1 - P_1 T - 2P_2 T + P_1 P_2 T^2) + (1 - P_1 T)$$

$$= (2 - P_1 T)(2 - P_2 T) - 2$$

$$= 4\left(1 - \frac{T}{r_1}\right)\left(1 - \frac{T}{r_2}\right) - 2$$

Proceeding exactly as in section I.11, we now seek a value for θ or t such that

either $(A + D) = 2\cos\theta = 4\cos^2(\theta/2) - 2;$

then $\lambda_1 = e^{i\theta}$, $\lambda_2 = e^{-i\theta}$

or $(A + D) = 2\cosh t = 4\cosh^2(t/2) - 2;$

then $\lambda_1 = e^t$, $\lambda_2 = e^{-t}$

or else $(A + D) = -2\cosh(-t) = -4\sinh^2(-t/2) - 2;$

then $\lambda_1 = -e^t$, $\lambda_2 = -e^{-t}$

To decide which of these three alternatives is to be used, all that we need to do is to find whether the product $(1 - T/r_1)(1 - T/r_2)$ lies between 0 and 1, or above this range, or below it.

As was shown in section I.12, if λ_1 and λ_2 are the eigenvalues for the unimodular matrix $\begin{bmatrix} A & B \\ C & D \end{bmatrix}$, then one of its eigenvectors has its components in the ratio $(\lambda_1 - D)/C$, the other in the ratio $(\lambda_2 - D)/C$.

INTERPRETED IN THE PRESENT CONTEXT, THESE RATIOS ARE THE y/V-VALUES, OR R-VALUES, WHICH ARE PROPAGATED UNCHANGED THROUGH THE RESONATOR. IF A WAVEFRONT WITH THIS CURVATURE EXISTS WITHIN THE RESONATOR, IT WILL REPRODUCE ITSELF. This is an important point which we shall be discussing later.

For those who prefer not to use the diagonalization methods described in chapter I, it should perhaps be pointed out that, in this instance, much the same results can be obtained by appealing directly to the $ABCD$ rule. If an R-value is to be propagated unchanged, we have the two equations

$$R_2 = \frac{AR_1 + B}{CR_1 + D} \text{ and } R_1 = R_2$$

Eliminating R_2 we obtain a quadratic equation in R_1:

$$CR_1^2 + (D - A)R_1 - B = 0$$

with the solutions

$$R_1 = \frac{(A - D) \pm \sqrt{(A - D)^2 + 4BC}}{2C}$$

$$= \frac{(A + D \pm \sqrt{(A + D)^2 - 4}) - 2D}{2C}$$

It is not difficult to verify that these two solutions for R coincide with the eigenvector ratios $(\lambda_1 - D)/C$ and $(\lambda_2 - D)/C$ obtained by diagonalization.

Before we consider the significance of these ratios, let us consider some alternative forms of resonator geometry for which it may be necessary to formulate a ray-transfer matrix.

III.4.1 Alternative reflectors

Fig.III.6 shows two different kinds of retroreflecting assembly that are sometimes used instead of a fully-reflecting end-mirror. It will be recalled that for

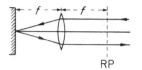

(a) Cat's-eye retroreflector

$$M = \begin{bmatrix} -1 & 0 \\ 0 & -1 \end{bmatrix}$$

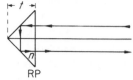

(b) Cube-corner retroreflector

$$M = \begin{bmatrix} -1 & \dfrac{-2t}{n} \\ 0 & -1 \end{bmatrix}$$

Figure III.6

this mirror the matrix was $\begin{bmatrix} 1 & 0 \\ -P & 1 \end{bmatrix}$, which, in the case of a flat mirror, reduces to a unit matrix.

(a) In order to describe the operation of a 'cat's-eye' reflector, we choose a reference plane which coincides with the first focal plane of the lens. If we consider the transfer of ray data from this plane to the flat mirror located in the second focal plane and back again, we obtain the overall matrix

$$M = \begin{bmatrix} 0 & f \\ -1/f & 0 \end{bmatrix} \begin{bmatrix} 1 & 0 \\ 0 & 1 \end{bmatrix} \begin{bmatrix} 0 & f \\ -1/f & 0 \end{bmatrix} = \begin{bmatrix} -1 & 0 \\ 0 & -1 \end{bmatrix}$$

In contrast with a plane mirror, therefore, the 'cat's-eye' *reverses* both the y- and the V-values. The change in the value of $V = n\sin v$ is caused by the change in the sign of n - the geometrical direction at which the ray returns is at 180° to its original direction.

(b) As is well known, exactly the same property is possessed by the cube-corner reflector, which reverses all three direction-cosines of a ray in whatever Cartesian coordinate system we care to employ. Provided that we choose a reference plane passing through the apex of the cube corner, its ray-transfer matrix will be of the form $\begin{bmatrix} -1 & 0 \\ 0 & -1 \end{bmatrix}$ for the (y,V)-values and also for the corresponding (x,U)-values when measured in the (x,z)-plane normal to the diagram.

If a cube-corner reflector is made in the form of a glass prism, then the effect of the glass thickness t measured from the apex to the hypotenuse face should also be included, producing an overall matrix referred to the hypotenuse face, which is

$$M = \begin{bmatrix} 1 & t/n \\ 0 & 1 \end{bmatrix} \begin{bmatrix} -1 & 0 \\ 0 & -1 \end{bmatrix} \begin{bmatrix} 1 & t/n \\ 0 & 1 \end{bmatrix} = \begin{bmatrix} -1 & -2t/n \\ 0 & -1 \end{bmatrix}$$

Exactly the same matrix applies when (y,V)-values are transferred by a 90° roof reflector whose roof axis is aligned normal to the (y,z)-plane, but in this case, the matrix for the corresponding (x,U)-values will be that of an ordinary plane end-mirror. The calculations for the (x,z)-plane and those for the (y,z)-plane must then be made on a separate basis.

III.4.2 *Bent and folded systems* (see Figure III.7)

There are several reasons why the axis of a laser

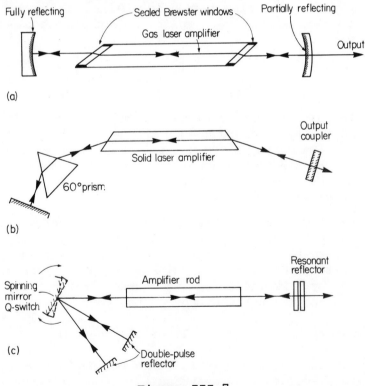

(a)

(b)

(c)

Figure III.7

resonator may not exist as a single straight line between the two end-mirrors:

(a) To avoid loss of light by reflection and to ensure that the laser beam is linearly polarized, the end surfaces, in the case of a glass amplifier, and the windows, in the case of a gas discharge tube, are often arranged at the Brewster angle.

(b) Again using refraction at the Brewster angle, prisms are often incorporated in order to assist the process of selecting the right laser wavelength.

(c) Folding by reflection may be introduced in order to save space, or to allow the inclusion of a rapidly rotating prism which acts as a Q-switch to generate one or more giant pulses.

(d) As has already been mentioned, the axis of the resonator may be bent round by several reflections and/or refractions so that it forms a complete cyclic path. Occasionally, this is arranged together with an isolator system to ensure that the radiation circulates in only one direction and no standing-wave patterns are produced. In most ring lasers (see Figure III.8), however, both cir-

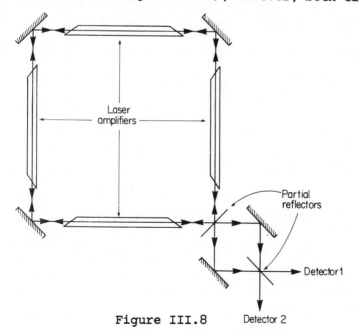

Figure III.8

culations are allowed to build up and the difference between their optical frequencies produces a
heterodyne beat note, which measures the absolute
rotation of the platform on which the resonator· is
carried. The result is a 'laser gyroscope', which
provides a very sensitive alternative to an inertia
gyroscope.

For all such systems, the procedure that we must
follow is to imagine a single ray travelling down the
optical axis of one section of the resonator and to
follow the course of this ray round other sections of
the resonator, inserting matrices as appropriate to
represent the various reduced gaps T and focal powers
P that it encounters. If one of the obliquely pres-
ented surfaces is a curved one, then astigmatism will
be generated, and the focal power P_y affecting the
propagation of rays in the (y,z)-plane will be greater
than the focal power P_x for the (x,z)-plane. Separate
matrices, M_y and M_x, will then be needed in order to
describe how the (y,V)-values and the (x,U)-values are
changed by a round trip through the resonator.

III.4.3 Misalignment effects

In following this procedure, it has to be assumed
that all the components of the resonator have been
positioned and oriented so that they conform to the
nominal geometry. In that case, the $\begin{bmatrix} 0 \\ 0 \end{bmatrix}$ ray that we
launch down the optical axis of one section of the
resonator will indeed repeat the same path after a
single traversal of the resonator has been accomplished
In practice, however, residual misalignment of each
component may cause small changes in either the ray
height or the ray angle, and the effect of all these
errors will be to produce a small lateral shift Δy and
a small angular shift ΔV such that the ray no longer
repeats its path along the same axis.

Because the treatment of these misalignments involves the use of an augmented 3×3 version of the ray-
transfer matrix, we shall discuss this matter separately in appendix B. The results that emerge,
however, are briefly as follows:

(a) If we are given for each ith intermediate reference plane the residual lateral and angular

misalignments that are generated, it is quite easy to calculate from these a pair of resultant values Δy and ΔV which can be associated with the overall ray-transfer matrix and which fully represent all the individual misalignments.

(b) Provided that the spur $(A + D)$ of this matrix is not equal to $+ 2$, we can find a *new effective axis* such that a ray launched along it does indeed return along exactly the same path after one resonator traversal. For this new effective ray, the ray height y_0 and ray angle V_0 are given by

$$y_0 = \frac{(1 - D)\Delta y + B\Delta V}{(2 - A - D)} \quad \text{and} \quad V_0 = \frac{C\Delta y + (1 - A)\Delta V}{(2 - A - D)}$$

Applying these results to the simple two-mirror resonator whose matrix spur we have already calculated, we find that it will be possible to find a new effective axis except when the geometry is such that

$$\left(1 - \frac{T}{r_1}\right)\left(1 - \frac{T}{r_2}\right) = 1$$

This implies that *either* $T = r_1 + r_2$ (in which case the mirrors are concentric), *or* r_1 and r_2 are both infinite (Fabry-Perot type system with flat mirrors) *or else* $T = 0$ (a situation that leaves no room for an amplifier!)

For all other systems the quantity $(2 - A - D)$ will not vanish and the centres of curvature of the two end-mirrors will be axially displaced from each other. It follows that, even if they do not quite coincide with the z-axis, the line joining them will still fall somewhere within the paraxial region of the system and will constitute the new effective axis. Even though the laser constructor may not know its location, if an effective axis exists, it will be found automatically by the laser process of regenerative build-up. It must be remembered, however, that if the misalignments of the mirrors are large and the étendue of the system is limited, as it usually is, then vignetting will occur and there will be an additional loss of energy from the resonator as the light encounters the stops of the system.

For continuously operating lasers, it is usually advantageous to avoid the condition $(A + D) = 2$, so that the optic axis is to some extent self-aligning. If a rotating mirror is employed to generate giant

pulses, however, it may be preferable to choose a resonator whose adjustment is critical; this helps to
ensure that the output pulse is a short one, with
oscillation confined to the moment when the rotating
prism sweeps through the *only* position for optical
feedback.

III.5 THE DISTINCTION BETWEEN STABLE AND UNSTABLE RESONATORS

As was shown in the last section, it is usually not
difficult to form a ray-transfer matrix that represents adequately the geometry of an optical resonator.
Having calculated the spur $(A + D)$, we can then find
solutions for eigenvalues λ_1 and λ_2 and for the
corresponding eigenvector ratios $(\lambda_1 - D)/C$ and
$(\lambda_2 - D)/C$.

It will be remembered that for values of $(A + D)$
between $+ 2$ and $- 2$ the eigenvalues are of the form
$e^{i\theta}$ and $e^{-i\theta}$ - complex numbers located on the unit
circle. To save confusion we specify that $\pi > \theta > 0$.
Outside this range, however, the eigenvalues are of
the form e^{t} and e^{-t} or, in the negative branch $- e^{t}$
and $- e^{-t}$ - real numbers which differ from unity. How
can we interpret these mathematical differences in
terms of light propagation inside the resonator?

Let us first consider a numerical example. The two-
mirror resonator shown in Figure III.9 consists of a
slightly convex mirror of 8 metres radius of curvature, spaced 1 metre away from a plane output mirror.

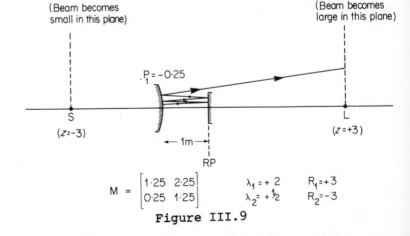

$$M = \begin{bmatrix} 1.25 & 2.25 \\ 0.25 & 1.25 \end{bmatrix} \qquad \begin{aligned} \lambda_1 &= +2 \\ \lambda_2 &= +\tfrac{1}{2} \end{aligned} \qquad \begin{aligned} R_1 &= +3 \\ R_2 &= -3 \end{aligned}$$

Figure III.9

Using the round-trip matrix of section III.4, and working in terms of metres and dioptres, we insert the values $T = 1 \cdot 0$, $P_1 = - 0 \cdot 25$ and $P_2 = 0$, and obtain for the round-trip from output mirror back to output mirror the matrix

$$M = \begin{bmatrix} 1 \cdot 25 & 2 \cdot 25 \\ 0 \cdot 25 & 1 \cdot 25 \end{bmatrix}$$

the spur of which is $(A + D) = 2 \cdot 5$. In order to solve for the eigenvalues, we specify that t is a positive real number and find that $\lambda_1 = 2$ and $\lambda_2 = 0 \cdot 5$. The corresponding eigenvector ratios are therefore $(\lambda_1 - D)/C = + 3$ and $(\lambda_2 - D)/C = - 3$. Interpreted as R-values, the former ratio indicates a spherical wave diverging from a point 3 metres to the left of the output mirror and the second indicates a converging wave directed to a point 3 metres to its right.

Consider now one of the rays in the ray pencil diverging from S, for example, the ray $\begin{bmatrix} 0 \cdot 03 \\ 0 \cdot 01 \end{bmatrix}$, which is quite close to the axis. Because it is one of the eigenvectors of the matrix M, we shall find that after one round trip it becomes $\begin{bmatrix} 0 \cdot 03\lambda_1 \\ 0 \cdot 01\lambda_1 \end{bmatrix} = 2 \begin{bmatrix} 0 \cdot 03 \\ 0 \cdot 01 \end{bmatrix}$, since 2 is the relevant eigenvalue. The new ray still belongs to the same pencil, but it is twice as far from the axis as before. After n traversals of the resonator, it will be 2^n times as far away and eventually it will encounter one of the stops of the system.

To visualize what is happening more clearly, let us convert our matrix to a new 'pair' of reference planes, both located at S, the point from which the ray pencil is diverging. Premultiplying and postmultiplying M by the appropriate $\mathcal{T}$-matrices, we obtain

$$\begin{bmatrix} 1 & -3 \\ 0 & 1 \end{bmatrix} \begin{bmatrix} 1 \cdot 25 & 2 \cdot 25 \\ 0 \cdot 25 & 1 \cdot 25 \end{bmatrix} \begin{bmatrix} 1 & 3 \\ 0 & 1 \end{bmatrix} = \begin{bmatrix} 1 & -3 \\ 0 & 1 \end{bmatrix} \begin{bmatrix} 1 \cdot 25 & 6 \\ 0 \cdot 25 & 2 \end{bmatrix}$$

$$= \begin{bmatrix} 0 \cdot 5 & 0 \\ 0 \cdot 25 & 2 \end{bmatrix}$$

Because the top right-hand element is zero, this new matrix represents an imaging relationship. For each resonator traversal, whatever distribution exists in the S-plane will be compressed in y-value and expanded in V-value by the magnification factor 2 (or, in the general case, $\lambda_1 = e^t$).

If, on the other hand, we perform the corresponding calculation for a reference plane containing L, we find a matrix $\begin{bmatrix} 2 & 0 \\ 0 \cdot 25 & 0 \cdot 5 \end{bmatrix}$ which indicates the opposite effect - in this plane the y-values are expanded and the V-values are compressed.

On a geometrical optics basis, therefore, whatever shape of beam is launched initially into the system (for example by spontaneous emission), the effect of shuttling backwards and forwards will cause the beam cross-section to shrink to a very Small central region in the S-plane and to become very Large in the L-plane. Eventually, the stops in the L-plane set a limit to this expansion, but the contraction in the S-plane continues until it reaches a diffraction limit determined by the maximum V-value that the system will handle without obstruction.

Instead of using a partially transmitting output mirror, laser designers often use a fully reflecting central region, the output being the annular portion of the beam that expands radially and escapes round the edge of the mirror after its last pass through the resonator (see Figure III.9). Although this beam is diverging slightly, it has diffraction-limited directionality, so it can be collimated, or focused to a small spot again, by means of an external system.

The resonator that we have been considering is of the inherently-lossy type that is usually referred to as 'unstable'. This is in some ways a misleading term, since the *shape* of the wavefronts being propagated converges smoothly and rapidly onto the self-repeating R-values specified by the eigenvector; nevertheless, with each round trip there occurs a radial expansion of the associated ray pencil and there is a progressive 'walk-off' of each individual ray, until it is lost from the system. Because of the radial expansion in both the x- and the y-directions, the energy returning to the central region of the output mirror is diluted

fourfold after each traversal, so the laser amplifier
needs to provide a double-passed gain of at least four,
(or, in the general case, $\lambda_1^2 = e^{2\mathcal{L}}$), in order to build
up or maintain laser oscillation.

Because of the need for a high-gain amplifier, the
unstable type of resonator did not at first find
favour with laser designers. Its great advantage is
that, even if constructed on a large scale, with a
very large Fresnel number, it produces a highly-
directional output without being too sensitive to
small changes in its adjustment. Unstable resonators
of one form or another are likely to dominate the
future design of many high-power industrial laser
systems. For carbon dioxide lasers, which operate in
the infrared at 10·6 μm wavelength, it may be preferable to use mirrors rather than lenses for any
external optics, so the gap in the annular output
beam is not disadvantageous.

Several other forms of unstable resonator are
possible, and it is not difficult to design them to
give a collimated output. Figure III.10 illustrates
two possibilities in which a pair of mirrors is
arranged to form an afocal system whose magnification differs from unity.

For the two concave mirrors shown in Fig.III.10a, we

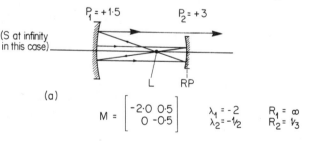

(a)

$$M = \begin{bmatrix} -2\cdot0 & 0\cdot5 \\ 0 & -0\cdot5 \end{bmatrix} \qquad \begin{matrix} \lambda_1 = -2 \\ \lambda_2 = -\frac{1}{2} \end{matrix} \qquad \begin{matrix} R_1 = \infty \\ R_2 = \frac{1}{3} \end{matrix}$$

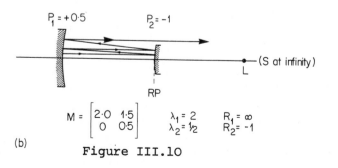

$$M = \begin{bmatrix} 2\cdot0 & 1\cdot5 \\ 0 & 0\cdot5 \end{bmatrix} \qquad \begin{matrix} \lambda_1 = 2 \\ \lambda_2 = \frac{1}{2} \end{matrix} \qquad \begin{matrix} R_1 = \infty \\ R_2 = -1 \end{matrix}$$

(b)

Figure III.10

have $P_1 = + 1 \cdot 5$, $P_2 = + 3$ and $T = 1$, so that

$M = \begin{bmatrix} -2 & 0 \cdot 5 \\ 0 & -0 \cdot 5 \end{bmatrix}$. The spur of this matrix is $- 2 \cdot 5$ and

its eigenvalues are evidently $\lambda_1 = - 2$ and $\lambda_2 = - 0 \cdot 5$. Of the eigenvector ratios, the first $(\lambda_1 - D)/C$ is evidently infinite, since C vanishes for this afocal system. For the second ratio, the formula $(\lambda_2 - D)/C$ gives an indeterminate result, but the alternative expression $B/(\lambda_2 - A)$ (see section I.12.1) gives $1/3$ as the second possible R-value. The two planes which are repeatedly imaged back on themselves are the plane at infinity and the plane passing through the common focus of the two mirrors. With successive traversals of the resonator, the beam cross-section will expand to fill the latter, and it will shrink to the diffraction limit in the plane at infinity. The gain needed for laser oscillation will be the same as in the previous example.

For high-power operation, at any rate, this system has a serious disadvantage: whenever the radiation is travelling in the $- z$-direction it is brought to focus inside the resonator at L and the sheer concentration of power in this region will almost certainly create a plasma by dielectric breakdown. It is therefore better to use a telescopic system with positive magnification as shown in Figure III.10b. In this case we have

$P_1 = + 0 \cdot 5$, $P_2 = - 1$ and $T = 1$. Hence $M = \begin{bmatrix} 2 \cdot 0 & 1 \cdot 5 \\ 0 & 0 \cdot 5 \end{bmatrix}$

and the spur $(A + D)$ has the same positive value, $2 \cdot 5$, as in our first example. The dominant R-value, representing the output, is

$$\frac{\lambda_1 - D}{C} = \frac{2 - 0 \cdot 5}{0} = \infty$$

The other eigenvector ratio is

$$\frac{B}{\lambda_2 - A} = \frac{1 \cdot 5}{0 \cdot 5 - 2 \cdot 0} = - 1$$

so the L-plane is located 1 metre to the right of the output mirror - at the common focus of the two mirrors in fact. As far as our output beam travelling in the $+ z$-direction is concerned, it will concentrate itself

into a diffraction-limited spot in the S-plane at
infinity, while in the L-plane its cross-section will
become as large as the stops of the system allow.

All of the examples considered so far have been of
the inherently-lossy, or unstable, type; the dominant
eigenvalue e^t represents the transverse magnification
(and the square root of the round trip gain required),
while the real eigenvector $(e^t - D)/C$ or, alternatively,
$B/(e^t - A)$ predicts the geometrical curvature radius
of the generated wavefront.

Let us now consider the geometry of a stable resonator. This is the type invariably used for a low-
power helium-neon gas laser which has a very low gain
coefficient. Two examples are shown in Figure III.11.

The resonator in Figure III.11a uses two concave
mirrors, each of 10 metres radius of curvature, separated by 34 cm. Working in metres and dioptres to
three decimal places, we have $P_1 = P_2 = + 0 \cdot 200$ and
$T = 0 \cdot 340$. We then calculate $M = \begin{bmatrix} 0 \cdot 800 & 0 \cdot 657 \\ -0 \cdot 386 & 0 \cdot 932 \end{bmatrix}$
and the spur $(A + D) = 1 \cdot 732 = 2\cos\theta$ (evidently
$\theta = 30°$ in this case). Since θ is restricted to be
in the range from $0°$ to $180°$, the eigenvalues are

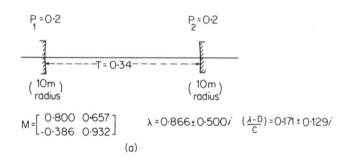

(a)

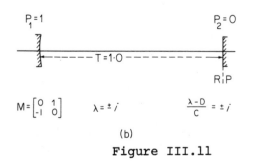

(b)

Figure III.11

then $e^{\pm i\theta}$ = cosθ ± $i$sinθ = 0·866 ± 0·500i, and the (complex) eigenvector ratios are

$$\frac{0·866 \pm 0·500i - 0·932}{-0·386} = 0·171 \pm 1·295i$$

For the particular geometry we have chosen, the angle θ is (almost exactly) 30° or $\pi/6$ radians. One immediate consequence is that if we consider *twelve* complete round trips in this resonator, then the over-all transfer matrix will collapse into the unit matrix Using F as in section I.11 to represent a diagonalizing matrix, we have

$$M^{12} = F\Lambda^{12}F^{-1}$$

$$= FIF^{-1} = I$$

where

$$\Lambda^{12} = \begin{bmatrix} e^{12i\theta} & 0 \\ 0 & e^{-12i\theta} \end{bmatrix} = \begin{bmatrix} 1 & 0 \\ 0 & 1 \end{bmatrix}$$

It follows that whatever paraxial ray is launched into the resonator will find itself repeating its original trajectory after twelve round trips have been completed. In this sense, therefore, we now have a stable resonator which does not allow individual rays to walk away from the axis and lose themselves.

As we shall see presently, it is by no means necessary to design a resonator so that a particular value of θ is exactly realised - *any* value will give stable oscillation in the sense of freedom from loss due to 'walk-off'. But the second example, shown in Figure III.11b, may help to emphasize the need to interpret these complex eigenvector ratios. It represents an ideally stable 'hemi-confocal' arrangement, in which

$P_1 = 1$, $P_2 = 0$ and $T = 1$. We then find $M = \begin{bmatrix} 0 & 1 \\ -1 & 0 \end{bmatrix}$ and $(A + D) = 0 = 2\cos\theta$ (evidently $\theta = 90°$ in this case). The eigenvalues are then $e^{\pm i\pi/2} = \pm i$ and the eigenvector ratios are $(\pm i - 0)/-1 = \mp i$ (wholly imaginary). It is not difficult to verify that for this matrix $M^2 = \begin{bmatrix} -1 & 0 \\ 0 & -1 \end{bmatrix}$ and $M^4 = I$. In double-pass the resonator is equivalent to the well-known confocal resonator and in quadruple-pass each paraxial ray repeats its original trajectory.

As we shall see in later sections, it is the complex eigenvector ratio rather than the actual eigenvalue which is usually of importance to the laser designer. For completeness, however, we now give a formula which enables us to calculate the effect of an n-fold traversal of any stable resonator.

If M is a unimodular matrix, with eigenvalues $e^{\pm i\theta}$, then

$$
M^n = \begin{bmatrix} A & B \\ C & D \end{bmatrix}^n = \begin{bmatrix} \dfrac{\sin(n+1)\theta - D\sin n\theta}{\sin\theta} & \dfrac{B\sin n\theta}{\sin\theta} \\[2ex] \dfrac{C\sin n\theta}{\sin\theta} & \dfrac{D\sin n\theta - \sin(n-1)\theta}{\sin\theta} \end{bmatrix}
$$

This result, which is known as Sylvester's theorem, can be used for any value of θ and for any integral value of n. (If trigonometric functions of θ are replaced by corresponding hyperbolic functions of t, then the same result can be used for n-fold traversals of an unstable resonator.)

The student who needs practice with de Moivre's theorem may like to verify that the above formula can be derived directly from the diagonalization procedure discussed in section I.12:

$$M^n = F\Lambda^n F^{-1}$$

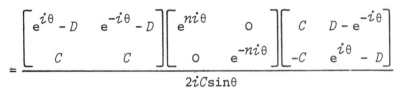

$$
= \frac{\begin{bmatrix} e^{i\theta} - D & e^{-i\theta} - D \\ C & C \end{bmatrix}\begin{bmatrix} e^{ni\theta} & 0 \\ 0 & e^{-ni\theta} \end{bmatrix}\begin{bmatrix} C & D - e^{-i\theta} \\ -C & e^{i\theta} - D \end{bmatrix}}{2iC\sin\theta}
$$

He may also like to check the following points:

(a) If we insert $n = 1$ in Sylvester's formula, then, since $(2\cos\theta) = (A + D)$, we obtain $M = \begin{bmatrix} A & B \\ C & D \end{bmatrix}$ as required.

(b) If we insert $n = -1$, we obtain the reciprocal matrix $M^{-1} = \begin{bmatrix} D & -B \\ -C & A \end{bmatrix}$.

(c) If $n\theta$ is an integral multiple of 2π, then we obtain the unit matrix.

III.6 PROPAGATION OF A GAUSSIAN BEAM AND ITS COMPLEX CURVATURE PARAMETER

At this stage in our discussion of stable resonators, we need to introduce the notion of a 'Gaussian beam'. This is a term used to describe a diffraction-limited beam of coherent radiation, whose energy remains concentrated near the axis of propagation and falls off rapidly according to a smooth Gaussian function. Such a beam is really the closest approximation that diffraction allows to a single ray or to a pencil of paraxial rays; as science marches on, we find that a concept which derives from Gauss only in a mathematical sense is now doing much to refine and enrich the ideas of 'Gaussian optics'.

As has been shown by Kogelnik and others, the details of how a Gaussian beam is propagated in free space are readily derived from the wave equation. Near the optic axis, the amplitude distribution $A(r,z)$ of a 'fundamental Gaussian mode' is described by

$$A(r,z) = A_0\left(\frac{w_0}{w}\right) \exp\left[i\left(\frac{2\pi z}{\lambda} + \phi\right) - r^2\left(\frac{1}{w^2} - \frac{2\pi i}{2\lambda R}\right)\right]$$

In this expression, the term $2\pi z/\lambda$ describes the phase-change which occurs along the axis of propagation; the term ϕ represents a small additional phase variation which depends on z according to the equation

$$\tan\phi = \left(\frac{\lambda z}{\pi w_0^2}\right)$$ (Note that λ as used here represents the wavelength of light, *not* an eigenvalue.

The coefficient of r^2 contains a real and imaginary component, both of which are vital to this chapter. The real component $1/w^2$ indicates that, in the radial direction, the amplitude modulus varies according to the Gaussian function $\exp(-r^2/w^2)$. The radius w is thus a 'spot radius', at which the light amplitude falls to $1/e$ and the irradiance or intensity to $1/e^2$

of its central value. The imaginary component $-2\pi i/2\lambda R$ describes the quadratic variation of the phase of the wavefield with radius and the term R represents in the usual way the radius of curvature of the surfaces of constant phase, the wavefronts travelling in the $+z$-direction.

As the Gaussian beam is propagated in space, the effects of diffraction cause it to expand and diverge slowly, so that both the spot radius w and the curvature radius R vary slowly with the z-coordinate. As derived from the wave equation, the laws governing these two parameters are

$$w(z)^2 = w_0^2\left[1 + \left(\frac{\lambda z}{\pi w_0^2}\right)^2\right]$$

and

$$R(z) = z\left[1 + \left(\frac{\pi w_0^2}{\lambda z}\right)^2\right]$$

See Figure III.12 for an illustration of the behaviour of these parameters when plotted in the yz-plane. It will be seen that the curve representing the locus of the $1/e^2$ radius is a hyperbola, whose closest approach to the z-axis is w_0 at $z = 0$ and whose

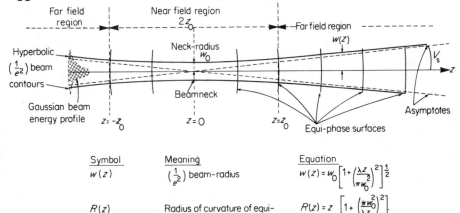

Symbol	Meaning	Equation
$w(z)$	$(\frac{1}{e^2})$ beam-radius	$w(z) = w_0\left[1 + \left(\frac{\lambda z}{\pi w_0^2}\right)^2\right]^{\frac{1}{2}}$
$R(z)$	Radius of curvature of equi-phase surfaces	$R(z) = z\left[1 + \left(\frac{\pi w_0^2}{\lambda z}\right)^2\right]$
$q(z)$	Complex curvature parameter	$\frac{1}{q(z)} = \frac{1}{R(z)} + \frac{i\lambda}{\pi[w(z)]^2}$
z_0	Confocal beam parameter	$z_0 = \frac{\pi w_0^2}{\lambda}$
V_s	Asymptotic angle of beam divergence	$V_s = \pm\left(\frac{\lambda}{\pi w_0}\right)$

Figure III.12

asymptotes are at an angle $V_S = \pm\ (\lambda/\pi w_0)$ to the
z-axis.

The surfaces of constant phase are planes near the 'neck' of the Gaussian beam and they acquire their strongest curvature at a distance $\pm\ z_0 = \pm\ \pi w_0^2/\lambda$ from the centre. The central region of length $2z_0$, over which the beam cross-section remains nearly constant, is sometimes referred to as the 'near field' and the diverging asymptote region as the 'far field'.

The initial term w_0/w represents the fact that the electric field strength is greater near the neck or focus of the beam than in the far-field region where the beam has expanded. But the total energy within the beam remains the same for all values of z (in the absence of an absorbing or amplifying medium); 86 per cent of the energy remains inside the $1/e^2$ contours.

Cumbersome though they may be, the equations given above for the spot radius $w(z)$ and the curvature radius $R(z)$ enable us to calculate how a Gaussian beam is propagated in free space or across a gap. It is intuitively obvious that if a Gaussian beam encounters a thin lens or a surface with converging power P, then its spot radius will be unchanged, but its divergence $(1/R)$ will be reduced to $(1/R) - P$.

A neat method of keeping track of both these beam parameters is obtained if we combine them into a single 'complex curvature parameter' $q(z)$. Instead of writing the coefficient of r^2 in the amplitude expression as $(1/w^2) - (2\pi i/2\lambda R)$, we write it simply as $-\ 2\pi i/2\lambda q$, where q is understood to be complex.

From this definition, we have immediately

$$\frac{1}{q} = \frac{1}{R} + \frac{i\lambda}{\pi w^2}$$

The real part of $1/q$ represents the divergence of the surfaces of constant phase and the imaginary part is a measure of $1/w^2$, the degree of power concentration in the axial region of the beam.

By substituting in the formulae already given for $R(z)$ and $w(z)$, it is not difficult to verify that for propagation in free space $q(z) = q_0 + z$, where $q_0 = (i\lambda/\pi w_0^2)^{-1}$ is the q-value obtained at the neck of the beam ($z = 0$). For propagation across a gap of width T, we shall therefore have $q_2 = q_1 + T$, exactly as if the q-value were a real R-value.

For refraction by a thin lens or surface of power P, we shall have four equations. By definition,

$$\frac{1}{q_2} = \frac{1}{R_2} + \frac{i\lambda}{\pi w_2^2} \quad \text{and} \quad \frac{1}{q_1} = \frac{1}{R_1} + \frac{i\lambda}{\pi w_1^2}$$

Thirdly, $1/R_2 = 1/R_1 - P$ represents the change of divergence and, finally, $w_2 = w_1$ represents the constancy of spot radius. Combining these equations, we find $1/q_2 = 1/q_1 - P$, so that once again the equation we need to transform the complex q-value is exactly the same as we should use for a real R-value.

Although the foregoing depends on formulae that we have quoted without proof, it indicates that we can transform a complex curvature parameter q just as if it were an R-value, using either a $\cdot\mathcal{T}$-matrix or an $\mathcal{R}$-matrix, or, indeed, any combination thereof. Furthermore, if the overall matrix representing a complete system has been calculated, we can use the $ABCD$ rule to obtain

$$q_2 = \frac{Aq_1 + B}{Cq_1 + D}$$

as our output value for the complex curvature parameter.

In section III.7, we shall show that the hitherto puzzling complex eigenvector ratio obtained for a stable resonator can be identified with the q-value of the fundamental Gaussian beam that the resonator will generate, given suitable apertures and sufficient laser amplification.

III.6.1 *Verification of the ABCD rule for Gaussian beams*

Before we do that, however, let us use the diffraction formula discussed in section III.2 to verify that if q_1 specifies the Gaussian beam existing in RP_1, then the output in RP_2 will be another Gaussian beam specified by

$$q_2 = \frac{Aq_1 + B}{Cq_1 + D}$$

Starting with the general diffraction formula for one dimension

$$A_2(y_2) = \left(\frac{-i}{\lambda B}\right)^{\frac{1}{2}} \int_{-\infty}^{\infty} A_1(y_1) \, \exp\left[\frac{2\pi i}{\lambda} W(y_1, y_2)\right] dy_1$$

we insert the Gaussian amplitude

$$A_1(y_1) = \exp\left[\frac{2\pi i y_1^2}{2\lambda q_1}\right]$$

and the eikonal formula

$$W(y_1, y_2) = \left(\frac{Ay_1^2 + Dy_2^2 - 2y_1 y_2}{2B}\right)$$

Hence

$$A_2(y_2) = \left(\frac{-i}{\lambda B}\right)^{\frac{1}{2}} \int_{-\infty}^{+\infty} \exp\left[\frac{2\pi i}{\lambda}\left(\frac{y_1^2}{2q_1} + \frac{Ay_1^2}{2B} - \frac{2y_1 y_2}{2B} + \frac{Dy_2^2}{2B}\right)\right] dy_1$$

$$= \left(\frac{-i}{\lambda B}\right)^{\frac{1}{2}} \exp\left[\frac{2\pi i D y_2^2}{2\lambda B}\right] \int_{-\infty}^{+\infty} \exp\left[y_1^2\left(\frac{\pi i}{\lambda q_1} + \frac{\pi i A}{\lambda B}\right) - y_1\left(\frac{2\pi i y_2}{\lambda B}\right)\right] dy_1$$

To evaluate the integral, we make two substitutions:

$$v = -\frac{y_2}{\lambda B} \quad \text{and} \quad \frac{i}{\lambda q_1} + \frac{iA}{\lambda B} = -\frac{1}{s^2}$$

The integral then becomes

$$\int_{-\infty}^{+\infty} \exp\left[-\frac{\pi y_1^2}{s^2}\right] \exp\left[2\pi i y_1 v\right] dy_1$$

This is a known Fourier transform, whose value is $\left[s\exp(-\pi v^2 s^2)\right]$. On substituting back, therefore, we obtain

$$A_2(y_2) = \left(\frac{-i}{\lambda B}\right)^{\frac{1}{2}} \exp\left[\frac{2\pi i D y_2^2}{2\lambda B}\right] \left[\frac{-1}{(i/\lambda q_1 + iA/\lambda B)}\right]^{\frac{1}{2}}$$

$$\times \exp\left[\frac{-\pi y_2^2}{\lambda^2 B^2} \frac{(-1)}{(i/\lambda q_1 + iA/\lambda B)}\right]$$

$$= \left[\frac{q_1}{Aq_1 + B}\right]^{\frac{1}{2}} \exp\left[\frac{2\pi i D y_2^2}{2\lambda B}\right] \exp\left[\frac{-\pi y_2^2}{\lambda^2 B^2} \frac{i\lambda B q_1}{(Aq_1 + B)}\right]$$

$$= \left[\frac{q_1}{Aq_1 + B}\right]^{\frac{1}{2}} \exp\left[\frac{\pi i y_2^2}{\lambda B}\left(D - \frac{q_1}{Aq_1 + B}\right)\right]$$

Rearranging

$$\left(D - \frac{q_1}{(Aq_1 + B)}\right) \quad \text{as} \quad \frac{ADq_1 + BD - q_1}{Aq_1 + B} \quad \text{or} \quad \frac{B(Cq_1 + D)}{(Aq_1 + B)}$$

we finally obtain

$$A_2(y_2) = \left[\frac{q_1}{Aq_1 + B}\right]^{\frac{1}{2}} \exp\left[\frac{2\pi i y_2^2}{\lambda} \frac{(Cq_1 + D)}{2(Aq_1 + B)}\right]$$

It will be shown in section III.8 that the complex quantity $\left[q_1/(Aq_1 + B)\right]^{\frac{1}{2}}$ which appears in front of the exponential term can be re-expressed as $\left[(w_1/w_2)\exp(i\phi_{12})\right]^{\frac{1}{2}}$ where w_2 is the real spot radius of the beam and ϕ_{12} is a real phase shift introduced between the two reference planes. (There will also, of course, be a very large phase shift of the whole beam introduced by the con-

stant term $\int_{z_1}^{z_2} n \, dz$, which we have omitted from the eikonal.)

We now turn our attention to the exponent in square brackets, and confirm that it exhibits the quadratic dependence on y_2 that characterizes a Gaussian beam. Identifying the exponent with $\left[(2\pi i/\lambda)(y_2^2/2q_2)\right]$, we can now verify that q_2, the complex curvature parameter of the Gaussian beam in the output plane, is indeed given by

$$q_2 = \frac{Aq_1 + B}{Cq_1 + D}$$

The above calculation has been done on a one-dimensional basis - effectively for cylindrical waves. Conveniently enough, however, any Gaussian function which represents the amplitude of a spherical wave can be regarded as the product of two separate one-dimensional cylindrical wave-functions.

Thus, if we multiply $A_2(y_2)$ by the corresponding expression for the x-direction we obtain, for spherical waves,

$$A_2(r_2) = A_2(x_2)A_2(y_2) = \left[\frac{q_1}{Aq_1 + B}\right] \exp \frac{2\pi i}{\lambda}\left[\frac{x_2^2 + y_2^2}{2q_2}\right]$$

$$= \left(\frac{w_1}{w_2}\right) \exp\left[i\phi_{12} + \frac{2\pi i}{\lambda}\left(\frac{r_2^2}{2q_2}\right)\right]$$

where w_1 and w_2 are the two spot radii and

$$q_2 = \left(\frac{Aq_1 + B}{Cq_1 + D}\right)$$

is the q-value for the output reference plane.

III.7 PREDICTING THE OUTPUT OF A LASER OSCILLATOR

We shall now interpret the complex eigenvector ratio for a stable resonator as a q-value which specifies the geometry of the Gaussian beam that will be generated if the system is made to oscillate in a fundamental mode.

According to the results of section III.5, if $\begin{bmatrix} A & B \\ C & D \end{bmatrix}$ is the resonator matrix and if $\exp(\pm i\theta)$ are its two eigenvalues, then the q-values which remain unchanged after a round trip are given by the equations

$$q = \frac{\exp(\pm i\theta) - D}{C} \quad \text{or, alternatively,} \quad q = \frac{B}{\exp(\pm i\theta) - }$$

where $\pi > \theta > 0$. Inverting the second equation and remembering that $\cos\theta = (A + D)/2$, we have

$$\frac{1}{q} = \frac{\exp(\pm i\theta) - A}{B} = \frac{\cos\theta - A}{B} \pm \frac{i\sin\theta}{B}$$

$$= \frac{(D - A)}{2B} \pm \frac{i\sin\theta}{B}$$

According to our defining equation, however, we must have

$$\frac{1}{q} = \frac{1}{R} + \frac{i\lambda}{\pi w^2}$$

Identifying real and imaginary parts, therefore, we have

$$\frac{1}{R} = \frac{D - A}{2B}$$

and, depending on which eigenvalue is used,

$$\frac{\lambda}{\pi w^2} = \frac{\pm \sin\theta}{B}$$

It can be shown that the negative value for w^2 generated by the second eigenvalue corresponds to an impossible situation for which the beam energy would have to increase enormously away from its centre. Discarding this solution and retaining only the first eigenvalue, $\lambda_1 = \exp(+ i\theta)$, we obtain a 'fundamental' Gaussian beam, whose spot radius is given by $w = (\lambda B/\pi\sin\theta)^{\frac{1}{2}}$.and whose surfaces of constant phase have a curvature given by $R = 2B/(D - A)$. The quantity $\sin\theta$ appearing above can, of course, be calculated directly from the matrix elements as

$$\sin\theta = \left[1 - \left(\frac{A + D}{2}\right)^2\right]^{\frac{1}{2}}$$

As an alternative way of characterizing the Gaussian beam, we can decompose the equation for q into its real and imaginary components. On the one hand, we have

$$q = \frac{\exp(+ i\theta) - D}{C} = \frac{A - D}{2C} + \frac{i\sin\theta}{C}$$

On the other hand, as was stated in the last section, if $q_0 = - i\pi w_0^2/\lambda$ represents the q-value of a Gaussian beam measured at its neck, then for a reference plane at a distance z to the right of the neck, we shall obtain (by application of the $ABCD$ rule for a gap z)

$$q = z + q_0 = z - \frac{i\pi w_0^2}{\lambda}$$

Comparing these two equations for the q-value at the output reference plane, we conclude that the neck of the beam is located at a distance $(A - D)/2C$ to the left of the reference plane and that the minimum spot radius which obtains there is

$$w_0 = \left(\frac{-\lambda\sin\theta}{\pi C}\right)^{\frac{1}{2}}$$

The distance $\pi w_0^2/\lambda$, which is sometimes referred to as the confocal beam parameter z_0, is given by

$$z_0 = \frac{-\sin\theta}{C}$$

Table 2. Relations between the matrix of a resonator and its optical properties

Let the matrix representing a round trip be $\begin{bmatrix} A & B \\ C & D \end{bmatrix}$ where $(AD - BC) = 1$ and all the elements are real.

Unstable systems	Property considered	Stable systems
$\frac{(A + D)}{2} > 1$ gives unstable resonator (positive branch) $\frac{(A + D)}{2} < -1$ gives unstable resonator (negative branch)	Matrix spur $(A + D)$	$1 > \frac{(A + D)}{2} > -1$ gives a stable resonator.
$\lambda_1 = \pm \exp(t) = \pm (\cosh t + \sinh t)$ * where $\cosh t = \pm \frac{(A + D)}{2}$ * $\sinh t = + \left[\left(\frac{A + D}{2} \right)^2 - 1 \right]^{\frac{1}{2}}$	Main eigenvalue λ_1 (t is taken positive, and θ in the range from π to O)	$\lambda_1 = \exp(i\theta) = \cos\theta + i\sin\theta$ where $\cos\theta = \frac{(A + D)}{2}$ $\sin\theta = \left[1 - \left(\frac{A + D}{2} \right)^2 \right]^{\frac{1}{2}}$
	Eigenvector ratios $\left(\frac{u}{V} \right) = \left(\frac{\lambda_1 - D}{C} \right)$ $\left(\frac{u}{v} \right) = \left(\lambda_1 - A \right)$	
Radius of curvature R $R = \frac{(\lambda_1 - D)}{C} = \frac{(A - D)}{2C} \pm \frac{\sinh t}{C}$ * $\frac{1}{} = (D - A) + \sinh t$ *		Complex curvature parameter q $q = \frac{(A - D)}{2C} + \frac{i\sin\theta}{C} = z - iz_0$ $\frac{1}{q} = \frac{(D - A)}{2B} + \frac{i\sin\theta}{B} = \frac{1}{R} + \frac{i\lambda}{}$ +

	parameters	equation for 1/q:
needed for the round trip (two-dimensional case) will be exp(2t). The above equations specify the radius of curvature of the output wavefront.	1. Radius of curvature	$R = \dfrac{2B}{(D-A)}$ ⎫ measured at
	2. Divergence of wavefront	$\dfrac{1}{R} = \dfrac{(D-A)}{2B}$ the output reference
	3. Beam radius	$w = \left(\dfrac{\lambda B}{\pi\sin\theta}\right)^{\frac{1}{2}}$ ⎭ plane
	4. Location of neck	From equation for q we obtain: $z = \dfrac{(A-D)}{2C}$ (to the left of the reference plane)
	5. Neck radius	$w_0 = \left(\dfrac{-\lambda\sin\theta}{\pi C}\right)^{\frac{1}{2}}$
	6. Confocal beam parameter	$z_0 = \dfrac{\pi w_0^2}{\lambda} = \dfrac{-\sin\theta}{C}$
	7. Far-field semi-angle (in radians)	$\dfrac{\lambda}{\pi w_0} = \dfrac{w_0}{z_0} = \left(\dfrac{-\lambda C}{\pi\sin\theta}\right)^{\frac{1}{2}}$
Good provided that the system operates with a large Fresnel number and appreciable gain.	Mode discrimination	Very poor unless the Fresnel number is kept small.

*Where two alternative signs are given, the upper one should be used for the positive branch and the lower one for the negative branch.

†Care must be taken to distinguish the *wavelength* λ from the *dominant eigen-value* represented by the symbol λ₁

The relations that we have just discussed, together with comparable relations that we obtained for an unstable resonator, are summarized in Table 2. Before we consider examples of their use, however, a caution will be sounded about the need to consider questions of mode discrimination.

First, unless a stable resonator is constructed with rather small apertures, so that the Fresnel number is comparable with unity, the loss of energy by diffraction at the edges will be extremely small, and there is no guarantee that only a centred Gaussian beam will be generated when laser gain is applied. Indeed, in some gas lasers, the gain produced by the discharge increases slightly near the walls of the tube, so that a higher-order transverse mode may actually be preferred.

Secondly, the mathematician might object that, if the Fresnel number is small enough to produce appreciable energy loss with each round trip, the modes should be described in terms of 'prolate spheroidal functions'. In practice, however, the amplitude distribution that results can still be represented very closely by a Gaussian function, multiplied for higher modes by a polynomial of appropriate order (Hermite polynomial for rectangular aperture and generalized Laguerre polynomial for circular). Provided that the polynomials are scaled correctly, the $ABCD$ rule can still be used to transform the q-values that specify the Gaussian function.

Thirdly, even if a resonator is restricted so that it generates only a centred Gaussian beam, difficulties can still be caused if the design is degenerate. We shall illustrate this by considering the two resonators described in section III.5.

For the first resonator (Figure III.11a), we calcul-ated a system matrix $M = \begin{bmatrix} 0 \cdot 800 & 0 \cdot 657 \\ -0 \cdot 386 & 0 \cdot 932 \end{bmatrix}$ with $\theta = \pi/6$, so that $\sin\theta = 0 \cdot 5$ in this case. Inserting these figures in the formulae of Table 2, we find

$$R = \frac{2(0 \cdot 657)}{0 \cdot 932 - 0 \cdot 800} = 10 \cdot 0 \text{ metres}$$

(The output wavefront has the same curvature as the output mirror surface.)

$$\text{Spot radius } w = \left[\frac{(0 \cdot 633) \, 10^{-6} \, (0 \cdot 657)}{\pi \, (0 \cdot 5)}\right]^{\frac{1}{2}}$$

$$= 5 \cdot 15 \times 10^{-4} \text{ metres,}$$

$$\text{or } 0 \cdot 52 \text{ mm}$$

(a suitable value for a small laser).

$$\text{Location of neck } z = \frac{(0 \cdot 800 - 0 \cdot 932)}{- \, 0 \cdot 386 \times 2}$$

$$= 0 \cdot 171 \text{ metres}$$

(the neck of the beam is 17 cm to the left, at the centre of the resonator).

$$\text{Neck radius } w_0 = \left[\frac{(-0 \cdot 633) \, 10^{-6} \, (0 \cdot 5)}{\pi \, (-0 \cdot 386)}\right]^{\frac{1}{2}}$$

$$= 5 \cdot 11 \times 10^{-4} \text{ metres,}$$

$$\text{or } 0 \cdot 51 \text{ mm}$$

(This is nearly equal to the spot radius, so the Gaussian beam has an almost constant cross-section inside the resonator.)

$$\text{Confocal beam parameter } z_0 = \frac{\pi \, (5 \cdot 11 \times 10^{-4})^2}{0 \cdot 633 \times 10^{-6}}$$

$$= 1 \cdot 30 \text{ metres}$$

The near-field region of the beam extends from its neck 17 cm to the left of the output mirror to a distance 1·13 metres to the right. At this point, it will have increased its spot radius to $\sqrt{2} w_0$ (or about 0·7 mm) and the radius of curvature of the wavefronts will be $R = 2 z_0 = 2 \cdot 6$ metres (see Figure III.13a).

$$\text{Far-field semi-angle } \frac{\lambda}{\pi w_0} = \frac{0 \cdot 633 \times 10^{-6}}{\pi \, (0 \cdot 51 \times 10^{-3})}$$

$$= 0 \cdot 395 \times 10^{-3} \text{ radians}$$

$$\text{or } 0 \cdot 40 \text{ milliradians}$$

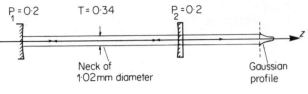

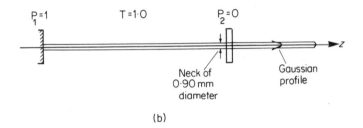

Figure III.13

The above figures correspond to a feasible design
for a small helium-neon gas laser, and they represent
that particular Gaussian beam which exactly repeats
itself with every single resonator traversal.

Let us suppose that into the same resonator we
launch a centred Gaussian beam whose spot radius and
R-value are slightly different. Perfect repetition
will no longer occur, but after a sixfold round trip,
because the eigenvalue that we have chosen corresponds
to $\theta = \pi/6$ (and because this is a symmetric mode),
the beam will again start to repeat itself.

If the resonator is constructed with a large Fresnel
number, then the latter type of Gaussian beam will be
able to circulate indefinitely. But as the apertures
of the two end-mirrors are reduced, the beam which
suffers least losses at the edges will be the one for
which the spot radius remains exactly the same for
each traversal. In practice, stable oscillation in
the fundamental mode is usually achieved even in situations where θ is an exact submultiple of π or 2π.

Another example of this situation is provided by the
hemi-confocal resonator which was the second stable
system discussed in section III.5. For this system

the matrix representing a round trip was $\begin{bmatrix} 0 & 1 \\ -1 & 0 \end{bmatrix}$ and
the main eigenvalue was $e^{+i\theta} = e^{i\pi/2} = i$, so that $\sin\theta = 1$.

Since $A = D$ in this case, we can see immediately that the radius of curvature R of the output wavefront will be infinite and, since $z = 0$, the neck of the Gaussian beam is located at the output mirror. The confocal parameter z_0 is exactly 1 metre and the formulae for both w and w_0 give as the beam radius

$$\left(\frac{\lambda}{\pi}\right)^{\frac{1}{2}} = \left[\frac{0 \cdot 633 \times 10^{-6}}{\pi}\right]^{\frac{1}{2}} = 4 \cdot 49 \times 10^{-4} \text{ metres,}$$

or   <u>0·45 mm</u> (see Figure III.13b)

Since this neck radius is much the same as in the previous example, we shall again have a semi-angle of divergence in the far field of 0·4 milliradians. If we assume that the laser tube is three times longer than before, we may expect perhaps three times more power in the beam, but this is liable to be distributed over more axial modes than in the shorter laser.

We shall not attempt here to discuss the problems of axial mode selection. There is, however, a simple formula according to which the optical frequency of a given mode depends not only on the spacing between the mirrors and the axial mode number but also on the transverse mode number and on the resultant eigenvalue $e^{i\theta}$. For coverage of such matters, the reader must refer to the work of Kogelnik and others given in the bibliography.

But before going on to consider mode-matching problems, we should perhaps return to a point that was mentioned in the last section - namely, the phase shift ϕ_{12} that a Gaussian beam experiences when it is transferred from one reference plane to another by application of the $ABCD$ rule. Associated with this phase change, in general, is a change in brightness, and therefore in field strength of the central region of the beam.

Let us consider first the case of a Gaussian beam whose neck radius w_0 is so small as to be comparable with λ. Over a considerable range of angles ($\pm \lambda/\pi w_0$ radians), we can regard the light reaching RP_1 as nearly equivalent to a spherical wavefront of radius R_1. We can represent this geometrically as a pencil of rays also diverging from the neck region. For a typical ray of this pencil, travelling at an angle V_1 to the axis, we can assign a ray vector $\begin{bmatrix} R_1 V_1 \\ V_1 \end{bmatrix}$ which will be transformed in the output plane into

$$\begin{bmatrix} R_2 V_2 \\ V_2 \end{bmatrix} = \begin{bmatrix} AR_1 V_1 + BV_1 \\ CR_1 V_1 + DV_1 \end{bmatrix}$$

Provided that the optical system is free from absorption, all the light arriving at RP_1 will pass through RP_2; in particular, that portion of the energy which passed through a circle of radius $y_1 = R_1 V_1$ in RP_1 must pass through a circular patch in RP_2 whose radius is $y_2 = AR_1 V_1 + BV_1$. Evidently, therefore, if we consider the problem in two dimensions, the energy concentration (irradiance) in the central region will be multiplied by an 'area factor' $\left(\dfrac{R_1}{AR_1 + B} \right)^2$ and the amplitude of the electric field will be multiplied by $\left(\dfrac{R_1}{AR_1 + B} \right)$, or, for a one-dimensional situation, by $\left(\dfrac{R_1}{AR_1 + B} \right)^{\frac{1}{2}}$.

As we have seen, when the neck region of a Gaussian beam is not negligibly small, we can obtain a successful formal description provided that we interpret each y/V ratio as representing not a real R-value but a complex q-value. It becomes very difficult, however, to find any meaning in the idea that either the y-value or the V-value should itself be complex. Nevertheless, as we found in section III.6, the multiplying factor $q_1/(Aq_1 + B)$ certainly makes its appearance when we make a diffraction calculation: we stated then that the modulus of this complex factor represents the ratio (w_1/w_2) corresponding to the ratio (y_1/y_2) considered above, and the argument ϕ_{12} represents a phase change of the whole beam (in addition to the much larger phase change $2\pi K/\lambda$).

To verify that

$$\left| \frac{q_1}{Aq_1 + B} \right| = \left(\frac{w_1}{w_2} \right)^2$$

we use the $ABCD$ rule in its inverted form:

$$\frac{1}{R_2} + \frac{i\lambda}{\pi w_2^2} = \frac{1}{q_2} = \frac{Cq_1 + D}{Aq_1 + B} = \frac{C + D/q_1}{A + B/q_1} = \frac{C + D/R_1 + i\lambda D/\pi w_1^2}{A + B/R_1 + i\lambda B/\pi w_1^2}$$

Multiplying above and below by the complex conjugate of the denominator, and re-expressing the denominator in its original form $(A + B/q_1)$, we obtain

$$\frac{1}{R_2} + \frac{i\lambda}{\pi w_2^2} = \frac{(C + D/R_1 + i\lambda D/\pi w_1^2)(A + B/R_1 - i\lambda B/\pi w_1^2)}{|A + B/q_1|^2}$$

Equating the imaginary parts of this equation, we have

$$\frac{i\lambda}{\pi w_2^2} = \frac{i\lambda}{\pi w_1^2} \frac{(DA + DB/R_1 - CB - DB/R_1)}{|A + B/q_1|^2}$$

$$= \frac{i\lambda}{\pi w_1^2} \frac{q_1^2}{|Aq_1 + B|^2}$$

(since $(AD - BC) = 1$). Hence, finally, we obtain the 'spot size transformation formula':

$$(w_1/w_2)^2 = |q_1/(Aq_1 + B)|^2 \quad \text{or, equivalently,}$$

$$(w_2/w_1) = |A + B/q_1|$$

In the particular case where a Gaussian beam is launched on a round trip inside an optical resonator, the complex factor $q_1/(Aq_1 + B)$ still enables us to find out what transformation of spot size or phase change is produced. But if we choose q_1 to coincide with our resonator eigenvalue $(e^{i\theta} - D)/C = B(e^{i\theta} - A)$, then the expression $q_1/(Aq_1 + B)$ reduces simply to $e^{-i\theta}$. Although as we should expect there is no change of spot size, there remains a phase change, which is determined by the resonator eigenvalue $e^{i\theta}$. (For the corresponding situation with an unstable resonator, there is no phase change, but the y-value is multiplied by the real eigenvalue e^{t}, the transverse magnification of the system.)

In the majority of optical instruments, a small phase change affecting the whole beam is of little importance. Inside a laser resonator, however, the fact that it occurs repeatedly causes a small shift in the optical frequency of the output. A correction for this kind of phase shift may also be needed in very accurate interferometric determinations of length, especially when infrared or microwave radiation is involved.

We should insert here two minor cautions concerning optical effects that can be produced by an output mirror and its substrate. In nearly every case the partially reflecting surface faces inwards; if we assume that it is curved, then, unless the other side of the transparent substrate is given a compensating curvature, the whole element will act as a weak lens - usually a diverging one. This lens will have no effect on the wave-shape that is generated inside the cavity, but it will produce a small change in the R-value of the beam that actually emerges. By making the final surface of the output mirror slightly convex, however, it is possible to produce a Gaussian beam that is slightly converging. In that case, the whole of the near-field region of the beam can be made available for external use (without any additional mode-matching optics of the type to be discussed in the next section).

Secondly, unless the final surface of the output mirror is wedged considerably, or is given a very efficient antireflection coating, interference effect-between this surface and the main partially reflecting surface are likely to produce an unwanted stripe or ring pattern across the output beam. For high-gain laser systems it may be better to design the resonator so that the output mirror is plane; the required optical feedback can then be obtained from both surfaces acting as a resonant reflector, and a 'smooth' output is produced.

It is, of course, possible to use a more rigorous approach to this problem of a laser oscillating in a single mode. If, for example, we use Fresnel-Huygens theory, we find that the amplitude distribution $A(y)$ of the light waves that emerge must satisfy an integral equation of the form

$$\lambda A(y) = \int A(y')K(y,y')\,dy'$$

In this equation the kernel $K(y,y')$ is a (complex) symmetric function which contains not only the changes of phase which occur as a result of a round trip inside the resonator but also the truncating effects produced by the finite apertures of the system. The losses introduced by the latter must, of course, be made good by the gain of the laser amplifier.

The simple solutions which have been discussed in this section correspond, in effect, to ignoring the effects of the apertures and replacing $K(y,y')$ by $\exp\left[(2\pi i/\lambda)W(y,y')\right]$, where W is the usual eikonal function for a round trip. This approximation gives useful results (a) for stable resonators operating with Fresnel number N close to unity, and (b) for unstable resonators operating with a large Fresnel number and with uniform gain.

Its predictions should, however, be used with caution if it is necessary to consider resonators which lie on the border between stability and instability (for example plane-parallel or concentric mirror arrangements for which $(A + D)/2$ has a nominal value of unity). For such cases a full diffraction calculation may be needed; and even this is likely to prove a useless exercise unless care is also taken to include any optical aberrations or pump-induced phase distortion and to consider any dynamic or non-linear effects that may be at work in an actual laser system. In many cases it is quite difficult to obtain a single-mode output.

III.8 APPLICATION OF THE $ABCD$ RULE TO MODE-MATCHING PROBLEMS

Although lasers are very specialized light sources, they are already used for a wide variety of applications; sooner or later, therefore, the owner of a laser may find that he needs to transform the shape of the Gaussian beam that emerges into something more suitable. In some cases he will be well advised to purchase a commercially available unit such as a beam-expander; but in others a simple lens may be all that is needed. In this section, we show how this sort of problem of beam transformation can be tackled.

Let us consider first, however, the problems that arise with classical imaging systems, if one of them has to be operated in cascade with another. If we assume, for example, that an approaching comet or a supernova has aroused great public interest, then a broadcasting engineer might need to couple a television camera system temporarily into an existing astronomical telescope.

For such a coupling to be fully satisfactory, the second optical system must possess an *étendue* at least as large as that provided by the first, and if resolution of detail is not to be lost its *space-bandwidth product* must also reach the same value. If these conditions are met, and if the exit window of the telescope together with the entrance window of the camera are both located at infinity, then all that is necessary is to ensure that the entrance pupil of the camera coincides in position and size with the exit pupil of the telescope. (The vignetting that is liable to occur if these conditions are not met is discussed in appendix A.)

Let us assume a more difficult situation in which, although the two optical systems both have the same étendue, their windows and pupils have different diameters and are located at different z-positions (although not at infinity).

In order to specify the aperture properties of the first system, we shall take RP_1 to lie just outside its last surface, but not so that it coincides with either its exit pupil or its exit window. We shall assume that the exit pupil lies at a distance R_1 to the left of RP_1 and we shall use V_1 to denote the maximum angle for a ray passing from the centre of the exit pupil to the edge of the exit window. The associated ray vector in RP_1 is evidently $\begin{bmatrix} V_1 R_1 \\ V_1 \end{bmatrix}$.

If the exit window lies at a distance R_1' to the left of RP_1 and V_1' denotes the maximum ray angle for a ray passing from its centre to the edge of the exit pupil, the associated ray vector is similarly $\begin{bmatrix} V_1' R_1' \\ V_1' \end{bmatrix}$.

As was discussed in section III.3, the étendue of the first system is given by the determinant of the matrix which is formed by these two extreme ray vectors. We shall call this matrix

$$S_1 = \begin{bmatrix} V_1R_1 & V_1'R_1' \\ V_1 & V_1' \end{bmatrix}.$$

Turning now to the second optical system, we specify its properties in similar fashion, choosing a reference plane RP_2 in some convenient location in front of its first component. If the entrance pupil lies at a distance R_2 to the left of RP_2, then a ray passing from its centre to the edge of the entrance window will be specified by the ray vector $\begin{bmatrix} V_2R_2 \\ V_2 \end{bmatrix}$, where V_2 is the maximum angle that the entrance window can accommodate. In like fashion $\begin{bmatrix} V_2'R_2' \\ V_2' \end{bmatrix}$ specifies a ray passing from the centre of the entrance window (located at a distance R_2' to the left of RP_2) to the edge of the entrance pupil. (It will be appreciated that both R_2 and R_2' may be negative in value.)

The output ray vectors thus specified can be combined into an output ray matrix

$$S_2 = \begin{bmatrix} V_2R_2 & V_2'R_2' \\ V_2 & V_2' \end{bmatrix}$$

Once again, the determinant of this matrix represents the étendue of the second optical system. For our present calculation, we shall assume that $\det(S_2)$ is exactly equal to $\det(S_1)$. (In practice, if the second optical system is a camera with an adjustable iris diaphragm, it might be good practice eventually to open it slightly *beyond* the point where exact matching occurs.)

In order to achieve exact matching of these two systems, we must now interpose between RP_1 and RP_2 a subsidiary optical system whose matrix M is such as to satisfy the following matrix equation:

$$MS_1 = S_2$$

where S_1 and S_2 are the matrices defined above.

Given that the determinant of S_1 is not a vanishing quantity, we can now postmultiply both sides of this matrix equation by the reciprocal matrix S_1^{-1}, and we then obtain an explicit equation for M:

$$MS_1S_1^{-1} = M = S_2S_1^{-1} = S_2 \frac{\text{adj}(S_1)^T}{\det(S_1)}$$

or, writing the elements in full,

$$\begin{bmatrix} A & B \\ C & D \end{bmatrix} = \frac{\begin{bmatrix} R_2V_2 & R_2'V_2' \\ V_2 & V_2' \end{bmatrix} \begin{bmatrix} V_1' & -V_1'R_1' \\ -V_1 & V_1R_1 \end{bmatrix}}{V_1V_1'(R_1 - R_1')}$$

In order to eliminate the V-values from the above expression it will be convenient to use the symbol m to represent the angular magnification (V_2/V_1) with which the ray leaving the centre of the exit pupil of system 1 is imaged on to the centre of the entrance pupil of system 2; since linear magnification is the reciprocal of angular magnification, m is also equal to the ratio $\dfrac{\text{(radius of exit pupil of system 1)}}{\text{(radius of entrance pupil of system 2)}}$.

In the same way, we use the symbol $m' = (V_2'/V_1')$ as a 'window magnification factor'. But we should note here that, in order for the two étendues to be equal, there must exist a connection between the values m and m'. We have

$$\det(S_1) = \begin{vmatrix} V_1R_1 & V_1'R_1' \\ V_1 & V_1' \end{vmatrix} = V_1V_1'(R_1 - R_1')$$

and

$$\det(S_2) = \begin{vmatrix} mV_1R_2 & m'V_1'R_2' \\ mV_1 & m'V_1' \end{vmatrix} = mm'V_1V_1'(R_2 - R_2')$$

It follows that

$$mm' = (R_1 - R_1')/(R_2 - R_2')$$

Bearing in mind this connection between m and m', we can now write

$$M = \begin{bmatrix} A & B \\ C & D \end{bmatrix} = \frac{\begin{bmatrix} mV_1R_2 & m'V_1'R_2' \\ mV_1 & m'V_1' \end{bmatrix} \begin{bmatrix} V_1' & -V_1'R_1' \\ -V_1 & V_1R_1 \end{bmatrix}}{V_1V_1'(R_1 - R_1')}$$

If we multiply the two matrices together and divide each element of the result by the scalar quantity $V_1V_1'(R_1 - R_1')$ in the denominator, we obtain, eventually,

$$M = \begin{bmatrix} A & B \\ C & D \end{bmatrix} = \begin{bmatrix} \dfrac{(mR_2 - m'R_2')}{(R_1 - R_1')} & \dfrac{(m'R_1R_2' - mR_1'R_2)}{(R_1 - R_1')} \\ \dfrac{(m - m')}{(R_1 - R_1')} & \dfrac{(m'R_1 - mR_1')}{(R_1 - R_1')} \end{bmatrix}$$

Each of the elements A, B, C and D is thus fully determined, but there still remains the task of devising an optical system which is characterized by such a matrix. We shall discuss here only the case where the required pupil and window magnification factors m and m' both have the same value. In that event the element C evidently vanishes, and we need an afocal system of angular magnification m; the matrix required now takes the form

$$\begin{bmatrix} A & B \\ C & D \end{bmatrix} = \begin{bmatrix} 1/m & \dfrac{m(R_1R_2' - R_1'R_2)}{(R_1 - R_1')} \\ 0 & m \end{bmatrix}$$

For an afocal system of angular magnification m, referred to a *conjugate* pair of reference planes, the basic matrix (listed in section III.1) is of the form $\begin{bmatrix} 1/m & 0 \\ 0 & m \end{bmatrix}$. However, if we premultiply and postmultiply this by suitable $\mathcal{T}$-matrices, we can control the B-value as required:

$$\begin{bmatrix} 1 & t_2 \\ 0 & 1 \end{bmatrix} \begin{bmatrix} 1/m & 0 \\ 0 & m \end{bmatrix} \begin{bmatrix} 1 & t_1 \\ 0 & 1 \end{bmatrix} = \begin{bmatrix} 1/m & (t_2m + t_1/m) \\ 0 & m \end{bmatrix}$$

In practical terms, we slide either the afocal unit or possibly the second optical system to such a position along the optical axis that

$$(t_2 m + t_1/m) = \frac{m(R_1 R_2' - R_1' R_2)}{(R_1 - R_1')}$$

(Since we have two adjustable parameters t_1 and t_2 at our disposal, this should not be too difficult.)

We give below a schematic representation of such a situation:

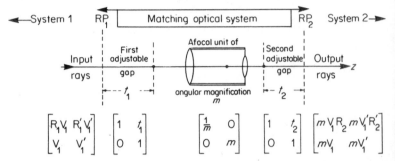

(*Note*. If calculation shows that negative values of t_1 or t_2 will be required, then it may help to use an afocal unit of *negative* magnification; the required t-values will still be negative, but the conjugate planes of the unit will be more accessible, so that there is less danger of RP_1 or RP_2 being 'buried' inside the matching system. But the user must be prepared to accept the resultant inversion of the final image!

Secondly, although matrix methods may be helpful in tackling such a problem, they are not a panacea. The engineer involved needs also to build on all the experience that has already been acquired in this area; he should, for example, consider the use of field lenses and he should study how the relaying of images is handled in a submarine periscope or in an optical information processing system.)

We turn now to what is in many ways an easier problem, that of transforming and shaping a laser beam so that it has the right geometry to match some external system. This is a problem that can arise at almost any kind of power level.

The power of a helium-neon gas laser, for example, is measured in milliwatts, and it is likely to be used as a measuring tool for alignment and interferometry, sometimes over great distances. It is also used as a light source for testing optical systems or for recording and reconstructing holograms. Medium-power lasers, with power measured in watts, are used in holography, communication and eye surgery, as dye-laser pumping sources and as rapidly scanned reading or writing devices. When we enter the 'high'-power range, the applications include welding and machining, pulsed holography and laser radar (including lunar ranging); more powerful still are certain pulsed lasers used for experiments in non-linear optics and in plasma studies, where the focused laser beam produces temperatures of a few million degrees, sufficient for some nuclear fusion to occur. (For some of the latter applications the laser pulse may last only 10^{-11} seconds, but its instantaneous power may reach 10^{13} watts!)

When the power in a laser beam reaches the megawatt level, its propagation may well provoke a catastrophic breakdown of the dielectric medium (air, glass, liquid, etc.) through which it is passing. At these intensities non-linear effects may produce self-focusing phenomena, and even at the kilowatt power level the heating effects of the beam may need to be considered unless the medium is perfectly transparent. We shall confine our discussion here to low-power and medium-power beams for which such effects are unimportant.

But we must point out here that even a one-milliwatt laser generates some slight degree of eye hazard. *The laser safety codes and regulations that have been devised provide a sensible degree of protection. Every owner of a laser should see that they are strictly observed.* Looking into a narrow laser beam is just as foolish as staring at the sun without proper eye protection.

In most mode-matching problems the Gaussian beam emerging from the laser has a circular cross-section, and this has to be sent into an external system which also possesses circular symmetry. Occasionally, however, it is required to transform a circular beam into one whose cross-section is highly elliptical, or vice versa. Provided that the x- and y-axes are chosen to coincide with the major and minor axes of such an

ellipse, it is quite easy to handle such transforma-
tions by specifying, for the x-direction as well as
the y-direction, separate q-values and separate ray-
transfer matrices. We conduct, in effect, two independent one-dimensional calculations, and we can then
determine the amplitude of the electric field at any
point (x,y,z) as the product of two functions
$\exp\left[\dfrac{2\pi i}{\lambda}\dfrac{(y^2)}{2q_y}\right]$ and $\exp\left[\dfrac{2\pi i}{\lambda}\dfrac{(x^2)}{2q_x}\right]$. The real portions
of $1/q_x$ and $1/q_y$ then represent the principal curva-
tures of the wavefronts, and if they are not identical
then astigmatism.must be present. Likewise, if the
imaginary portions are not identical then the equi-
energy contours of the beam are not circular but
elliptical.

It will be sufficient to consider here only the one-
dimensional problem or the related problem with
cylindrical symmetry. In that case, for any plane of
constant z, the geometry of the Gaussian beam is completely specified by its complex q-value, which has
just two degrees of freedom.

Let us now try to determine an optical system which
will transform one given q-value into another. Remembering that the determinant of a ray-transfer matrix
must be unity, we have one real and one complex
equation which the four real elements A, B, C and D
must satisfy:

$$q_2 = \frac{(Aq_1 + B)}{(Cq_1 + D)} \quad \text{and} \quad (AD - BC) = 1$$

Since the q-value of a Gaussian beam always contains
an imaginary element, the former equation can be separated into its real and imaginary components. In
effect, therefore, we have three equations, and the
matrix elements are not fully determined (except in
the unlikely event that we need to specify also the
auxiliary phase shift $\phi_{12} = \arg\left[(Aq_1 + B)/q_1\right]$ that
occurs between the two reference planes).

The fact that we have one parameter still at our
disposal can be a considerable asset in finding a suitable matching system; if, for example, only a few
lenses are available, it is useful to be able to
tailor the design so as to accommodate one of them.

III.8.1 Beam-transforming optical systems

We begin by reviewing the two simplest kinds of transformation - those that can be accomplished by a single $\mathscr{T}$-matrix or a single $\mathscr{R}$-matrix.

If RP_2 is separated from RP_1 by an empty gap of thickness T, we have immediately from the *ABCD* rule

$$q_2 = \frac{1.q_1 + T}{0.q_1 + 1} = q_1 + T$$

In other words, the real part of q, the z-value which describes the location of the beam-neck, is increased by T, but the imaginary part ($- iz_0$) is unchanged. It will be remembered that z_0 is the 'confocal beam parameter' and has the value $\pi w_0^2/\lambda$, where w_0 is the neck radius.

Let us now consider the effect of a thin lens of power P. Using the inverted form of the *ABCD* rule we obtain

$$1/q_2 = \frac{Cq_1 + D}{Aq_1 + B} = \frac{- Pq_1 + 1}{q_1} = 1/q_1 - P$$

In this case the real part of $1/q$, representing the divergence $1/R$, is reduced by the converging power of the lens; but the imaginary part ($i\lambda/\pi w^2$), which specifies the local beam radius w, remains unchanged. (A transparent lens changes the shape of a wavefield but not its energy.)

In rare cases the input beam already has the correct neck radius or the correct beam radius, and the only matching system required is either a simple gap or a single lens of the right focal power. More usually, however, our mode-matching system will need to contain at least one $\mathscr{T}$-matrix and one $\mathscr{R}$-matrix.

If the gap precedes the lens, then its value needs to be adjusted so that, by the time the beam reaches the lens, it has already spread to the desired diameter; the lens is then chosen to produce the desired divergence or curvature of the wavefronts. Alternatively, if the lens is used first, it must be chosen to generate a new Gaussian beam with the required final neck radius, and the gap is then adjusted so that the above-mentioned neck is located at the desired distance from RP_2.

To succeed with these methods we need to be able to find a lens with the right focal length, and a set of trial-case lenses of the kind used by ophthalmologists will prove a useful investment. Difficulty sometimes arises, however, because the prescribed gap-value turns out to be either negative or awkwardly large. In some cases a second lens may help to achieve a more compact arrangement, and in others a neat solution can be found by choosing a lens of about the right focal length and then sandwiching it between *two* gaps, each of which can be adjusted as required.

In specifying the two gaps of such a system, we shall use dimensionless '*g*-parameters', which measure, in terms of the focal length, the distances from RP_1 to the first focal plane and from the second focal plane to RP_2. The system matrix is thus

$$
\begin{bmatrix} A & B \\ C & D \end{bmatrix} = \begin{bmatrix} 1 & g_2 f \\ 0 & 1 \end{bmatrix} \begin{bmatrix} 0 & f \\ -1/f & 0 \end{bmatrix} \begin{bmatrix} 1 & g_1 f \\ 0 & 1 \end{bmatrix}
$$

$$
= \begin{bmatrix} 1 & g_2 f \\ 0 & 1 \end{bmatrix} \begin{bmatrix} 0 & f \\ -1/f & -g_1 \end{bmatrix} = \begin{bmatrix} -g_2 & f(1-g_1 g_2) \\ -1/f & -g_1 \end{bmatrix}
$$

In discussing the geometry of the input and output beams it is often convenient to think in terms of the size and location of the beam neck. Let us consider here the case where the two reference planes are actually coincident with the two beam necks (see Figure III.14). It then follows that both input and

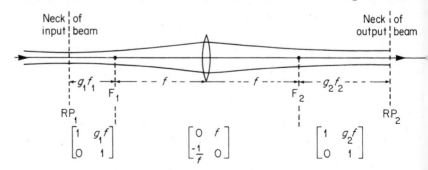

(Note: as in Figure III.13, the radii of the Gaussian beam are shown exaggerᵉ

Figure III.14

output beams have a q-value that is purely imaginary. We can therefore write $q_1 = -iz_{01}$ and $q_2 = -iz_{02}$, where the positive real quantities z_{01} and z_{02} denote the corresponding confocal beam parameters. Inserting these values in the $ABCD$ rule, we obtain

$$-iz_{02} = \frac{(-g_2)(-iz_{01}) + f(1 - g_1g_2)}{(-1/f)(-iz_{01}) - g_1}$$

$$= \frac{ig_2z_{01} + f(1 - g_1g_2)}{iz_{01}/f - g_1}$$

Multiplying by the denominator $(iz_{01}/f - g_1)$, which is non-zero, we obtain finally, after separating the real and imaginary components,

$$z_{02}z_{01} = f^2(1 - g_1g_2)$$

and

$$g_1z_{02} = g_2z_{01}$$

Since z_{01} and z_{02} are both necessarily positive, the second equation shows that g_1 and g_2 must both have the same sign, so that the product (g_1g_2) must also be positive. It follows from the first equation that, when the beam neck is transferred in this way from RP_1 to RP_2, the geometric mean of z_{01} and z_{02} can never exceed the focal length of the lens employed. In a symmetrical beam-relaying system for which $g_1 = g_2 = g$, we shall have $z_{01} = z_{02} = z_0 = f(1 - g^2)^{\frac{1}{2}}$.

We have already seen that, if the neck radius of a Gaussian beam is made very small, the asymptotic angle of divergence $(\lambda/\pi w_0)$ becomes appreciable, and the beam begins to resemble a family of spherical wavefronts diverging from the neck region. As this geometrical optics limit is approached, both z_{01} and z_{02} become very small, and the first equation then states that the term $(1 - g_1g_2)$ tends towards zero. Eventually, therefore, g_2 becomes the reciprocal of g_1, in agreement with Newton's equation for the focal distances of a classical object and image.

The second equation can be rewritten in terms of the neck radii to give

$$w_{02}^2/w_{01}^2 = z_{02}/z_{01} = g_2/g_1$$

In the geometrical limit, where $g_2 = 1/g_1$, the factor g_2 can be regarded as the transverse magnification (w_{02}/w_{01}) with which the neck region has been imaged.

Except in this limiting case, however, it would be quite wrong to regard any transformation of the beam neck as a conventional imaging process. Consider, for example, the simple case where the neck of the input beam is located in the first focal plane, so that g_1 equals zero. We might perhaps expect the 'image of the beam neck' to appear at infinity in the output space. In fact, however, the second equation shows that g_2 must also vanish, so that the neck of the output beam is located in the second focal plane. The confocal beam parameters in this case are given by $z_{01}z_{02} = f^2$.

(In optical terms, the disturbance in the second focal plane is really the far-field or Fraunhofer pattern of the input beam; this would normally appear at infinity, but the lens focuses it on to the second focal plane. In mathematical terms, the amplitude distributions in these two planes are always related as a pair of Fourier transforms, and because the first function is a real Gaussian function with no quadratic phase shift, the second function must also be wholly real.)

If the confocal parameter of the input beam z_{01} is equal to the focal length f, then z_{02} is also equal to f, and we have a self-repeating system which could be used as an optical relay system, as illustrated in Figure III.15. If made sufficiently free from absorption and reflection losses, such systems offer one

Four consecutive elements of an optical relay system

Figure III.15

possibility for point-to-point optical communication. For a laser beam of 800 nm wavelength transmitted with a minimum neck radius of 10 mm, the confocal beam parameter would be equal to $\pi 10^{-4}/8.10^{-7}$, or just under 400 metres. A relay system would therefore need a series of very weak (1/400 diopter) lenses, each about 100 mm in diameter and spaced (in a straight line!) at 800 metre intervals.

As it happens, there is one other case, easily identifiable by the matrix method, for which the neck of the output beam coincides *exactly* with the geometrical image of the neck of the input beam.

We know here that both q_1 and q_2 must be wholly imaginary, and for imaging between the reference planes the ray-transfer matrix must be of the form $\begin{bmatrix} m & 0 \\ -1/f & 1/m \end{bmatrix}$, where m is the lateral magnification factor. The *ABCD* rule then states

$$- iz_{02} = \frac{- mi z_{01} + 0}{(iz_{01}/f + 1/m)}$$

We can see immediately that this equation will be satisfied only if the denominator is wholly real; and since z_{01} is necessarily real and positive the focal length f must be infinite. It is therefore an *afocal* system which possesses this special property. We can, if we wish, regard the matrix of such a system as the product of two successive 'focal plane matrices' $\begin{bmatrix} 0 & f \\ -1/f & 0 \end{bmatrix}$ and $\begin{bmatrix} 0 & mf \\ -1/mf & 0 \end{bmatrix}$. These generate, in effect, a double Fourier transformation process, in the first stage of which the neck of the beam is transferred from the input focal plane to the intermediate focal plane, and thence to the output focal plane (see Figure III.16).

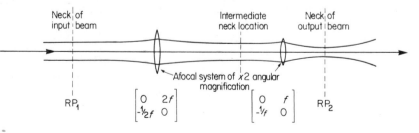

(Neck-radius is halved but divergence is doubled)

Figure III.16

In this section so far we have been considering situations where both the input and output q-values are known and the *ABCD* rule enables us to find out what kind of optical system is needed to effect the transformation. Let us now discuss two slightly different problems, in which the beam-transforming system is known, but there is something incomplete about our knowledge of the input and output q-values.

III.8.2 *Obtaining a narrow focus*

You are given (a) a laser whose output is a Gaussian beam of neck radius w_{01} and corresponding confocal beam parameter $z_{01} = (\pi w_{01}^2/\lambda)$, and (b) an optical system whose transfer matrix from RP_1 to RP_2 is
$\begin{bmatrix} A & B \\ C & D \end{bmatrix}$. The optical system is mounted on a slide so that its spacing from the laser can be varied. What will be the minimum spot radius w_{02} (measured at the neck) in the beam emerging from RP_2, and how will it depend on the spacing z_1 between the neck of the input beam and the location of RP_1?

Solution

Clearly the input q-value for this situation will be given by $q_1 = z_1 - iz_{01}$, and the output q-value will be

$$q_2 = \frac{(Aq_1 + B)}{(Cq_1 + D)} = \frac{(Aq_1 + B)}{(Cq_1 + D)} \frac{(Cq_1^* + D)}{(Cq_1^* + D)}$$

(* denotes a complex conjugate). If we now retain only the imaginary part of this equation, we obtain

$$Im(q_2) = -iz_{02} = \frac{-iAz_{01}(Cz_1 + D) + (Az_1 + B)iCz_{01}}{(Cz_1 + D)^2 + C^2z_{01}^2}$$

$$= \frac{-iz_{01}\left[ACz_1 + AD - ACz_1 - BC\right]}{(Cz_1 + D)^2 + C^2z_{01}^2}$$

$$= \frac{-iz_{01}}{(Cz_1 + D)^2 + C^2z_{01}^2}$$

(the expression in square brackets reducing to unity).

If we now replace z_{01} by $\pi w_{01}^2/\lambda$ and z_{02} by $\pi w_{02}^2/\lambda$, the final result is

$$w_{02}^2 = \frac{w_{01}^2}{(Cz_1 + D)^2 + C^2\pi^2 w_{01}^4/\lambda^2}$$

In the case where our optical system is a simple lens of focal length f, with RP_1 and RP_2 located at the two focal planes, we have

$$\begin{bmatrix} A & B \\ C & D \end{bmatrix} = \begin{bmatrix} 0 & f \\ -1/f & 0 \end{bmatrix}$$

and hence

$$w_{02}^2 = \frac{w_{01}^2 f^2}{z_1^2 + \pi^2 w_{01}^4/\lambda^2}$$

If we wish to minimize this neck radius, it will usually help if the given lens has a short focal length; but it will also help, paradoxically, if we slide the lens away from the laser so that the distance z_1 is considerably greater than the confocal beam parameter $z_{01} = \pi w_{01}^2/\lambda$. By doing this we obtain an increase in the spot radius of the beam which enters the aperture of the lens, and provided this is not overfilled we obtain a more strongly convergent output beam. Ultimately it is the numerical aperture (in this case $\lambda/\pi w_{02}$) which determines the neck radius.

The general equation that we have just obtained can be regarded as a 'neck-size transformation formula':

$$z_{02}/z_{01} = w_{02}^2/w_{01}^2 = \frac{1}{(Cz_1 + D)^2 + C^2 z_{01}^2} = \frac{1}{|Cq_1 + D|^2}$$

This equation, although different, is comparable with the spot-size transformation formula, which we obtained in section III.7 by taking only the imaginary parts of the equation for $1/q$. In its inverted form, the spot-size transformation formula reads

$$w_2^2/w_1^2 = |A + B/q_1|^2$$

III.8.3 *Finding a Gaussian beam with two curvatures prescribed*

A Gaussian beam is propagated across an empty gap of width d between RP_1 and RP_2. All you are told is that when the wavefronts cross RP_1 they have radius of curvature R_1 and when they cross RP_2 they have radius of curavure R_2. Find the other parameters of the beam, its neck size and location and the local spot radii in RP_1 and RP_2.

Solution

If we put $q_1 = z_1 - iz_0$, we have from the *ABCD* rule $q_2 = q_1 + d = (z_1 + d) - iz_0$. If both these equations are inverted we then obtain

$$\frac{1}{q_1} = \frac{1}{z_1 - iz_0} = \frac{z_1 + iz_0}{z_1^2 + z_0^2} = \frac{1}{R_1} + \frac{i\lambda}{\pi w_1^2} \qquad (\text{III.1})$$

and

$$\frac{1}{q_2} = \frac{1}{(z_1 + d) - iz_0} = \frac{(z_1 + d) + iz_0}{(z_1 + d)^2 + z_0^2} = \frac{1}{R_2} + \frac{i\lambda}{\pi w_2^2} \quad (\text{III.2})$$

Taking only the real portions of these two equations, we obtain from equation (III.2)

$$z_1^2 + 2z_1 d + d^2 + z_0^2 = z_1 R_2 + dR_2 \qquad (\text{III.3})$$

and from equation (III.1)

$$z_1^2 \qquad\qquad + z_0^2 = z_1 R_1 \qquad (\text{III.4})$$

Subtracting equation (III.4) from equation (III.3) give

$$2z_1 d + d^2 = dR_2 + z_1(R_2 - R_1)$$

or

$$z_1 = \frac{d(R_2 - d)}{(2d + R_1 - R_2)} \qquad (\text{III.5})$$

To obtain an expression for z_0 we now substitute for z_1 in equation (III.4) and obtain, after a little algebra,

$$z_0^2 = R_1 z_1 - z_1^2$$

$$= \frac{dR_1(R_2 - d)(2d + R_1 - R_2) - d^2(R_2 - d)^2}{(2d + R_1 - R_2)^2}$$

$$= \frac{d(R_2 - d)(2dR_1 + R_1^2 - R_1R_2 - dR_2 + d^2)}{(2d + R_1 - R_2)^2}$$

This can be simplified to give

$$z_0^2 = \pi^2 w_0^4 / \lambda^2 = \frac{d(R_2 - d)(R_1 + d)(d + R_1 - R_2)}{(2d + R_1 - R_2)^2} \qquad \text{(III.6)}$$

To obtain a value for w_1 we use the imaginary part of equation (III.1) to obtain

$$\frac{\lambda^2}{\pi^2 w_1^4} = \frac{z_0^2}{(z_1^2 + z_0^2)^2} = \frac{z_0^2}{(z_1 R_1)^2}$$

(by virtue of equation III.4). Inverting this equation, and substituting from equations (III.5) and (III.6), we find

$$\frac{\pi^2 w_1^4}{\lambda^2} = \frac{d^2(R_2 - d)^2 R_1^2 (2d + R_1 - R_2)^2}{(2d + R_1 - R_2)^2 d(R_2 - d)(R_1 + d)(d + R_1 - R_2)}$$

Hence

$$w_1^4 = \frac{\lambda^2 d(R_2 - d)R_1^2}{\pi^2(R_1 + d)(d + R_1 - R_2)} \qquad \text{(III.7)}$$

To obtain w_2 we proceed in a similar way with the imaginary part of equation (III.2):

$$\frac{\lambda^2}{\pi^2 w_2^4} = \frac{z_0^2}{\left((z_1 + d)^2 + z_0^2\right)^2} = \frac{z_0^2}{R_2^2(z_1 + d)^2}$$

(by virtue of equation III.3). Inverting this equation, and substituting from equations (III.5) and (III.6), we find eventually

$$w_2^4 = \frac{\lambda^2 d(R_1 + d)R_2^2}{\pi^2(R_2 - d)(d + R_1 - R_2)} \qquad \text{(III.8)}$$

Finally, for the distance z_2 (measured from the neck to RP$_2$) we have

$$z_2 = (z_1 + d) = \frac{d(R_2 - d)}{(2d + R_1 - R_2)} + d$$

$$= \frac{d(R_1 + d)}{(2d + R_1 - R_2)} \qquad \text{(III.}$$

The solution that we have just obtained will be realizable physically only if equations (III.6), (III.7 and (III.8) all indicate positive values for w_0^2, w_1^2 and w_2^2 respectively. In empty space, for example, it is not possible for a beam to change from one that is diverging into one that is subsequently converging.

Since the geometry of a Gaussian beam is the same even if the direction of propagation is reversed, these equations can also be used to solve a corresponding resonator problem. If inward-facing spherical mirrors, having radii of curvature $r_1 = -R_1$ and $r_2 = R_2$ respectively, are positioned in RP$_1$ and RP$_2$, then the wavefronts of the Gaussian beam will coincide with either mirror surface; they will therefore be reflected backwards and forwards indefinitely (see Figure III.17). The beam shape that we have calculated is that of the fundamental mode of the corresponding optical resonator. (Notice, however, that because the mirror in RP$_1$ will be facing to the right its curva-

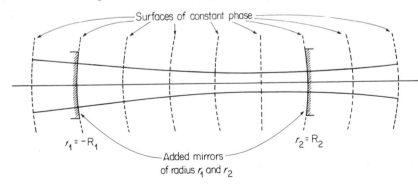

If mirrors are added to fit the equi-phase surfaces of a Gaussian beam, then a standing wave pattern becomes possible

Figure III.17

ture r_1 must be given the opposite sign to that which
is needed for the waves travelling in the positive
z-direction.)

III.9 RAY-TRANSFER MATRICES FOR DISTRIBUTED
LENS-LIKE MEDIA

The centred optical systems considered so far in
this book have all been formed from just two kinds of
basic element - the gap and the refracting or reflecting surface. The latter can be regarded as a boundary discontinuity between two media of different but
uniform refractive index. Each of these elements is
easily identified and easily specified by the appropriate $\mathcal{T}$-matrix or $\mathcal{R}$-matrix.

Situations sometimes arise, however, where the
refractive index of a medium is not constant but
varies smoothly with the position coordinate (x,y,z).
Perhaps the simplest case, which we shall discuss
briefly in this section, is a 'lens-like distribution'
in which the index n varies quadratically with distance from the optic axis Oz and shows cylindrical
symmetry about that axis. We shall see that, when
light is propagated paraxially through such a distribution, the effect of the 'parabolic' profile is
to confine the radiation so that it remains in the
vicinity of the optic axis.

We might remark here that this subject is not a new
one - it was discussed half a century ago in R. W.
Wood's *Physical Optics* - but it has acquired importance recently for two main reasons.

The first is the growing need for efficient relay
systems in optical communication. There have been
several studies of how a lens-like distribution can
be deliberately generated, for example in a gaseous
medium, by contriving suitable temperature gradients
inside a tube; solid state versions involve carefully controlling an impurity diffusion process inside
an optical fibre.

The second reason is that, whenever a powerful laser
beam is used, this sort of distribution is liable to
arise automatically. At the megawatt power levels
reached by pulsed lasers, the fact that the refractive
index of a medium *rises* very slightly means that 'self-
focusing effects' are likely to occur, either inside
or outside the laser resonator. These may lead to

expensive breakdown or fracture, and they also make it difficult to know with what geometry the laser beam is being propagated.

In order to represent such a distribution, we shall use the equation

$$n(x,y,z) = n_0(1 - \tfrac{1}{2}\alpha^2 r^2) = n_0 \left[1 - \tfrac{1}{2}\alpha^2 (x^2 + y^2)\right]$$

Here n_0 represents the on-axis value of the index n and the quadratic coefficient α can be regarded as the reciprocal of the radius at which (hypothetically) the index n would fall to half its central value. (In practical situations only the central regions of this parabolic profile really exist, and even at the edge of the system the quantity $\alpha^2 r^2$ will usually be very small - less than 0·01.)

It is, of course, also possible for the refractive index to increase rather than decrease as the square of the distance off-axis; to represent this situation we may have to assume that the parameter α takes on an imaginary value. (Quadratic variation of absorption or gain can also be represented by taking α to be complex.)

In order to build up a ray-transfer matrix which represents the optical properties of such a distribution, we begin by considering an infinitesimally thin slice contained between the planes $z = z_0$ and $z = z_0 + \delta z$. We can regard such a slice as introducing not only a gap of reduced thickness $\delta z/n_0$ but also a non-uniform retardation given by

$$W(x,y) = \left[n(x,y) - n_0\right]\delta z = -\tfrac{1}{2}n_0\alpha^2 (x^2 + y^2)\delta z$$

If we compare this non-uniform retardation with that which would be generated by a very weak thin lens of power P, namely $- P(x^2 + y^2)/2$, we find that the power of the 'equivalent infinitesimal lens' must be given by $P_e = n_0\alpha^2\delta z$.

It follows that, in order to represent the optical effect of this thin slice, we must use *two* matrices

$$\mathscr{T}(\delta z) = \begin{bmatrix} 1 & \delta z/n_0 \\ 0 & 1 \end{bmatrix} \quad \text{and} \quad \mathscr{R}(\delta z) = \begin{bmatrix} 1 & 0 \\ -n\alpha^2\delta z & 1 \end{bmatrix}$$

There are perhaps four different ways in which these two matrices could be combined. We could form

(1) $\quad M(\delta z) = \mathscr{T}(\delta z) \, \mathscr{R}(\delta z)$

or (2) $\quad M(\delta z) = \mathscr{R}(\delta z) \, \mathscr{T}(\delta z)$

Alternatively, we could split one of the two matrices in half and use these two halves as premultiplier and postmultiplier respectively. We thus obtain

(3) $\quad M(\delta z) = \mathscr{T}(\delta z/2) \, \mathscr{R}(\delta z) \, \mathscr{T}(\delta z/2)$

or (4) $\quad M(\delta z) = \mathscr{R}(\delta z/2) \, \mathscr{T}(\delta z) \, \mathscr{R}(\delta z/2)$

As will become clear later, it makes little difference in the end which of these four alternatives is chosen; we shall restrict ourselves here to considering only the first and third possibility.

Case (1)

We have

$$M(\delta z) = \begin{bmatrix} 1 & \delta z/n_0 \\ 0 & 1 \end{bmatrix} \begin{bmatrix} 1 & 0 \\ -n_0 \alpha^2 \delta z & 1 \end{bmatrix} = \begin{bmatrix} 1 - (\alpha \delta z)^2 & \delta z/n_0 \\ -n_0 \alpha^2 \delta z & 1 \end{bmatrix}$$

At this stage we make the substitutions

$$K = \alpha n_0 \quad \text{and} \quad \theta = 2\sin^{-1}(\alpha \delta z/2) \quad \text{or} \quad \alpha \delta z = 2\sin(\theta/2)$$

We then have

$$M(\delta z) = \begin{bmatrix} 2\cos\theta - 1 & \left(2\sin(\theta/2)\right)/K \\ -2K\sin(\theta/2) & 1 \end{bmatrix}$$

At this point we check that the determinant of this matrix is unity, and we note that its spur is $2\cos\theta$, so that its eigenvalues are $e^{\pm i\theta}$.

Having found this 'infinitesimal matrix' $M(\delta z)'$, let us now raise it to its pth power, where p is a very large number. We shall then obtain the ray-transfer matrix for a slice of finite thickness $t = p\delta z$. For this purpose we recall Sylvester's theorem, which was quoted in section III.5. If $M = \begin{bmatrix} A & B \\ C & D \end{bmatrix}$ is a unimodular matrix with eigenvalues $e^{\pm i\theta}$, then

$$M^P = \begin{bmatrix} \dfrac{\sin(p+1)\theta - D\sin p\theta}{\sin\theta} & \dfrac{B\sin p\theta}{\sin\theta} \\[2em] \dfrac{C\sin p\theta}{\sin\theta} & \dfrac{D\sin p\theta - \sin(p-1)\theta}{\sin\theta} \end{bmatrix}$$

On substituting for B, C and D the values that we have just derived, we obtain

$$M(t) = \left[M(\delta z)\right]^P$$

$$= \begin{bmatrix} \dfrac{\sin p\theta\cos\theta + \cos p\theta\sin\theta - \sin p\theta}{\sin\theta} & \dfrac{2\sin p\theta\sin(\theta/2)}{K\sin\theta} \\[2em] \dfrac{-2K\sin(\theta/2)\sin p\theta}{\sin\theta} & \dfrac{\sin p\theta - \sin p\theta\cos\theta + \cos p\theta\text{s}}{\sin\theta} \end{bmatrix}$$

$$= \begin{bmatrix} \cos p\theta - \dfrac{\sin p\theta(1-\cos\theta)}{\sin\theta} & \dfrac{\sin p\theta}{K\cos(\theta/2)} \\[2em] \dfrac{-K\sin p\theta}{\cos(\theta/2)} & \cos p\theta + \dfrac{\sin p\theta(1-\cos\theta)}{\sin\theta} \end{bmatrix}$$

If we now assume that $\delta z \to 0$ and $p \to \infty$ in such a way that their product $(p\delta z) = t$ remains constant, the behaviour of the parameter θ will be such that $\theta \to 0$ but $p\theta = \alpha t$ remains unaltered. In the above matrix, therefore, the term $(1 - \cos\theta)/\sin\theta$ will be negligibly small and $\cos(\theta/2)$ will be indistinguishable from unity. We thus obtain finally

$$M(t) = \begin{bmatrix} \cos p\theta & \dfrac{\sin p\theta}{K} \\[1.5em] -K\sin p\theta & \cos p\theta \end{bmatrix} = \begin{bmatrix} \cos\alpha t & \dfrac{\sin\alpha t}{\alpha n_0} \\[1.5em] -\alpha n_0\sin\alpha t & \cos\alpha t \end{bmatrix}$$

Before we discuss the optics of this solution, we shall show that the same matrix $M(t)$ is obtained even if we use one of the other possibilities noted above as our starting point.

Case (3)

In this case our infinitesimal matrix is the triple product

$$M(\delta z) = \begin{bmatrix} 1 & \frac{1}{2}\delta z/n_0 \\ 0 & 1 \end{bmatrix} \begin{bmatrix} 1 & 0 \\ -n_0\alpha^2\delta z & 1 \end{bmatrix} \begin{bmatrix} 1 & \frac{1}{2}\delta z/n_0 \\ 0 & 1 \end{bmatrix}$$

$$= \begin{bmatrix} 1 - \frac{1}{2}(\alpha\delta z)^2 & \frac{\delta z}{n_0}\left(1 - (\frac{1}{2}\alpha\delta z)^2\right) \\ -n_0\alpha^2\delta z & 1 - \frac{1}{2}(\alpha\delta z)^2 \end{bmatrix}$$

Changing to the same parameters θ and K as before, we re-express this matrix as

$$M(\delta z) = \begin{bmatrix} 1 - 2\sin^2(\theta/2) & \frac{2\sin(\theta/2)}{K}(1 - \sin^2(\theta/2)) \\ -2K\sin(\theta/2) & 1 - 2\sin^2(\theta/2) \end{bmatrix}$$

$$= \begin{bmatrix} \cos\theta & \frac{\sin\theta\cos(\theta/2)}{K} \\ -2K\sin(\theta/2) & \cos\theta \end{bmatrix}$$

(It will be noticed that the main diagonal elements of this matrix are now the same; as before, the determinant is unity and the eigenvalues are $e^{\pm i\theta}$.)

Again using Sylvester's theorem, we obtain finally

$$M(p\delta z) = M^p$$

$$= \begin{bmatrix} \dfrac{\sin p\theta\cos\theta + \cos p\theta\sin\theta - \sin p\theta\cos\theta}{\sin\theta} & \dfrac{\sin\theta\cos(\theta/2)\sin p\theta}{K\sin\theta} \\ -\dfrac{2K\sin(\theta/2)\sin p\theta}{\sin\theta} & \dfrac{\cos\theta\sin p\theta - \sin p\theta\cos\theta + \cos p\theta\sin\theta}{\sin\theta} \end{bmatrix}$$

$$= \begin{bmatrix} \cos p\theta & \cos(\theta/2)\dfrac{\sin p\theta}{K} \\ \dfrac{-K\sin p\theta}{\cos(\theta/2)} & \cos p\theta \end{bmatrix}$$

Once again, in the limit where $\theta \to 0$, $\cos(\theta/2)$ is indistinguishable from unity, so our final matrix becomes, as before,

$$M(t) = \begin{bmatrix} \cos p\theta & \dfrac{\sin p\theta}{K} \\ \\ -K\sin p\theta & \cos p\theta \end{bmatrix} = \begin{bmatrix} \cos\alpha t & \dfrac{\sin\alpha t}{n_0\alpha} \\ \\ -n_0\alpha\sin\alpha t & \cos\alpha t \end{bmatrix}$$

From now on we shall use this matrix $M(t)$ to represent the optical effect of *any* arbitrary length of the medium, whether long or infinitesimally short. Apart from the multiplying factor αn_0 it resembles a 'rotation matrix', and it possesses the same property that $M(t_1)M(t_2) = M(t_2)M(t_1) = M(t_1 + t_2)$. It is also easily verified by means of Sylvester's theorem that $\big(M(t)\big)^N = M(Nt)$; this will be true even if N is not an integer.

Cursory inspection of this matrix shows that it has periodic properties; over a distance $t = 2\pi/\alpha$ it reduces to a unit matrix. Over half this distance it resembles that for an afocal system of magnification minus unity, and over a distance $t = \pi/2\alpha$ it has the same form as that for a lens of focal power $K = n_0\alpha$, referred to its two focal planes. We shall now use the matrix to examine in more detail the path of a geometric ray, or the progress of a Gaussian beam, through such a medium.

III.9.1 Propagation of paraxial rays

Let us use an input reference plane RP$_0$ located at $z = 0$ and an output reference plane RP$_z$, which we can move at will to any desired z-position. For any given input ray $\begin{bmatrix} y_0 \\ V_0 \end{bmatrix}$ we can then write

$$\begin{bmatrix} y(z) \\ V(z) \end{bmatrix} = M(z)\begin{bmatrix} y_0 \\ V_0 \end{bmatrix} = \begin{bmatrix} \cos\alpha z & \dfrac{\sin\alpha z}{\alpha n_0} \\ -\alpha n_0\sin\alpha z & \cos\alpha z \end{bmatrix}\begin{bmatrix} y_0 \\ V_0 \end{bmatrix}$$

If, for example, our input ray comes in parallel to the axis at height y_0, then since $V_0 = 0$ we find

$$y(z) = y_0\cos\alpha z$$

and
$$V(z) = -y_0 n_0\alpha\sin\alpha z$$

or, since $v(z) = V(z)/n_0$,

$$v(z) = -y_0 \alpha \sin \alpha z$$

(= $dy(z)/dz$, of course).

It will be noticed that, whenever z is an odd multiple of $\pi/2\alpha$, the y-coordinate of this ray vanishes, independent of the height y_0 at which it entered the input plane, and whenever z is an even multiple of $\pi/2\alpha$, the v-values vanish and the ray is travelling parallel to the optic axis again.

According to the geometric model, therefore, a parallel input beam is periodically focused to a point, and then recollimated, at intervals $\Delta z = \pi/\alpha$. This situation is illustrated in Figure III.18.

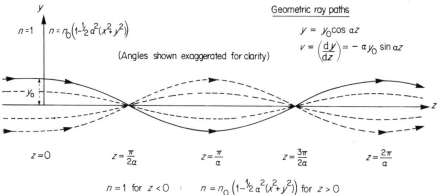

$$n = 1 \text{ for } z < 0 \quad : \quad n = n_0\left(1-\tfrac{1}{2}\alpha^2(x^2+y^2)\right) \text{ for } z > 0$$

Figure III.18

But, because of the effects of diffraction, the focusing of the beam at the 'crossover positions' will never be perfectly sharp. If y_m is the maximum radius of the beam at $z = 0$, then the maximum V-value in the 'focal region' $z = \pi/2\alpha$ will be $V_m = n_0\alpha y_m$; as indicated in section III.3, therefore, we must expect the beam diameter at this point to be at least $0\cdot61\lambda/V_m = 0\cdot61\lambda/(n_0\alpha y_m)$. The product of the beam diameters at $z = m\pi/\alpha$ and at $z = (2m + 1)\pi/2\alpha$ will therefore be approximately $(2y_m)(0\cdot61\lambda)/(n_0\alpha y_m)$, or $1\cdot22\lambda/n_0\alpha$. This is an important limitation, and the ray diagram shown in Figure III.18 is a geometric idealization.

III.9.2 Propagation of a Gaussian beam

In order to follow the progress of a Gaussian beam through a lens-like medium, all that we need do is to apply the *ABCD* rule, the coefficients A, B, C and D being the known functions of z that were obtained above. Thus if q_0 represents the complex curvature parameter of the input beam in the plane $z = 0$, then

$$q(z) = \frac{A(z)q_0 + B(z)}{C(z)q_0 + D(z)}$$

Equally, if we are interested only in how the spot radius w changes with distance through the medium, we can use the spot size transformation formula:

$$\left(\frac{w(z)}{w(0)}\right)^2 = \left| A(z) + B(z)/q_0 \right|^2$$

For the case where the input beam has its neck located at $z = 0$, we have immediately $w(0) = w_0$ and $1/q_0 = i\lambda/\pi w_0^2$, so that

$$w(z)^2 = w_0^2\left(A(z)^2 + \frac{B(z)^2\lambda^2}{\pi^2 w_0^4}\right)$$

$$= w_0^2\left(\cos^2\alpha z + \frac{\sin^2\alpha z\ \lambda^2}{\pi^2\alpha^2 n_0^2 w_0^4}\right)$$

$$= w_0^2\left[1 + \frac{\lambda^2 - \pi^2\alpha^2 n_0^2 w_0^4}{\pi^2\alpha^2 n_0^2 w_0^4}\ (\sin^2\alpha z)\right]$$

This formula shows that, unless $\lambda^2 = \pi^2\alpha^2 n_0^2 w_0^4$, the beam radius fluctuates between two extreme values: one of them occurs when $\sin\alpha z = 0$ and its value is just w_0, and the other occurs when $\sin^2\alpha z = 1$. In the latter case $w(z) = w_0(\lambda/\pi\alpha n_0 w_0^2) = (\lambda/\pi\alpha n_0 w_0)$.

If $\pi\alpha n_0 w_0^2 > \lambda$, then this second value is a minimum, as illustrated in Figure III.19a. If $\pi\alpha n_0 w_0^2 < \lambda$, then the second value is a maximum, as illustrated in Figure III.19b. Finally, if $\pi\alpha n_0 w_0^2 = \lambda$, the coefficient of $\sin^2\alpha z$ vanishes and the beam radius retains its initial value at all distances, as illustrated in Figure III.19c. In this case, evidently, the input beam is exactly matched to the 'fundamental Gaussian mode' of this lens like distribution. The beam radius of this fundamental mode is given by $w_f = (\lambda/\pi\alpha n_0)^{\frac{1}{2}}$ and the corresponding input q-value is given by $q_0 = -i\pi w_f^2/\lambda = -i/\alpha n_0$.

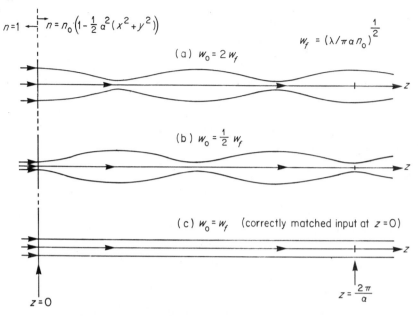

$$n = n_0 \left(1 - \tfrac{1}{2} a^2 (x^2 + y^2) \right)$$

$$W_f = (\lambda / \pi a n_0)^{\frac{1}{2}}$$

(a) $W_0 = 2 W_f$

(b) $W_0 = \tfrac{1}{2} W_f$

(c) $W_0 = W_f$ (correctly matched input at $z = 0$)

$$z = \frac{2\pi}{a}$$

$$z = 0$$

Figure III.19

[We could, of course, have derived this q-value more directly, calculating it as the eigenvector ratio of the matrix $M(z)$. We obtain, almost by inspection,

$$q_0 = \frac{\lambda_1 - D}{C} = \frac{e^{iaz} - \cos az}{- an_0 \sin az} = (-i/an_0)\,]$$

It will be noticed that, even if the input beam is not matched to this fundamental mode, the product of the maximum and minimum beam diameters between which it fluctuates remains always the same, being equal to

$$(2w_0)(2\lambda/\pi a n_0 w_0) = 4\lambda/\pi a n_0 = (2w_f)^2 = 1 \cdot 27 (\lambda/a n_0)$$

The last expression quoted agrees quite closely with that calculated previously for a uniform beam according to the Airy formula. But the agreement is slightly fortuitous, since different criteria are being used for specifying the beam diameter.

There are many ramifications of this topic. In gaseous lens systems, for example, it frequently happens that the coefficient α is not constant but varies along the direction of propagation. It is also possible for the coefficient α to be purely imaginary, in which case the medium behaves rather like a travelling-wave version of an unstable resonator. If we write $a = i\alpha$, the equation for the refractive index distribution becomes

$$n(z) = n_0(1 + \tfrac{1}{2}a^2 r^2)$$

and the matrix representing the medium becomes

$$M(z) = \begin{bmatrix} \cosh az & \dfrac{\sinh az}{an_0} \\ -an_0 \sinh az & \cosh az \end{bmatrix}$$

A 'negative' lens-like distribution of this kind can be generated, on a relatively slow time-scale, by the action of a continuous laser. If several kilojoules of energy are being transported, then the heating produced by any residual absorption tends to *reduce* the index of the medium in the central region. This process, which is the opposite of self-focusing, is usually called 'thermal lensing'. For a medium of given absorption, the initial rise of temperature depends not only on the intensity and duration of the beam but also on the beam radius; but where propagation through the atmosphere is concerned, the complicating effects of wind, turbulence and convection may all need to be considered.

As the last remark suggests, ray-transfer matrices cannot be expected to solve all the problems that arise with continuous distributions of refractive index. They are applicable only when the distribution is still lens-like and the beam geometry is still Gaussian.

There was perhaps an additional reason for including some mention of lens-like media in this chapter. We believe that it is an instructive example, which may encourage the student to look for the parallels that often exist between optics and other subjects. In this case we have in mind particularly the field of electrical engineering.

The transition that we have made, from a system containing a finite number of lenses to a continuous distribution, is closely akin to the transition from a multisection filter (with lumped inductances and capacitances) to a continuous transmission line. More generally, the eigenvalue, which indicates to the optician whether a resonator is stable or unstable, determines, in communication, the phase shift or attenuation of an electrical signal; and the eigenvector ratio, instead of representing the R-value or q-value of a self-repeating wave-pattern, becomes for the electrical engineer a voltage-current ratio - an iterative impedance. As with the invention of the laser itself, the search for such parallels is often fruitful, but care must be taken to consider the differences as well as the similarities!

In the last section of this chapter we shall consider a few problems; these may serve as recapitulation of some of the ground that has been covered. Suggestions for further reading on ray-transfer matrices are appended at the end of the book.

III.10 ILLUSTRATIVE PROBLEMS
Problem 1

You are given a sealed optical system and are asked to identify the elements of its ray-transfer matrix
$M_0 = \begin{bmatrix} A_0 & B_0 \\ C_0 & D_0 \end{bmatrix}$ The only alterations you can make are to attach either or both of two weak thin lenses, of known focal power P_1 and P_2, to the front and the back of the system. The only facility for measurement available is a distant source of known angular size together with a cathetometer, which enables you to measure the equivalent focal length $f = -1/C$ for each of the four possible combinations.

Show that, if C_0, C_1, C_2 and C_{12} represent the four measured C-values, the solution for the matrix must be

$$\begin{bmatrix} A_0 & B_0 \\ C_0 & D_0 \end{bmatrix} = \begin{bmatrix} \dfrac{C_0 - C_2}{P_2} & \dfrac{C_{12} + C_0 - C_1 - C_2}{P_1 P_2} \\ C_0 & \dfrac{C_0 - C_1}{P_1} \end{bmatrix}$$

Solution

For the case where both lenses have been added, the modified ray-transfer matrix is evidently

$$\begin{bmatrix} A_{12} & B_{12} \\ C_{12} & D_{12} \end{bmatrix} = \begin{bmatrix} 1 & 0 \\ -P_2 & 1 \end{bmatrix} \begin{bmatrix} A_0 & B_0 \\ C_0 & D_0 \end{bmatrix} \begin{bmatrix} 1 & 0 \\ -P_1 & 1 \end{bmatrix}$$

$$= \begin{bmatrix} 1 & 0 \\ -P_2 & 1 \end{bmatrix} \begin{bmatrix} A_0 - B_0 P_1 & B_0 \\ C_0 - D_0 P_1 & D_0 \end{bmatrix}$$

$$= \begin{bmatrix} A_0 - B_0 P_1 & B_0 \\ -P_2 (A_0 - B_0 P_1) + (C_0 - D_0 P_1) & -P_2 B_0 + D_0 \end{bmatrix}$$

Hence, when both lenses are present, the measured C-value is given by

$$C_{12} = B_0 P_1 P_2 - D_0 P_1 - A_0 P_2 + C_0$$

If only the front lens has been added, we must put $P_2 = 0$, and obtain

$$C_1 = \quad - D_0 P_1 \quad + C_0$$

If only the back lens has been added, we must put $P_1 = 0$, and obtain

$$C_2 = \quad - A_0 P_2 + C_0$$

Hence we obtain the required solutions

$$A_0 = \frac{C_0 - C_2}{P_2}, \quad D_0 = \frac{C_0 - C_1}{P_1} \quad \text{and} \quad B_0 = \frac{C_{12} + C_0 - C_1 - C}{P_1 P_2}$$

In the experimental procedure that we have just discussed, only two auxiliary lenses are needed. If, on the other hand, several known thin lenses are available, then it becomes possible to make a whole series of measurements of C as a function of the varying lens powers P_1 and P_2. Since in each case we have $C = B_0 P_1 P_2 - D_0 P_1 - A_0 P_2 + C_0$, the overall matrix can be expressed in the partial derivative form

$$
\begin{bmatrix} A_0 & B_0 \\ C_0 & D_0 \end{bmatrix} = \begin{bmatrix} \dfrac{-\partial C}{\partial P_2} & \dfrac{\partial^2 C}{\partial P_1 \partial P_2} \\ C_0 & \dfrac{-\partial C}{\partial P_1} \end{bmatrix}
$$

(This result is nearly a century old. It was obtained, in a slightly different form, by L. Pendlebury in his book *Lenses and Systems of Lenses Treated after the Manner of Gauss*, Deighton and Bell, Cambridge, 1884, pp.55 and 91.)

Problem 2

Consider two bounded spherical wavefronts and their associated ray pencils radiating from two separate axial image points O and O'. Let the extreme rays of these two pencils be characterized in a given reference plane by the ray vectors $\begin{bmatrix} y \\ V \end{bmatrix}$ and $\begin{bmatrix} y' \\ V' \end{bmatrix}$

Show that, in the plane where these two marginal rays intersect (so that the area irradiated by the two wavefronts is the same), the difference between the sagittae of the two waves is equal to one-half the determinant of $\begin{bmatrix} y & y' \\ V & V' \end{bmatrix}$

Show also that, if one marginal ray vector $\begin{bmatrix} y \\ V \end{bmatrix}$ is replaced by its diametrically opposite ray vector $\begin{bmatrix} -y \\ -V \end{bmatrix}$, a second plane can be found where the two wavefronts occupy the same area. Evaluate the sagittal difference for this second plane of intersection, and find a physical interpretation for the change in its sign.

Solution

(a) Figure III.20a illustrates a ray vector $\begin{bmatrix} y \\ V \end{bmatrix}$ emanating from O and a second ray vector $\begin{bmatrix} y' \\ V' \end{bmatrix}$ emanating from O'. In order to locate the plane of their intersection, we seek a $\mathscr{T}$-matrix such that the calculations

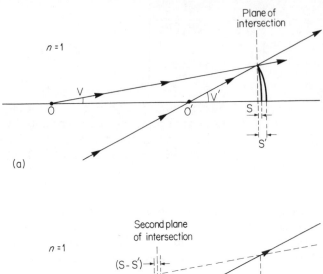

(a)

(b)

Figure III.20

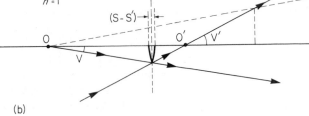 both produce the same y-value.

In other words, we require

$$y + TV = y' + TV'$$

Hence

$$T = \frac{(y - y')}{(V' - V)}$$

(Obviously T will be finite only if $V \neq V'$.) At the plane of intersection, located a distance T to the right of the given reference plane, the shared y-value will be

$$y + \frac{V(y - y')}{(V' - V)} = \frac{(yV' - y'V)}{(V' - V)}$$

and the two ray vectors, referred to the plane of

intersection, will be $\begin{bmatrix} \dfrac{(yV' - y'V)}{(V' - V)} \\ V \end{bmatrix}$ and $\begin{bmatrix} \dfrac{(yV' - y'V)}{(V' - V)} \\ V' \end{bmatrix}$

Because the V-values are different, the two spherical wavefronts which are associated with these ray vectors must have different radii of curvature and different sagittae. Remembering from section III.2 that a sagitta can be calculated as half the product of the y-value and the V-value, we obtain for the sagittal difference the equation

$S' - S$ = Sagitta of O'-wave - Sagitta of O-wave

$$= \tfrac{1}{2} \frac{(yV' - y'V)}{(V' - V)} (V' - V) = \tfrac{1}{2}(yV' - y'V)$$

$$= \tfrac{1}{2} \det \begin{bmatrix} y & y' \\ V & V' \end{bmatrix}$$

In the example illustrated, it will be seen that it is the second wave, emanating from O', which has the greater sagitta or greater divergence; and since $V' > V$, the determinant must have a positive value (in *any* reference plane).

(b) Figure III.20b illustrates the situation where $\begin{bmatrix} y \\ V \end{bmatrix}$ has been replaced by the diametrically opposite ray vector $\begin{bmatrix} -y \\ -V \end{bmatrix}$. The new plane of intersection is located at a distance $(- y - y')/(V' + V)$ to the right of the reference plane, in other words at a distance $(y + y')(V + V')$ to its left.

The common y-value at this point of intersection is $(- yV' + y'V)/(V' + V)$ and the excess of the O'-wave sagitta is now

$$(S' - S) = \tfrac{1}{2} \frac{(- yV' + y'V)}{(V' + V)} (V' + V) = - \tfrac{1}{2} \det \begin{bmatrix} y & y' \\ V & V' \end{bmatrix}$$

the *reverse* of the previous value.

As will be seen from the figure, when the O'-wave reaches the plane of intersection it has not yet

reached the axial point O', and is obviously *convergent*
The divergence of the O-wave, on the other hand, is
slightly greater than at the intersection region shown
in Figure III.20a.

It is perhaps worth noting that for any bounded
spherical wave, which is being propagated in air in
the direction of the z-axis, the rate of change of
sagitta S with distance z is constant. If V represents the limiting ray angle (N.A.), this rate is given
by

$$\frac{(dS)}{(dz)} = \frac{d}{dz}(\tfrac{1}{2}yV) = \tfrac{1}{2}V(dy/dz) = \tfrac{1}{2}V^2$$

Hence, for the two waves of this problem,

$$\frac{d}{dz}(S' - S) = \tfrac{1}{2}(V'^2 - V^2)$$

The distance Δz between the two planes of intersection
is given by

$$\Delta z = \frac{(y - y')}{(V' - V)} + \frac{(y' + y)}{(V' + V)} = \frac{2 \det\begin{bmatrix} y & y' \\ V & V' \end{bmatrix}}{(V'^2 - V^2)}$$

Hence the change in sagittal difference is going
to be

$$\Delta(S' - S) = \tfrac{1}{2}(V'^2 - V^2)\Delta z = \det\begin{bmatrix} y & y' \\ V & V' \end{bmatrix}$$

As we have already calculated, the change occurs
symmetrically, from $-\tfrac{1}{2}$ to $+\tfrac{1}{2}$ the value of this
determinant.

The student may like to verify that a similar formula
for the increase of saggita applies to the propagation
of a Gaussian beam. If the sagitta is calculated with
respect to the $1/e^2$ contours, its value is given by
$S = (\lambda/2\pi)(z/z_0)$, where z is the distance to the neck
of the beam and $z_0 = \pi w_0^2/\lambda$ is the usual confocal beam
parameter. If we say that V for the geometric ray
pencil corresponds to $\lambda/\pi w_0$ - the asymptotic angle of
divergence - for the Gaussian beam, then the same formula $dS/dz = \tfrac{1}{2}V^2$ is obeyed, even in the vicinity of the
beam neck.

Problem 3

In a to-and-fro resonator system for a dye laser, the fully reflecting mirror on the left has converging focal power P_1 and the output mirror on the right has converging power P_2. The intervening region contains several lens elements, so that the matrix for transferring data from the mirror on the left to that on the right is no longer a simple translation matrix. If we denote this matrix by $M = \begin{bmatrix} A_0 & B_0 \\ C_0 & D_0 \end{bmatrix}$, then show that the round-trip matrix calculated with respect to the output mirror must take the form

$$\begin{bmatrix} A & B \\ C & D \end{bmatrix} = \begin{bmatrix} A_0 & B_0 \\ C_0 & D_0 \end{bmatrix} \begin{bmatrix} 1 & 0 \\ -P_1 & 1 \end{bmatrix} \begin{bmatrix} D_0 & B_0 \\ C_0 & A_0 \end{bmatrix} \begin{bmatrix} 1 & 0 \\ -P_2 & 1 \end{bmatrix}$$

Solution

What needs to be considered in this problem is the effect of reverse propagation through the intervening optical system. We can visualize this as a chain of $\mathcal{R}$-matrices and $\mathcal{T}$-matrices whose product *taken from left to right* yields the overall matrix $\begin{bmatrix} A_0 & B_0 \\ C_0 & D_0 \end{bmatrix}$.
We seek to prove that, if the same chain of matrices is written down and multiplied together in reverse order, the result will be $\begin{bmatrix} D_0 & B_0 \\ C_0 & A_0 \end{bmatrix}$, that is the two main diagonal elements are interchanged but none of the elements is changed in sign.

We shall show later that this can be proved by mathematical induction, making use of the fact that for each individual $\mathcal{R}$- and $\mathcal{T}$-matrix the A- and D-values are always the same. It seems preferable, however, to prove the proposition first by using an optical argument.

Let us assume that all the optical elements of our system are mounted together on a short length of optical bench. In Figure III.21a we show an input ray $\begin{bmatrix} y_1 \\ V_1 \end{bmatrix}$ travelling in the + z-direction, together with

168

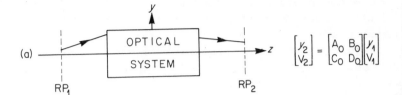

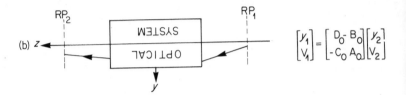

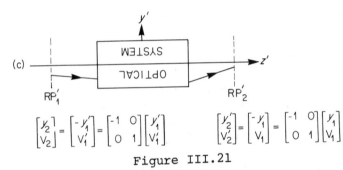

Figure III.21

the corresponding output ray $\begin{bmatrix} y_2 \\ V_2 \end{bmatrix} = \begin{bmatrix} A_0 & B_0 \\ C_0 & D_0 \end{bmatrix} \begin{bmatrix} y_1 \\ V_1 \end{bmatrix}$

also travelling in the + z-direction. In Figure III.21
we show the new geometry that results if the whole
optical bench is rotated about its centre by 180°, so
that the y-axis and the z-axis are both reversed, but
the x-axis remains the same.

If we now use ' suffices to represent the ray
vectors referred to new y'- and z'-axes, we can reverse
the direction of what was the original output ray and
use it as a new input ray (see Figure III.21c).
Because of the reversibility of light rays, the new
output ray will emerge in the right of the new diagram
travelling along the tracks of the old input ray. In
each case the V-values will be unchanged, but the
y-values will be reversed. We thus have

$$\begin{bmatrix} y_2 \\ V_2 \end{bmatrix} = \begin{bmatrix} -y_1' \\ V_1' \end{bmatrix} = \begin{bmatrix} -1 & 0 \\ 0 & 1 \end{bmatrix} \begin{bmatrix} y_1' \\ V_1' \end{bmatrix}$$

and

$$\begin{bmatrix} y_2' \\ V_2' \end{bmatrix} = \begin{bmatrix} -y_1 \\ V_1 \end{bmatrix} = \begin{bmatrix} -1 & 0 \\ 0 & 1 \end{bmatrix} \begin{bmatrix} y_1 \\ V_1 \end{bmatrix}$$

Remembering that, by inversion of the original matrix, we have

$$\begin{bmatrix} y_1 \\ V_1 \end{bmatrix} = \begin{bmatrix} D_0 & -B_0 \\ -C_0 & A_0 \end{bmatrix} \begin{bmatrix} y_2 \\ V_2 \end{bmatrix}$$

we combine all three equations and obtain finally as our new output-input relationship

$$\begin{bmatrix} y_2' \\ V_2' \end{bmatrix} = \begin{bmatrix} -1 & 0 \\ 0 & 1 \end{bmatrix} \begin{bmatrix} D_0 & -B_0 \\ -C_0 & A_0 \end{bmatrix} \begin{bmatrix} -1 & 0 \\ 0 & 1 \end{bmatrix} \begin{bmatrix} y_1' \\ V_1' \end{bmatrix}$$

$$= \begin{bmatrix} -D_0 & B_0 \\ -C_0 & A_0 \end{bmatrix} \begin{bmatrix} -1 & 0 \\ 0 & 1 \end{bmatrix} \begin{bmatrix} y_1' \\ V_1' \end{bmatrix}$$

$$= \begin{bmatrix} D_0 & B_0 \\ C_0 & A_0 \end{bmatrix} \begin{bmatrix} y_1' \\ V_1' \end{bmatrix}$$

This is the ray-transfer matrix for an optical system which has been turned back to front, so that the order in which all the lens surfaces and gaps are traversed is reversed. It resembles quite closely the reciprocal matrix, but, as we should expect, the element C_0 which represents the equivalent focal power still has the same value.

Alternative proof by mathematical induction

Let

$$M = \begin{bmatrix} A & B \\ C & D \end{bmatrix} = M_1 M_2 M_3 M_4 \ldots M_i \ldots M_n$$

where each individual matrix M_i is unimodular and such that its two main diagonal elements A_i and D_i are equal. Let M_B represent the 'backwards' product

$$M_B = M_n \cdots M_i \cdots M_4 M_3 M_2 M_1$$

Then it is required to prove that

$$M_B = \begin{bmatrix} D & B \\ C & A \end{bmatrix}$$

Consider the matrix chain

$$M_B \begin{bmatrix} -1 & 0 \\ 0 & 1 \end{bmatrix} M = M_n \cdots M_i \cdots M_3 M_2 M_1 \begin{bmatrix} -1 & 0 \\ 0 & 1 \end{bmatrix} M_1 M_2 M_3$$

$$\cdot \rightarrow \cdots M_i \cdots$$

If we evaluate first the product of the three central matrices, we obtain

$$M_1 \begin{bmatrix} -1 & 0 \\ 0 & 1 \end{bmatrix} M_1 = \begin{bmatrix} A_1 & B_1 \\ C_1 & D_1 \end{bmatrix} \begin{bmatrix} -1 & 0 \\ 0 & 1 \end{bmatrix} \begin{bmatrix} A_1 & B_1 \\ C_1 & D_1 \end{bmatrix}$$

$$= \begin{bmatrix} A_1 & B_1 \\ C_1 & D_1 \end{bmatrix} \begin{bmatrix} -A_1 & -B_1 \\ C_1 & D_1 \end{bmatrix}$$

$$= \begin{bmatrix} (-A_1^2 + B_1 C_1) & B_1(D_1 - A_1) \\ C_1(D_1 - A_1) & D_1^2 - B_1 C_1 \end{bmatrix}$$

But, since $A_1 = D_1$ and $(A_1 D_1 - B_1 C_1) = 1$, this matrix can be reduced to $\begin{bmatrix} -1 & 0 \\ 0 & 1 \end{bmatrix}$ — the same central matrix as before.

Proceeding similarly, we find that

$$M_2 \begin{bmatrix} -1 & 0 \\ 0 & 1 \end{bmatrix} M_2 = \begin{bmatrix} -1 & 0 \\ 0 & 1 \end{bmatrix}$$

and so on, so that in the end we obtain

$$M_B \begin{bmatrix} -1 & 0 \\ 0 & 1 \end{bmatrix} M = \begin{bmatrix} -1 & 0 \\ 0 & 1 \end{bmatrix}$$

If we now postmultiply the left side of this equation by $M^{-1} \begin{bmatrix} -1 & 0 \\ 0 & 1 \end{bmatrix}$ and the right side by the equivalent

expression $\begin{bmatrix} D & -B \\ -C & A \end{bmatrix} \begin{bmatrix} -1 & 0 \\ 0 & 1 \end{bmatrix}$, we obtain

$$M_B \begin{bmatrix} -1 & 0 \\ 0 & 1 \end{bmatrix} \cdot M \cdot M^{-1} \begin{bmatrix} -1 & 0 \\ 0 & 1 \end{bmatrix} = \begin{bmatrix} -1 & 0 \\ 0 & 1 \end{bmatrix} \begin{bmatrix} D & -B \\ -C & A \end{bmatrix} \begin{bmatrix} -1 & 0 \\ 0 & 1 \end{bmatrix}$$

This reduces to our required result

$$M_B = \begin{bmatrix} D & B \\ C & A \end{bmatrix}$$

Problem 4

Using the resonator considered in Problem 3, show that the round-trip matrix can also be represented by

$$\begin{bmatrix} A_r & B_r \\ \dfrac{(A_r^2 - 1)}{B_r} & A_r \end{bmatrix} \begin{bmatrix} 1 & 0 \\ -P_2 & 1 \end{bmatrix}, \text{ where } P_2 \text{ again represents}$$

the power of the output mirror.

If it is essential that this resonator operate in a stable configuration, what values of P_2 will be permissible? If one of these values is used, what will be the radius of curvature of the wavefront that emerges when the laser oscillates?

Solution

We observe at the outset that the matrix $\begin{bmatrix} A_r & B_r \\ \dfrac{(A_r^2 - 1)}{B_r} & A_r \end{bmatrix}$

can be regarded as the most general form of a unimodular matrix for which the two main diagonal elements are equal. In this case, it represents the transfer of a ray leaving RP_2 in the $-z$-direction back on to the same reference plane; this transfer is accomplished by the combined action of the optical system inside the resonator and the left-hand mirror - a complete catadioptric system. It was proved at the end of section II.11 that such matrices always have equal A- and D-values.

To confirm this, without appeal to section II.11, we multiply out the triple chain obtained in Problem 3 and obtain:

$$\begin{bmatrix} A_0 & B_0 \\ C_0 & D_0 \end{bmatrix} \begin{bmatrix} 1 & 0 \\ -P_1 & 1 \end{bmatrix} \begin{bmatrix} D_0 & B_0 \\ C_0 & A_0 \end{bmatrix}$$

$$= \begin{bmatrix} A_0 & B_0 \\ C_0 & D_0 \end{bmatrix} \begin{bmatrix} D_0 & B_0 \\ C_0 - P_1 D & A_0 - B_0 P_1 \end{bmatrix}$$

$$= \begin{bmatrix} A_0 D_0 + B_0 C_0 - B_0 D_0 P_1 & 2A_0 B_0 - B_0^2 P_1 \\ 2C_0 D_0 - D_0^2 P_1 & B_0 C_0 + A_0 D_0 - B_0 D_0 P_1 \end{bmatrix}$$

If we now rename the two main diagonal elements $A_0 D_0 + B_0 C_0 - B_0 D_0 P_1$ as A_r and rename the upper right-hand element $2A_0 B_0 - B_0^2 P_1$ as B_r, we obtain a matrix of the form stipulated. (We omit here the routine of checking that the above matrix is still unimodular, but it should always be done.)

Multiplying out the two matrices that remain, we obtain for the round-trip matrix:

$$\begin{bmatrix} A & B \\ C & D \end{bmatrix} = \begin{bmatrix} A_r & B_r \\ \dfrac{(A_r^2 - 1)}{B_r} & A_r \end{bmatrix} \begin{bmatrix} 1 & 0 \\ -P_2 & 1 \end{bmatrix}$$

$$= \begin{bmatrix} A_r - B_r P_2 & B_r \\ \dfrac{(A_r^2 - 1)}{B_r} - A_r P_2 & A_r \end{bmatrix}$$

In order that this shall represent a *stable* resonator, the spur $(A + D)$ must lie between 2 and -2. Rearranging the condition $2 > (2A_r - B_r P_2) > -2$, we find that P_2 must lie between $2(A_r + 1)/B_r$ and $2(A_r - 1)/B_r$.

If, for example, we have the simplest possible cavity, which contains only a plane mirror at a reduced distance T, then we have $A_r = 1$ and $B_r = 2T$, and P_2 must lie between $2/T$ and zero. In other words, the output mirror can be concave but not convex, and its radius of curvature must not be less than T ($r_2 = T$ gives a hemispherical resonator).

We revert now to the more general resonator. Provided that the condition for stability has been satisfied, the radius of curvature of the Gaussian beam emerging from the output mirror when the laser oscillates in its fundamental mode will be given by (see section III.7)

$$R = \frac{2B}{(D - A)} = \frac{2B_r}{(A_r) - (A_r - B_r P_2)} = 2/P_2 = r_2$$

(radius of curvature of the output mirror). In other words, as physical considerations would indicate, the pattern of waves that is generated is one which has constant phase across the output mirror.

For any to-and-fro system which is operating in a stable manner, this condition of constant phase is satisfied at both the end-mirrors, and a complete standing-wave pattern occupies the intervening region. The condition will not be satisfied in the case of unstable operation, for which some of the energy is always escaping in a radial direction.

If we wish to know not only the radius of curvature but also the spot radius of the Gaussian beam that will be generated, we could obtain the latter directly from the rather tedious approach employed in section III.8 (see Figure III.17). It is probably better, however, to form the round-trip matrix for the resonator and then determine its eigenfunctions. Proceeding in this way, we obtain two useful checks: firstly that the resonator is a stable one and secondly that the R-value obtained does agree with the known shape of the output mirror.

Problem 5

You are asked to help in designing a short sealed-off carbon dioxide laser for low-power operation at a single frequency. A preliminary specification suggests that this should be based on a stable resonator 50 cm long, and a parallel output beam is required. It is also suggested that θ (the phase angle in the eigen-value) should be nominally one radian. What mirror curvatures would you specify for this system and approx imately what tube diameter?

Solution

Since we are dealing with a stable resonator, the requirement for a collimated output tells us immediate that the output mirror must be plane. Working in metre therefore, we can insert $T = 0 \cdot 5$ and $P_2 = 0$ in the standard matrix for a two-mirror resonator and obtain (section III.4)

$$\begin{bmatrix} A & B \\ C & D \end{bmatrix} = \begin{bmatrix} 1 - P_1 T - 2P_2 T + P_1 P_2 T^2 & T(2 - P_1 T) \\ - P_1 - P_2 + P_1 P_2 T & 1 - P_1 T \end{bmatrix}$$

$$= \begin{bmatrix} 1 - \tfrac{1}{2}P_1 & 1 - \tfrac{1}{4}P_1 \\ - P_1 & 1 - \tfrac{1}{2}P_1 \end{bmatrix}$$

As we should expect, the two main diagonal elements are equal and the matrix is unimodular. We now deter-mine P_1 from the requirement that $(A + D) = 2 - P_1 = 2\cos\theta$, where the angle θ is one radian. Hence $P_1 = 0 \cdot 92$ and $r_1 = 2/P_1 = 2 \cdot 17$ (radius of curvature in metres).

With this value for the mirror curvature, the round-trip matrix has the numerical form

$$\begin{bmatrix} A & B \\ C & D \end{bmatrix} = \begin{bmatrix} 0 \cdot 54 & 0 \cdot 77 \\ -0 \cdot 92 & 0 \cdot 54 \end{bmatrix}$$

It should be noticed that up to this point no accoun has been taken of the wavelength at which this laser i to operate. In order to choose the tube diameter, however, we need to know the geometry of the fundamen-tal Gaussian mode that will be generated.

Using the formula for the neck radius w_0, we find

$$w_0 = \left(\frac{(-\lambda\sin\theta)}{\pi C} \right)^{\frac{1}{2}} = \left(\frac{(10 \cdot 6)(10^{-6})(0 \cdot 842)}{(0 \cdot 92)(\pi)} \right)^{\frac{1}{2}}$$

$$= 1 \cdot 76 \times 10^{-3} \text{ metres, or } \underline{1 \cdot 76 \text{ mm}}$$

For the confocal parameter z_0, we find

$$z_0 = (- \sin\theta/C) = \frac{0 \cdot 842}{0 \cdot 920} = 0 \cdot 915 \text{ metres}$$

It is worth noticing that, if the same resonator geometry were being used for a visible laser, the neck radius w_0 would be about four times smaller but the confocal parameter z_0 would still be the same.

In seeking a suitable value for the tube diameter, we need to remember that the beam will be slightly larger at the other end of the resonator, its spot radius being given by

$$w = w_0(1 + T^2/z_0^2)^{\frac{1}{2}} = 1 \cdot 14 w_0$$

As a result of this slight expansion the maximum beam diameter, measured at the $1/e^2$ points, is almost exactly 4 mm.

For most gas lasers, in order to allow for diffraction losses, and for some degree of misalignment, the diameter of the discharge tube is made at least twice the $1/e^2$ diameter. For a carbon dioxide laser discharge the tube walls may produce troublesome oblique reflections, and they will cool the gas slightly so as to produce a higher gain-coefficient. Furthermore, the total gas volume enclosed may be important in determining the life of a sealed-off system.

In this case a tube with 10 mm internal diameter might prove a reasonable choice, but it might be better to use a larger tube in conjunction with a 10 mm diaphragm.

Problem 6

As shown in Figure III.22, a collimated laser beam 2 cm in diameter is being focused by a simple glass plano-convex lens, whose focal power is 10 dioptres,

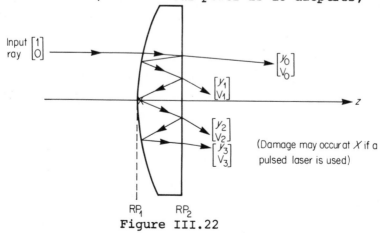

Figure III.22

central thickness 1 cm and refractive index 1·5.
Using paraxial optics, calculate the position of any
focused image which is formed as a result of internal
reflection by the surfaces of the lens and which falls
inside the volume of the glass.

If the lens surfaces are uncoated, and if spherical
aberration and self-trapping effects can be neglected
by how much will the light intensity at this focus
exceed that in the incident parallel beam?

Solution

Let $\mathcal{R}_1$ and $\mathcal{R}_2$ respectively denote the refraction
matrices of the first and second surface. Let $\mathcal{R}_1'$ and
$\mathcal{R}_2'$ denote the corresponding reflection matrices, for
reflection *from the glass side*. Finally, let the
translation matrix $\mathcal{T}$ represent the 1 cm thickness of
the glass.

For transfer of ray data *through* the first and second
lens surfaces, without any internal reflections, the
matrix will be

$$M_0 = \mathcal{R}_2 \, \mathcal{T} . \mathcal{R}_1$$

If, on the other hand, a ray emerges in the $+ z$-
direction after suffering n internal reflections, the
transfer matrix will be

$$M_n = \mathcal{R}_2 (\mathcal{T} \, \mathcal{R}_1' \, \mathcal{T} \, \mathcal{R}_2')^n \, \mathcal{T} \, \mathcal{R}_1$$

For this particular lens, the second surface is
plane, so both $\mathcal{R}_2$ and $\mathcal{R}_2'$ reduce to the unit matrix
The above matrix can therefore be simplified to the
form

$$M_n = (\mathcal{T} \, \mathcal{R}_1' \, \mathcal{T})^n \, \mathcal{T} \, \mathcal{R}_1$$

Working in centimetres, we see immediately that
$r_1 = 5$, so that

$$\mathcal{R}_1 = \begin{bmatrix} 1 & 0 \\ -0 \cdot 1 & 1 \end{bmatrix} \quad \text{and} \quad \mathcal{R}_1' = \begin{bmatrix} 1 & 0 \\ \dfrac{-2n}{r_1} & 1 \end{bmatrix} = \begin{bmatrix} 1 & 0 \\ -0 \cdot 6 & 1 \end{bmatrix}$$

while

$$\mathcal{T} = \begin{bmatrix} 1 & t/n \\ 0 & 1 \end{bmatrix} = \begin{bmatrix} 1 & 2/3 \\ 0 & 1 \end{bmatrix}$$

With these substitutions, we obtain

$$M_n = \left(\begin{bmatrix} 1 & 2/3 \\ 0 & 1 \end{bmatrix} \begin{bmatrix} 1 & 0 \\ -0 \cdot 6 & 1 \end{bmatrix} \begin{bmatrix} 1 & 2/3 \\ 0 & 1 \end{bmatrix} \right)^n \begin{bmatrix} 1 & 2/3 \\ 0 & 1 \end{bmatrix} \begin{bmatrix} 1 & 0 \\ -0 \cdot 1 & 1 \end{bmatrix}$$

$$= \left(\begin{bmatrix} 1 & 2/3 \\ 0 & 1 \end{bmatrix} \begin{bmatrix} 1 & 2/3 \\ -0 \cdot 6 & 0 \cdot 6 \end{bmatrix} \right)^n \begin{bmatrix} 0 \cdot 9333 & 2/3 \\ -0 \cdot 1 & 1 \end{bmatrix}$$

$$= \begin{bmatrix} 0 \cdot 6 & 1 \cdot 0667 \\ -0 \cdot 6 & 0 \cdot 6 \end{bmatrix}^n \begin{bmatrix} 0 \cdot 9333 & 2/3 \\ -0 \cdot 1 & 1 \end{bmatrix}$$

If only the nth image were of interest, then direct use could now be made of Sylvester's theorem. In this case, however, it will be better to make a progressive calculation. From each successive y_i and V_i, we can then determine the corresponding R_i, and if R_i lies between $0 \cdot 67$ and 0 this means that the ray being considered has been brought to a focus during its last passage through the thickness of the lens. On the other hand, if R lies between 0 and $-0 \cdot 67$, we know that it will do so during its next journey, after reflection from the plane surface.

Proceeding according to this scheme and operating on the marginal ray $\begin{bmatrix} 1 \\ 0 \end{bmatrix}$, we find first of all that

$$\begin{bmatrix} y_0 \\ V_0 \end{bmatrix} = \begin{bmatrix} 0 \cdot 933 \\ -0 \cdot 1 \end{bmatrix} \quad \therefore R_0 = -9 \cdot 33 \text{ (location of normal focused image)}$$

$$\begin{bmatrix} y_1 \\ V_1 \end{bmatrix} = \begin{bmatrix} 0 \cdot 6 & 1 \cdot 0667 \\ -0 \cdot 6 & 0 \cdot 6 \end{bmatrix} \begin{bmatrix} 0 \cdot 933 \\ -0 \cdot 1 \end{bmatrix} = \begin{bmatrix} 0 \cdot 4533 \\ -0 \cdot 62 \end{bmatrix}$$

$\therefore R_1 = -0 \cdot 731$ (only slightly less than $-0 \cdot 67$)

$$\begin{bmatrix} y_2 \\ V_2 \end{bmatrix} = \begin{bmatrix} 0 \cdot 6 & 1 \cdot 0667 \\ -0 \cdot 6 & 0 \cdot 6 \end{bmatrix} \begin{bmatrix} 0 \cdot 4533 \\ -0 \cdot 62 \end{bmatrix} = \begin{bmatrix} -0 \cdot 390 \\ -0 \cdot 644 \end{bmatrix}$$

$\therefore R_2 = 0 \cdot 606$ (*inside the glass*)

$$\begin{bmatrix} y_3 \\ V_3 \end{bmatrix} = \begin{bmatrix} 0\cdot6 & 1\cdot0667 \\ -0\cdot6 & 0\cdot6 \end{bmatrix} \begin{bmatrix} -0\cdot390 \\ -0\cdot644 \end{bmatrix} = \begin{bmatrix} -0\cdot921 \\ -0\cdot152 \end{bmatrix}$$

$\therefore R_3 = 6\cdot06$, etc.

(Notice that, when the ray that is being traced is brought to a focus, its y-value changes sign.)

Since the R-values that are calculated represent *reduced* radii, the position of the focused image or 'Boys point' that is produced after two internal reflections will be $(nR_2) = 0\cdot909$ cm to the left of the plane surface, that is about 1 mm in from the curved front surface.

For an uncoated surface of refractive index 1·5, th fraction of incident power reflected at normal incid- ence is 0·04. In the focused image that we are consi ering, therefore, the total power will be only $(0\cdot04)$ of that in the original beam. On the other hand, if the image size is diffraction-limited, its radius wil be reduced from 1 cm, its original value, to $(0\cdot61\lambda)/$ which in this case is $(0\cdot61\lambda)/0\cdot644$. (Notice again th the V-values calculated include the refractive index of the medium.)

If we assume a helium-neon laser wavelength, then t increase of intensity at the centre of this image, averaged over the area of the Airy disc, will be $(0\cdot04)^4 (V_2)^2/(0\cdot61\lambda)^2 = 710$ approximately. (For a giant pulse ruby or neodymium laser, the factor may b slightly less, because of the longer wavelength, but it may well be sufficient to cause dielectric breakdo with catastrophic results to an otherwise satisfactor lens.)

IV
Matrices in
Polarization Optics

IV.1 POLARIZED LIGHT - ITS PRODUCTION AND ANALYSIS

We assume that the reader has some knowledge of
polarized light such as is usually given in an elementary course on optics. He will know that for all
electromagnetic radiation the oscillating components
of the electric and magnetic fields are directed at
right angles to each other and to the direction of
propagation. In this chapter, as previously, we shall
assume a right-handed set of axes, with the Oz-axis
pointing along the direction of propagation. In
dealing with problems of polarization, we shall ignore
the magnetic field, and our interest will be mainly in
how the electric field vector is orientated with respect to the transverse xy-plane. We shall start by
considering some of the ways in which polarized light
is produced and how it may be specified or analysed.
Let us begin with two extreme cases.

Most ordinary sources of light, such as the sun or
an incandescent light bulb, produce light which we
describe as incoherent and unpolarized; light of this
sort is a chaotic jumble of almost innumerable independent disturbances, each with its own *direction* of
travel, its own optical *frequency* and its own state of
polarization. Just how innumerable are these 'states'
or 'modes of the radiation field'?

As far as the transverse modes are concerned, we
found in the last chapter that, for a source of area A
radiating into a solid angle Ω, the number of distinguishable *directions* of travel is given by $A\Omega/\lambda^2$ - for
a source of 1 cm^2 area radiating into a hemisphere,
this will be well over 10^8. Secondly, if we consider
a source which is observed over a period of one second

and ask how many optical *frequencies*, in theory at
least, can be distinguished, we obtain a multiplicity
factor of well over 10^{14}. By contrast with these
figures, if we consider the *polarization* of an electromagnetic wavefield, we find that the number of independent orthogonal states is just <u>two</u>!

Unpolarized light is sometimes said to be a random
mixture of 'all sorts' of polarization; but it would
be more correct to say that, whenever we try to analyse
it, in terms of any chosen orthogonal pair of states,
we can find no evidence for preference of one state
over the other. If we could make our observations fast
enough, we should find that the 'instantaneous' state o
polarization is passing very rapidly through all possible combinations of the two states we have chosen, in
a statistically random fashion. In most cases the rate
at which these changes occur is over 10^{12} per second,
so that what we observe is a smoothed average.

We have said enough, perhaps, to indicate that, while
unpolarized light is easy to produce, it is difficult
to describe, whether mathematically or in terms of a
model. (This point is discussed further in appendix C.

We turn now to the opposite extreme, namely the completely coherent light that man has recently learnt to
generate inside a single mode laser oscillator. Although this device represents a notable triumph of
human ingenuity, the light that it produces is about
as simple as one could specify. Let us assume that we
have a plane wave of angular frequency ω which is
travelling with velocity c in the direction Oz. Since
we know that the vibrations of the electric field
vector **E** are transverse, they can be specified in terms
of an x-component E_x, of peak amplitude H, and a y-
component E_y, of peak amplitude K. We thus have

$$E_x = H\cos\left[\omega\left(t - \frac{z}{c}\right) + \phi_x\right]$$

$$= \text{real part of } H\exp i\left[\omega\left(t - \frac{z}{c}\right) + \phi_x\right]$$

and

$$E_y = K\cos\left[\omega\left(t - \frac{z}{c}\right) + \phi_y\right]$$

$$= \text{real part of } K\exp i\left[\omega\left(t - \frac{z}{c}\right) + \phi_y\right]$$

(It is a matter of convention whether the time dependence in this complex exponential is displayed as $e^{i\omega t}$
or as $e^{-i\omega t}$. In this chapter we shall use the former
convention.)

If we use Δ to represent the phase difference
$(\phi_y - \phi_x)$, and if the symbols $\hat{\imath}$ and $\hat{\jmath}$ denote unit
vectors directed along the axes Ox and Oy, then these
two equations can be combined in the space vector form:

$$\mathbf{E}(x,y,z,t) = (H\hat{\imath} + Ke^{i\Delta}\hat{\jmath})\exp i\left[\omega\left(t - \frac{z}{c}\right) + \phi_x\right]$$

Alternatively, and for the purposes of this book
this will be preferable, we can use the column vector
form:

$$\begin{bmatrix} E_x \\ E_y \end{bmatrix} = \begin{bmatrix} H \\ Ke^{i\Delta} \end{bmatrix}\exp i\left[\omega\left(t - \frac{z}{c}\right) + \phi_x\right]$$

In the above expression there is no dependence on x
or y, since a plane wave of indefinite lateral extent
has been assumed. (A close approximation to such a
wave could be obtained by choosing the point of observation so that it is at a great distance from a laser
point source or, admittedly, at much lower frequencies,
from a continuous-wave radio transmitter.)

A column vector or ket such as $\begin{bmatrix} He^{i\phi_x} \\ Ke^{i\phi_y} \end{bmatrix}$ or $\begin{bmatrix} H \\ Ke^{i\Delta} \end{bmatrix}$ is

usually referred to as a Maxwell column, or as a Jones
vector. In this book we shall use the former description. As we shall see, the Maxwell column provides a
complete description of the state of polarization for
any light beam that is fully polarized. For the coherent plane wave now being considered, we can see immediately that, if either H or K vanishes, the transverse
vibrations must be *vertically* or *horizontally* polarized.
If the phase difference $(\phi_y - \phi_x) = \Delta$ vanishes, we
shall find that the polarization is *linear*, and if
$H = K$ while $\Delta = \pi/2$, we have light that is *circularly*
polarized. For the general case, we say that the light
is *elliptically* polarized.

If these geometrical descriptions of the transverse

182

vibrations are difficult to visualize, it may be helpful to think in terms of a model. Let us imagine that,
in a given xy-plane, we are following the motion of a
very small charged particle, which is simultaneously
attached to the origin by a weak spring and pulled
away from it by the oscillating electric field vector.
If we choose our time-origin so that $\phi_x = 0$ and measure
distance in suitable units, then the instantaneous x-
and y-coordinates of the test-charge, its *displacement*
from the origin, will vary with time according to the
equations

$\quad x = H\cos(\omega t)$

and

$\quad y = K\cos(\omega t + \Delta)$

It follows that, whenever ωt changes through a
range of 2π (which happens at least 10^{14} times every
second), the test-charge executes one cyclic traversal
of a simple Lissajous' figure. See Figure IV.1. (It

Lissajous figures produced by parametric equations $x = \cos \omega t$ and $y = \cos(\omega t + \Delta)$

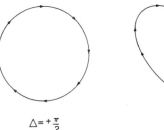

$\Delta = 0$

$\Delta = +\frac{\pi}{6}$

$\Delta = +\frac{\pi}{3}$

$\Delta = +\frac{\pi}{2}$

$\Delta = +\frac{2\pi}{3}$

$\Delta = +\frac{5\pi}{6}$

Figure IV.1

is, of course, very simple to produce such patterns at
much lower frequencies on a cathode ray oscilloscope;
all that is necessary is to apply two sinusoidal signals,
of nominally the same audio frequency, to the input
terminals of the X- and Y-deflection amplifiers. When
the X-deflection is controlled in this way, so as to
produce an 'XY-display', no time-base is required.)

Each of these patterns represents the locus of the
tip of the oscillating electric field vector. In order
to analyse their geometry, our first step will be to
eliminate the time-parameter (ωt) from the above pair
of equations. Before proceeding with this, however,
let us consider some of the types of light that fall
between the fully coherent and the fully incoherent
extremes.

There are two main ways in which 'partially coherent
light' can be produced.

(a) One can start with a laser. In some scientific
    experiments, the complete coherence of a laser beam
    can be an embarassment, because of the unwanted
    effects of interference or diffraction or 'speckle'
    that it produces. Occasionally, therefore, 'quasi-
    thermal' light is manufactured by sending a laser
    beam through a coherence-degrading device, such as
    a rapidly rotating ground-glass diffuser.

(b) In the vast majority of cases, however, partially
    coherent light is produced by sending incoherent
    light from an ordinary thermal source through some
    combination of *optical*, *spectroscopic* and *polariz-
    ing* filters.

The simplest type of optical filter is a pinhole. It
refines the directionality of a light beam by the in-
efficient process of rejecting nearly all the light
that falls upon it, transmitting only a central portion
with a greatly reduced étendue. Equally wasteful is
the spectroscopic filter. Whether it is a piece of
coloured glass, a monochromator using prisms or
gratings, or an interference device, it refines the
spectral purity of a beam only by rejecting all the
light outside a narrow band of wavelengths.

If one attempts to refine ordinary light so as to
approach the directionality and spectral purity provided by a laser, then the combined filtering losses
are such that practically none of the original energy

survives. Until the laser was invented, the light used in an optical laboratory could be *either* directional *or* monochromatic, but not both at once.

If all that we need for an experiment is light that is fully polarized, then much more efficient filtering becomes possible; since the original light is distributed between only *two* states of polarization, only half the energy has to be rejected.

One of the most efficient types of polarizing filter is that which is based on the double-refraction properties of a uniaxial crystal such as calcite. Early forms, such as the Nicol prism, have been replaced by more modern types such as the Glan-Foucault polarizer, which will transmit one preferred plane of polarization with an efficiency which is limited mainly by reflection losses; the unwanted orthogonal plane is rejected almost completely by total internal reflection. For such prisms, however, the incident light must be approximately collimated.

Also remarkably efficient, over a wide range of wavelengths and directions, are the various polarizing sheets produced by the Polaroid Corporation. One example, HN-32 sheet polarizer, transmits about 32 per cent of the visible light that falls upon it, and the fraction that remains in the unwanted state of polarization is about one part in 10,000.

In the laboratory, at any rate, the most convenient method of obtaining fully polarized light is either to start with a laser or to use one of the above forms of polarizing filter. There are, of course, several other ways in which some degree of polarization can arise naturally; this happens, for example, whenever light is reflected obliquely from a polished transparent surface or is scattered through something approaching 90° by an assembly of very small scattering particles. (There are also situations, both astronomically and in the laboratory, where light is generated in the presence of powerful oriented electric or magnetic fields, and exhibits either a general degree of polarization or one that varies sharply with the wavelength.)

In many of these situations the filtering action is incomplete and we obtain light that is only partially polarized. It will be shown in the next section that the 'degree of polarization' for such a beam can be defined in terms of its 'Stokes parameters'. For purposes of analysis, the light can be regarded as a mixture of two fully polarized but independent beams of unequal strength and opposite states of polarization.

IV.1.1 *Plane-polarized light*

Let us now imagine that we have converted unpolarized light into plane-polarized light by passing it through a sheet polarizer. The vibrations of the electric field vector now lie entirely in one direction in the transverse xy-plane, and we shall call the plane containing this direction and the propagation direction Oz the 'vibration-plane'. How can we determine its orientation?

For many polarizers, the vibration-plane of the transmitted light - the 'pass-plane' - is indicated by an engraved straight line or double-ended arrow; if a sheet polarizer has square edges, the pass-plane is likely to be parallel to one pair of edges (but 45° cuts also find favour in some laboratories!)

A useful first check is to hold the polarizer in the hand whilst looking through it at the light reflected obliquely from a glossy horizontal surface, for example a polished table. The polarizer is rotated in its own plane until the reflected light as nearly as possible disappears. We then know that, since the rejected reflections are mainly horizontally polarized, the pass-plane of our polarizer must be within a few degrees of vertical. We can therefore inscribe on it a tentative marking for the pass-plane line PP'.

To check more accurately the orientation of this preliminary line, or to verify one that has already been inscribed, we can proceed as follows. The polarizer is set up on a horizontal optical bench with the line PP' adjusted to be accurately vertical (or horizontal). A second polarizer is than inserted, either on the lamp side or on the observer side, and rotated until the transmitted light is as nearly as possible extinguished.

Without disturbing the second polarizer, the first is now turned back to front, and adjusted so that the line PP' has the same orientation as before. If PP' already coincides with the true pass-plane of the polarizer, then the field of view will again appear dark; but if appreciable light is transmitted, then the first polarizer will need to be rotated a few

degrees until a new 'extinction' position is found.
A second line is engraved upon it, and the correct
position of the pass-plane lies midway between the two
markings.

Having checked our polarizer in this way, we can now
specify the angle θ which the vibration-plane of the
emergent light makes with the horizontal x-axis. The
equations governing the transverse electric field components are then of the form:

$$E_x = A\cos\theta\cos(\omega t + \phi)$$

and

$$E_y = A\sin\theta\cos(\omega t + \phi)$$

(These equations are similar to those that we have
already used for describing a fully coherent laser
beam; here, however, we are using light which contains a whole range of different directions and frequencies, and the amplitude A and the phase angle ϕ
of the oscillating field are not really constant.
They are themselves functions of lateral position and
time, which fluctuate rapidly and irregularly about
their mean value. The distance and time over which
they remain roughly constant are sometimes called the
'coherence width' and the 'coherence time' of the beam
Strictly speaking, therefore, we should write, for
each position (x,y) in the transverse plane,

$$\begin{bmatrix} E_x \\ E_y \end{bmatrix}(x,y) = \begin{bmatrix} \cos\theta \\ \sin\theta \end{bmatrix} A(x,y,t)\cos\{\omega t + \phi(x,y,t)\}$$

So far as the Maxwell column is concerned, however,
the fluctuations occur only in a scalar multiplier
which affects both components equally. For most purposes, at any rate, the phase angle ϕ and its fluctuations are of no significance, and we can replace
$A(x,y,t)$ by a constant root-mean-square value which we
shall call A.)

Let us return now to consider the motion of our
hypothetical small test-charge. Under the influence of

this synchronized pair of field vectors, its displacement will be given by

$$x = A\cos\theta\cos(\omega t + \phi) \quad \text{and} \quad y = A\sin\theta\cos(\omega t + \phi)$$

Multiplying the first equation by $\sin\theta$ and the second
by $\cos\theta$, we obtain immediately, for the locus of the
transverse displacement, the equation $x\sin\theta = y\cos\theta$ –
a straight line motion in the plane of vibration.

IV.1.2 *Elliptically polarized light*

A convenient method of producing elliptically polarized light is to send a plane-polarized beam through a
'phase plate', that is a slice of uniaxial crystal. As
is shown in chapter V, such a slice introduces a phase
difference between the components of the vibration
parallel to and perpendicular to a special direction
in the crystal, which is known as the optic axis. The
vibration parallel to the optic axis is called the
'extraordinary' vibration (the E-vibration) and that
perpendicular to the optic axis is called the 'ordinary'
vibration (the O-vibration). For most phase plates (if
made from a *negative* uniaxial crystal), the refractive
index for the E-vibration is smaller than for the O-
vibration, and the optic axis is also termed the 'fast
axis'. In marking the orientation of a phase plate, it
is best to inscribe a line on the mounting which is
parallel to the fast axis. (To check this marking we
can use the same trial and error procedure as was described for the orientation of a polarizer – the phase
plate is tested both ways round in between a pair of
polarizers set to give extinction.)

We shall suppose that the phase plate which we are
using has its optic axis parallel to the x-axis and
that its thickness is such that it *advances* the O-
vibration by an angle Δ radians with respect to the
E-vibration. (In that case the 'fast' axis is at $90°$
to the x-axis.) The vibrations coming out of the phase
plate are then given by

$$x = A\cos\theta\cos\omega t \quad \text{and} \quad y = A\sin\theta\cos(\omega t + \Delta)$$

If we eliminate ωt between these two equations, we get
this equation linking the x- and y-vibrations in the
resultant beam:

$$\frac{x^2}{A^2\cos^2\theta} - \frac{2xy\cos\Delta}{A^2\sin\theta\cos\theta} + \frac{y^2}{A^2\sin^2\theta} = \sin^2\Delta$$

that is

$$\frac{x^2}{H^2} - \frac{2xy\cos\Delta}{HK} + \frac{y^2}{K^2} = \sin^2\Delta$$

where $H = A\cos\theta$ and $K = A\sin\theta$. Here, H is the component of the original vibration parallel to the x-axis and K is the component of the original vibration parallel to the y-axis. By squaring and adding, it will be seen that $H^2 + K^2 = A^2$, the square of the amplitude of the original vibration, which is therefore proportional to the energy flow in the original vibration. It will be called I – the intensity (or, more properly perhaps, the irradiance).

We can now see some special cases of this equation. If $\Delta = 0$, that is if there is no phase plate, then $\cos\Delta = 1$ and $\sin\Delta = 0$, and the equation becomes

$$x^2/H^2 - 2xy/HK + y^2/K^2 = 0$$

that is

$$(x/H - y/K)^2 = 0$$

that is

$$y/x = K/H$$

This just describes the plane-polarized condition of the original beam. Now suppose that we have a 'half-wave' plate, for which $\Delta = \pi$, so that $\sin\Delta$ is 0 but $\cos\Delta$ is now – 1. The resulting equation is similar except that there is now a change of sign, so that

$$y/x = - K/H$$

This means that the ratio of the displacement parall to the two axes is the same as before but with reverse sign, so that when x is at its maximum positive y is a its maximum negative, and vice versa. We still have plane-polarized vibration, but it is now vibrating at an angle θ on the other side of the optic axis, that i the half-wave plate has turned the vibration-plane through an angle of 2θ.

Now suppose that $\Delta = \pi/2$, that is we have a 'quarter-wave' plate. This means that $\cos\Delta = 0$ and $\sin\Delta = 1$, s that the equation linking x and y becomes

$$\frac{x^2}{H^2} + \frac{y^2}{K^2} = 1$$

This is a well-known equation of an ellipse lying with its major and minor axes parallel to the x- and y-axes. The semi-axis parallel to the x-axis is H and that parallel to the y-axis is K. If $\theta = 45^\circ$, then H and K are equal, and the equation becomes

$$x^2 + y^2 = A^2/2$$

and we say that the light is circularly polarized.

If a beam of fully polarized light is viewed through a polarizer, and the latter is gradually rotated through 360°, it is quite easy to distinguish experimentally between light that is linearly, elliptically or circularly polarized. For linear states there will occur two orientations for which the light is completely extinguished. For elliptical states there will be two maxima and two minima of intensity, but the minima will not be completely dark. For circularly polarized light, the brightness will remain constant.

Any state of polarization which is elliptical or circular is called either 'right-handed' or 'left-handed'. For a right-handed state, the phase angle Δ by which the y-vibration leads the x-vibration must lie between 0 and π. In that case, the tip of the electric field vector travels round its elliptical locus in a *clockwise* direction, as seen by an observer looking from the $+ z$-direction towards the light source. See figure IV.1 on page 182.

IV.2 THE STOKES PARAMETERS FOR SPECIFYING POLARIZATION

The original equation obtained by eliminating ωt from the equations describing the x- and y-vibrations represents an ellipse of a more general type, with its semi-axes not parallel to the x- and y-axes. To find the orientation and ratio of the axes of the ellipse in this case, it proves profitable to use a more sophisticated mathematical description than we have used so far. For a pure state of polarization, specified by H, K and Δ, we define the four Stokes parameters of the beam as follows:

$$I = H^2 + K^2 = A^2$$

$$Q = H^2 - K^2 = A^2\cos^2\theta - A^2\sin^2\theta = A^2\cos2\theta = I\cos2\theta$$

$$U = 2HK\cos\Delta = 2(A\cos\theta)(A\sin\theta)\cos\Delta = A^2\sin2\theta\cos\Delta$$
$$= I\sin2\theta\cos\Delta$$

$$V = 2HK\sin\Delta = I\sin2\theta\sin\Delta$$

It can be shown by quite elementary algebra that, for this fully polarized beam of light, $I^2 = Q^2 + U^2 + V^2$. We shall see later the physical significance of these quantities.

From the defining equations we see that

$$H^2 = (I + Q)/2$$

$$K^2 = (I - Q)/2$$

$$\sin^2\Delta = V^2/4H^2K^2 = V^2/(I^2 - Q^2)$$

so that the equation representing the vibration becomes

$$\frac{2x^2}{(I + Q)} - \frac{4Uxy}{(I^2 - Q^2)} + \frac{2y^2}{(I - Q)} = \frac{V^2}{(I^2 - Q^2)}$$

that is

$$\frac{2x^2(I - Q)}{V^2} - \frac{4Uxy}{V^2} + \frac{2y^2(I + Q)}{V^2} = 1$$

If we now put

$$P = \frac{2(I - Q)}{V^2}, \quad G = \frac{2U}{V^2} \quad \text{and} \quad F = \frac{2(I + Q)}{V^2}$$

we get the equation

$$Px^2 - 2Gxy + Fy^2 = 1$$

We can most easily find the orientation and ratio of axes of this ellipse as follows. In polar coordinates, if (R, ϕ) represents the point whose Cartesian coordinates are (x, y), then $x = R\cos\phi$ and $y = R\sin\phi$, so that the equation becomes

$PR^2\cos^2\phi - 2GR^2\cos\phi\sin\phi + FR^2\sin^2\phi = 1$

Using the well-known formulae for $\cos 2\phi$ and $\sin 2\phi$, this becomes

$PR^2(1 + \cos 2\phi)/2 - GR^2\sin 2\phi + FR^2(1 - \cos 2\phi)/2 = 1$

Writing $2\phi = \beta$ and $2/R^2 = W$, this becomes

$W = (P + F) - 2G\sin\beta + (P - F)\cos\beta$

Now, at the ends of the major and minor axes of the ellipse, R is respectively a maximum and a minimum and therefore W is respectively a minimum and a maximum. Thus, the values of β corresponding to the axes of the ellipse are given by $dW/d\beta = 0$:

$dW/d\beta = - 2G\cos\beta - (P - F)\sin\beta$

Therefore, if α is a value of β corresponding to an axis of the ellipse, then

$$\tan\alpha = \frac{\sin\alpha}{\cos\alpha} = \frac{2G}{F - P}$$

This gives two possible values of α.
Successive values of an angle having a fixed tangent differ by π, so, if α_1 is the smaller and α_2 the larger value of α between 0 and 2π, then $\alpha_2 = \alpha_1 + \pi$, so that $\sin\alpha_2 = - \sin\alpha_1$ and $\cos\alpha_2 = - \cos\alpha_1$. Therefore, if W_1 and W_2 are the corresponding values of $2/R^2$, this gives

$$\frac{R_2^2}{R_1^2} = \frac{W_1}{W_2} = \frac{(P + F) - 2G\sin\alpha_1 + (P - F)\cos\alpha_1}{(P + F) - 2G\sin\alpha_2 + (P - F)\cos\alpha_2}$$

Substituting back in terms of the Stokes parameters, we find that the smaller angle $\alpha/2$ which one of the axes of the ellipse makes with the x-axis is given by the equation

$\tan\alpha = 2G/(F - P) = U/Q = \tan 2\theta\cos\Delta$

while the ratio of the squares of the lengths of the minor and major axes becomes, after some reduction:

$$\frac{I - \sqrt{(Q^2 + U^2)}}{I + \sqrt{(Q^2 + U^2)}} = \frac{1 - \sqrt{(1 - \sin^2 2\theta\sin^2\Delta)}}{1 + \sqrt{(1 - \sin^2 2\theta\sin^2\Delta)}}$$

These relations can both be checked experimentally, using two polaroids, a phase plate and a photoelectric cell.

The equations which we have just obtained relate the Stokes parameters for a single fully polarized disturbance to the corresponding Maxwell column. Whereas the elements H and K represent amplitudes of the oscillating electric field vector, which at optical frequencies cannot be directly observed, the Stokes parameters are linearly related to the actual energy measurements that we can make with a photocell - they are real measures of intensity or intensity difference (Such measurements will be discussed in section IV.4.)

Arising from this distinction, there are two important consequences which apply whenever we are going to *combine* two or more beams, each of known polarization, and we wish to predict a state of polarization for the mixture that results.

(a) If the disturbances to be combined are 'mutually coherent' then it is their Maxwell columns which must be added to predict the Maxwell column of the result. (This case, which corresponds to 'additio of amplitudes', arises mainly where the beams have come from a fully compensated interferometer or have been derived from a single laser source.)

(b) On the other hand, if the disturbances to be combined are 'mutually incoherent', then it is their Stokes columns which must be added in order to calculate the Stokes column for the result. This case, which corresponds to 'addition of intensities', arises much more frequently, for example whenever we are dealing with beams that are unpolarized or only partially polarized. In that case the effective values of H, K and Δ are obtained as statistical averages over a finite time of observation. For a mathematical analysis of this situati see appendix C.

In this appendix it is shown that, whereas for a ful polarized beam we must have $I^2 = Q^2 + U^2 + V^2$, for a completely unpolarized beam the three components Q, U and V must all vanish, and only the intensity paramete I remains.

For a beam that is partially polarized, we define a 'degree of polarization' P which is equal to the positive square root of the ratio $(Q^2 + U^2 + V^2)/I^2$. For any beam which is physically possible, P will lie somewhere between unity and zero.

If we wish, we can decompose a partially polarized beam into two independent beams, one of which is fully polarized while the other is completely unpolarized. Evidently, for this decomposition we shall have

$$
\begin{bmatrix} I \\ Q \\ U \\ V \end{bmatrix} = \begin{bmatrix} PI \\ Q \\ U \\ V \end{bmatrix} + \begin{bmatrix} (1-P)I \\ 0 \\ 0 \\ 0 \end{bmatrix}
$$

In many cases, however, it is more convenient to use an alternative decomposition into two fully polarized beams, characterized by exactly opposite state of polarization. Provided that P does not vanish, we obtain

$$
\begin{bmatrix} I \\ Q \\ U \\ V \end{bmatrix} = \frac{(1 + P)}{2P} \begin{bmatrix} PI \\ Q \\ U \\ V \end{bmatrix} + \frac{(1 - P)}{2P} \begin{bmatrix} PI \\ -Q \\ -U \\ -V \end{bmatrix}
$$

where

$$PI = (Q^2 + U^2 + V^2)^{\frac{1}{2}}$$

If P vanishes, then we are dealing with completely unpolarized light, and it can be regarded as a 50%-50% mixture of any oppositely polarized pair of states that we care to choose.

It should be remarked here that, for a fully polarized beam of *unit* intensity, the parameters Q, U and V can assume any real value between 1 and $-$ 1, provided that the sum of their squares remains equal to unity. If we regard these three values as coordinates referred to a set of Cartesian axes, then we obtain a spatial representation in which the space point (Q,U,V) corresponding to any state of polarization lies somewhere on a sphere of unit radius - the Poincaré sphere. On such a sphere, evidently, the points (Q,U,V) and $(-Q,-U,-V)$

representing opposite states lie on opposite ends of a diameter. (There are many aspects of polarization optics for which the Poincaré sphere provides an elegant geometrical representation, but we shall not pursue them here.)

IV.3 USE OF THE MUELLER CALCULUS FOR TRANSFORMING A STOKES COLUMN

We now move on to the application of matrix methods to these ideas. The four Stokes parameters associated with a beam can be regarded as the elements of a 4×1 matrix, which we shall call the Stokes column of the beam, representing it by a bold symbol **S**:

$$\mathbf{S} = \begin{bmatrix} I \\ Q \\ U \\ V \end{bmatrix}$$

(In some books, the parameters are labelled I, M, C, S or P_0, P_1, P_2, P_3. In this chapter a bold symbol is used for each Stokes column in order to distinguish it from S used as an abbreviation for a sine-function. Bold symbols will also be used later to denote a Hermitian conjugate.)

Now, we shall show that for any type of device, that is for a polaroid sheet with its pass-plane at any angle with the axis, for a phase plate of any retardation and any orientation of the fast axis, and for a rotator which merely twists the plane of polarization like certain organic liquids, the Stokes parameters of the beam coming out of the device are linear functions of the four Stokes parameters of the input beam. Thus:

$$I_2 = M_{11}I_1 + M_{12}Q_1 + M_{13}U_1 + M_{14}V_1$$

$$Q_2 = M_{21}I_1 + M_{22}Q_1 + M_{23}U_1 + M_{24}V_1$$

$$U_2 = M_{31}I_1 + M_{32}Q_1 + M_{33}U_1 + M_{34}V_1$$

$$V_2 = M_{41}I_1 + M_{42}Q_1 + M_{43}U_1 + M_{44}V_1$$

Here, the Stokes parameters with subscript 1 refer to the beam before entering the device, those with subscript 2 refer to the beam after leaving the device and the Ms with double subscripts are characteristic only of the device and its orientation. These four equations we write in matrix form thus:

$$\begin{bmatrix} I_2 \\ Q_2 \\ U_2 \\ V_2 \end{bmatrix} = \begin{bmatrix} M_{11} & M_{12} & M_{13} & M_{14} \\ M_{21} & M_{22} & M_{23} & M_{24} \\ M_{31} & M_{32} & M_{33} & M_{34} \\ M_{41} & M_{42} & M_{43} & M_{44} \end{bmatrix} \begin{bmatrix} I_1 \\ Q_1 \\ U_1 \\ V_1 \end{bmatrix}$$

that is

$$S_2 = M S_1$$

Here, S_1 is the Stokes column of the beam entering the device, S_2 is the Stokes column of the beam leaving the device and M is a 4×4 matrix which is characteristic of the device and its orientation and which is called, after its inventor, the Mueller matrix of the device. Thus, if a beam passes through a series of devices in succession, we can find the nature of the emerging beam, knowing only the nature of the original beam and the devices through which it passes.

Listed below in Table 3 are the Mueller matrices that describe the operation of various frequently employed devices. For lists of the Stokes columns and Maxwell columns for various types of polarized light reference should be made to appendix C.

Using the data of Table 3, we shall now work through some simple calculations. These will be based on the photoelastic effect, which is widely used in analysing the strains in model structures in both civil and mechanical engineering. The physical facts of photo-elasticity are briefly these. If a sheet of transparent substance, such as glass or plastic, is subjected to strain in directions parallel to its surface, it becomes a phase plate, the optic axis being parallel to the surface. The effect was investigated by Kerr, Brewster, Filon and Coker, among others. They found the laws of photoelasticity which, in their simplest form, state (a) that the direction of the optic axis at any point in the plate is that of the strain there, and (b) that the retardation of the ordinary ray com-

Table 3. Mueller matrices for ideal linear polarizers, linear retarders, rotation of axes and circular retarders

The angle θ specifies how the pass-plane of a polarizer or the fast axis of a linear retarder is oriented with respect to the x-axis.

Type of device	$\theta = 0$	$\theta = \pm\pi/4$	$\theta = \pi/2$	θ takes any value
Ideal linear polarizer at angle θ	$\frac{1}{2}\begin{bmatrix}1&1&0&0\\1&1&0&0\\0&0&0&0\\0&0&0&0\end{bmatrix}$	$\frac{1}{2}\begin{bmatrix}1&0&\pm1&0\\0&0&0&0\\\pm1&0&1&0\\0&0&0&0\end{bmatrix}$	$\frac{1}{2}\begin{bmatrix}1&-1&0&0\\-1&1&0&0\\0&0&0&0\\0&0&0&0\end{bmatrix}$	$\frac{1}{2}\begin{bmatrix}1&C_2&S_2&0\\C_2&C_2^2&C_2S_2&0\\S_2&C_2S_2&S_2^2&0\\0&0&0&0\end{bmatrix}\begin{matrix}C_2=\\S_2=\end{matrix}$
Quarter-wave linear retarder with fast axis at angle θ	$\begin{bmatrix}1&0&0&0\\0&1&0&0\\0&0&0&1\\0&0&-1&0\end{bmatrix}$	$\begin{bmatrix}1&0&0&0\\0&0&0&\mp1\\0&0&1&0\\0&\pm1&0&0\end{bmatrix}$	$\begin{bmatrix}1&0&0&0\\0&1&0&0\\0&0&0&-1\\0&0&1&0\end{bmatrix}$	$\begin{bmatrix}1&0&0&0\\0&C_2^2&C_2S_2&-S_2\\0&C_2S_2&S_2^2&C_2\\0&S_2&-C_2&0\end{bmatrix}$
Half-wave linear retarder with fast axis at angle θ	$\begin{bmatrix}1&0&0&0\\0&1&0&0\\0&0&-1&0\\0&0&0&-1\end{bmatrix}$	$\begin{bmatrix}1&0&0&0\\0&-1&0&0\\0&0&1&0\\0&0&0&-1\end{bmatrix}$	$\begin{bmatrix}1&0&0&0\\0&1&0&0\\0&0&-1&0\\0&0&0&-1\end{bmatrix}$	$\begin{bmatrix}1&0&0&0\\0&C_4&S_4&0\\0&S_4&-C_4&0\\0&0&0&-1\end{bmatrix}\begin{matrix}C_4\\S_4\end{matrix}$
Linear retarder with retardation δ and with fast axis at angle θ	$\begin{bmatrix}1&0&0&0\\0&1&0&0\\0&0&\beta&\mu\\0&0&-\mu&\beta\end{bmatrix}$	$\begin{bmatrix}1&0&0&0\\0&\beta&0&\mp\mu\\0&0&1&0\\0&\pm\mu&0&\beta\end{bmatrix}$	$\begin{bmatrix}1&0&0&0\\0&1&0&0\\0&0&\beta&-\mu\\0&0&\mu&\beta\end{bmatrix}$	$\begin{bmatrix}1&0&0&0\\0&C_2^2+S_2^2\beta&C_2S_2(1-\beta)&-S_2\mu\\0&C_2S_2(1-\beta)&S_2^2+C_2^2\beta&C_2\mu\\0&S_2\mu&-C_2\mu&\beta\end{bmatrix}$

Note: for this row we use as abbreviated symbols $\beta = \cos\delta$ and $\mu = \sin\delta$.

Rotation of x-axis to new angle θ with respect to old x-axis: *or*, action of circular retarder which retards *left* circular state by 2θ radians

$$\begin{bmatrix}1&0&0&0\\0&C_2&S_2&0\\0&-S_2&C_2&0\\0&0&0&1\end{bmatrix}$$

If we denote this 'rotator' matrix by $R(\theta)$, th can use it to transform the Mueller matrix for polarizing device or system at a given orienta into the corresponding matrix for the same dev system at a new orientation $(\theta + \phi)$. The tran tion formula reads

$$M(\theta + \phi) = R(-\theta)\,M(\phi)\,R(\theta)$$

If a complete polarizing system, compounded from several devices, is to be rotated, then it ma labour to use the above unitary transformation. For a single device, however, it is usually qui use a matrix taken from the appropriate column in the above table. (For derivation of these mat see appendix D.)

pared with the extraordinary ray emerging at any point is proportional to the magnitude of the strain at that point. If this retardation measured in radians is called δ, then the second law can be put in the form

$$\delta = csd$$

where s is the strain at the point, d is the thickness of the plate and c is a constant called the strain optical coefficient, which depends only upon the material of the plate and on the wavelength. Thus, if we can find the direction of the optic axis and the

phase difference at every point in a loaded plate, we
can determine the strain distribution completely.
Knowing the strain pattern which is produced in a
plastic model by a known set of stresses, we can
deduce by simple proportion the strains produced in
a geometrically similar full-size structure by a set
of stresses proportional to those in the model, thus
avoiding a great deal of calculation.

We shall now develop the theory of this method using
the Mueller matrices.

The model is placed between two Polaroid sheets, or
'polaroids', whose pass-planes are normally 'crossed',
that is at $90°$, so that if the model were not there no
light would get through. We shall, to begin with, consider instead quite general orientations of the pass-
planes of the polaroids. We shall consider, for simplicity, that the optic axis of the phase plate is
horizontal, but this will not mean any loss of generality. We suppose that the phase plate introduces a
phase difference δ between the ordinary and the extraordinary beams, the former being retarded. We suppose
also that the first polaroid through which the light
passes has its pass-plane at an angle α above the
horizontal direction Ox, whereas the second, which
follows the phase plate, has its pass-plane at an
angle γ below the horizontal, that is at an angle of
$-\gamma$ in the usual convention. We put a beam of unpolarized light of intensity I_1 into the first polaroid.
By the use of Mueller matrices, we now calculate the
intensity of the beam of light emerging from the final
polaroid. We shall do this in three steps. First,
the first polaroid. If the light emerging from the
first polaroid has a Stokes column S_2, then, using the
formula for the Mueller matrix of a polaroid, we find
that

$$\mathbf{S}_2 = \tfrac{1}{2}\begin{bmatrix} 1 & \cos 2\alpha & \sin 2\alpha & 0 \\ \cos 2\alpha & \cos^2 2\alpha & \cos 2\alpha \sin 2\alpha & 0 \\ \sin 2\alpha & \sin 2\alpha \cos 2\alpha & \sin^2 2\alpha & 0 \\ 0 & 0 & 0 & 0 \end{bmatrix}\begin{bmatrix} I_1 \\ 0 \\ 0 \\ 0 \end{bmatrix}$$

$$= \frac{1}{2}\begin{bmatrix} I_1 \\ I_1\cos2\alpha \\ I_1\sin2\alpha \\ 0 \end{bmatrix} = \frac{I_1}{2}\begin{bmatrix} 1 \\ \cos2\alpha \\ \sin2\alpha \\ 0 \end{bmatrix}$$

We now pass the beam of light from the first polaroid through the phase plate, where its Stokes column becomes S_3, that is S_2 multiplied by the Mueller matrix of the phase plate. This is

$$S_3 = \frac{I_1}{2}\begin{bmatrix} 1 & 0 & 0 & 0 \\ 0 & 1 & 0 & 0 \\ 0 & 0 & \cos\delta & \sin\delta \\ 0 & 0 & -\sin\delta & \cos\delta \end{bmatrix}\begin{bmatrix} 1 \\ \cos2\alpha \\ \sin2\alpha \\ 0 \end{bmatrix}$$

$$= \frac{I_1}{2}\begin{bmatrix} 1 \\ \cos2\alpha \\ \sin2\alpha\cos\delta \\ -\sin2\alpha\sin\delta \end{bmatrix}$$

(Notice that, as these matrix calculations proceed, any scalar multiplier such as I_1 can be transferred at will from one side of the expression to the other.) The beam emerging from the phase plate is now passed into the second polaroid, where it becomes S_4, that is S_3 multiplied by the Mueller matrix of the second polaroid. This is

$$S_4 = \frac{1}{2}\begin{bmatrix} 1 & \cos2\gamma & -\sin2\gamma & 0 \\ \cos2\gamma & \cos^2 2\gamma & -\sin2\gamma\cos2\gamma & 0 \\ -\sin2\gamma & -\sin2\gamma\cos2\gamma & \sin^2 2\gamma & 0 \\ 0 & 0 & 0 & 0 \end{bmatrix}\begin{bmatrix} 1 \\ \cos2\alpha \\ \sin2\alpha\cos\delta \\ -\sin2\alpha\sin\delta \end{bmatrix}$$

$$= \frac{I_1}{4} \begin{bmatrix} 1 + \cos2\gamma\cos2\alpha - \sin2\gamma\sin2\alpha\cos\delta \\[2mm] \cos2\gamma + \cos^2 2\gamma\cos2\alpha - \sin2\gamma\cos2\gamma\sin2\alpha\cos\delta \\[2mm] -\sin2\gamma - \sin2\gamma\cos2\gamma\cos2\alpha + \sin^2 2\gamma\sin2\alpha\cos\delta \\[4mm] 0 \end{bmatrix}$$

The first element of S_4 is I_4, the intensity of the final emerging beam. Now consider the special case where the first and second polaroids are crossed, that is their pass-planes are perpendicular to one another, so that if there were no model present no light at all would get through. This means that $\alpha + \gamma = 90°$, whence $2\alpha + 2\gamma = 180°$, so that $\sin2\gamma = \sin2\alpha$ and $\cos2\gamma$ is $-\cos2\alpha$. The expression for the intensity of the emerging light then becomes, after a little manipulation,

$$I_4 = \frac{1}{2}(1 - \cos^2 2\alpha - \sin^2 2\alpha\cos\delta)\frac{I_1}{2} = \tfrac{1}{2}I_1\sin^2 2\alpha\sin^2\frac{\delta}{2}$$

This expression vanishes first if $\sin2\alpha = 0$. This will happen if $\alpha = 0$ or $90°$, which means that the optic axis of the phase plate is parallel to the pass-plane of one or other of the polaroids. This gives the curves called the *isoclinics*, from which the direction of the strain at each point can be deduced.

The expression also vanishes if $\sin(\delta/2) = 0$, that is if $\delta/2 = 0$, π, 2π, or any other integral multiple of π. From this, we can find the value of the magnitude of the strain at each point, since this is proportional to δ. Curves along which $\sin(\delta/2)$ vanishes are called *isochromatics*. When interpreting the isochromatics, the presence of the isoclinics on the screen at the same time leads to some confusion.

It is often desirable to eliminate the isoclinics, and this can be done by the use of two quarter-wave plates, as follows. One quarter-wave plate is placed between the first polarizer and the model, with its fast axis at $+ 45°$ to the horizontal, that is in the first and third quadrants. The other quarter-wave plate is placed just after the model, with its fast axis again at $45°$ to the axis, but in the second and fourth quadrants, so that conventionally the angle is regarded as - 45°.

We shall now show that, when these quarter-wave plates have been inserted, the intensity on the screen depends only on the phase-shift introduced by the phase plate and not on the orientation of the phase plate. We shall assume, for simplicity, that the pass-plane of the first polaroid is horizontal, while that of the second polaroid is vertical. This can always be arranged in practice while looking for isochromatics

We put into the first polaroid a beam of unpolarized light, for which all of the Stokes parameters except I_1 are zero. Using the formula for the Mueller matrix of a polaroid, we find that the Stokes column of the beam emerging from the first polaroid is

$$S_2 = \tfrac{1}{2} \begin{bmatrix} 1 & 1 & 0 & 0 \\ 1 & 1 & 0 & 0 \\ 0 & 0 & 0 & 0 \\ 0 & 0 & 0 & 0 \end{bmatrix} \begin{bmatrix} I_1 \\ 0 \\ 0 \\ 0 \end{bmatrix} = \tfrac{1}{2} \begin{bmatrix} I_1 \\ I_1 \\ 0 \\ 0 \end{bmatrix}$$

The beam now passes through the quarter-wave plate whose fast axis is at + 45° with respect to the x-axis. Using the formula for the appropriate Mueller matrix listed in Table 3, we obtain a Stokes column which represents right-handed circularly polarized light, namely

$$S_3 = \tfrac{1}{2} \begin{bmatrix} 1 & 0 & 0 & 0 \\ 0 & 0 & 0 & -1 \\ 0 & 0 & 1 & 0 \\ 0 & 1 & 0 & 0 \end{bmatrix} \begin{bmatrix} I_1 \\ I_1 \\ 0 \\ 0 \end{bmatrix} = \tfrac{1}{2} \begin{bmatrix} I_1 \\ 0 \\ 0 \\ I_1 \end{bmatrix}$$

We now pass the beam through a phase plate whose retardation is δ and whose fast axis is oriented at some unknown angle θ. In Table 3 the Mueller matrix for such a device is given, for the sake of brevity, with the terms $\cos 2\theta$ and $\sin 2\theta$ replaced by C_2 and S_2 and with $\cos \delta$ and $\sin \delta$ replaced by β and μ respectively. In terms of these symbols, we obtain for the Stokes column of the emergent beam

$$
S_4 = \frac{1}{2}
\begin{bmatrix}
1 & O & O & O \\
O & C_2 + S_2^2\beta & C_2 S_2 (1-\beta) & -S_2\mu \\
O & C_2 S_2 (1-\beta) & S_2^2 + \beta C_2 & C_2\mu \\
O & S_2\mu & -C_2\mu & \beta
\end{bmatrix}
\begin{bmatrix}
I_1 \\
O \\
O \\
I_1
\end{bmatrix}
= \frac{1}{2}
\begin{bmatrix}
I_1 \\
-S_2\mu I_1 \\
C_2\mu I_1 \\
\beta I_1
\end{bmatrix}
$$

After the beam passes through the second quarter-wave plate with its fast axis at $-45°$, its Stokes column becomes

$$
S_5 = \frac{1}{2}
\begin{bmatrix}
1 & O & O & O \\
O & O & O & 1 \\
O & O & 1 & O \\
O & -1 & O & O
\end{bmatrix}
\begin{bmatrix}
I_1 \\
-S_2\mu I_1 \\
C_2\mu I_1 \\
\beta I_1
\end{bmatrix}
= \frac{1}{2}
\begin{bmatrix}
I_1 \\
\beta I_1 \\
C_2\mu I_1 \\
S_2\mu I_1
\end{bmatrix}
$$

Finally, after passing through the polaroid with its pass-plane vertical, the output beam has the following Stokes column:

$$
S_6 = (\tfrac{1}{2})(\tfrac{1}{2})
\begin{bmatrix}
1 & -1 & O & O \\
-1 & 1 & O & O \\
O & O & O & O \\
O & O & O & O
\end{bmatrix}
\begin{bmatrix}
I_1 \\
\beta I_1 \\
C_2\mu I_1 \\
S_2\mu I_1
\end{bmatrix}
= \frac{1}{4}
\begin{bmatrix}
I_1 - \beta I_1 \\
-I_1 + \beta I_1 \\
O \\
O
\end{bmatrix}
= \frac{I_1}{4}
\begin{bmatrix}
1-\beta \\
-1+\beta \\
O \\
O
\end{bmatrix}
$$

The final intensity is represented by the first element of this Stokes column, that is

$$
I_6 = \tfrac{1}{4} I_1 (1 - \beta) = I_1 \frac{(1 - \cos\delta)}{4} = \tfrac{1}{2} I_1 \sin^2\left(\frac{\delta}{2}\right)
$$

We now see that the final intensity depends only on the original intensity I_1 and on δ, the phase-shift introduced by the phase plate, *not* on the angle θ describing the orientation of the phase plate. Thus, the isoclinics which depend on the orientation of the plate have vanished, and we are left only with the isochromatics.

Sometimes a linear polarizer and a $\lambda/4$ phase plate at $\pm$ 45° are sandwiched together on the same sheet, and the combination is called a right-handed or left-handed circular polarizer. In using such a filter, great care must be taken not to mount it 'back to front'. (The student will find it instructive, in the calculation that we have just done, to try the effect of interchanging the first polaroid and the first $\lambda/4$ plate.)

The method of calculation followed above was to evaluate successively each of the Stokes columns. Although it would be possible to proceed in a different way by telescoping together all of the 4×4 Mueller matrices, the number of calculations required would be significantly greater. If the computer programs of Appendix G are employed, this disadvantage is not serious. (It is worth noticing that, if the initial

unpolarized beam $\begin{bmatrix} I_1 \\ 0 \\ 0 \\ 0 \end{bmatrix}$ is replaced by a polarized beam $\begin{bmatrix} I \\ Q \\ U \\ V \end{bmatrix}$

most of the effects of this input state of polarization are obliterated by the first polarizer; the only trace that remains is that, in succeeding stages, the quantity I_1 is replaced by $I_1 + Q_1$.)

IV.4 EXPERIMENTAL DETERMINATION OF THE ELEMENTS OF A MUELLER MATRIX OR A STOKES COLUMN

We now interrupt the discussion of problems to describe the techniques by which the Stokes column of any beam of light and the Mueller matrix of any optical device can be found experimentally. (See also section II.8, where we were able to find experimentally the elements of a ray-transfer matrix for an optical imaging system.)

First, we need to be able to find the Stokes parameters of any beam. We assume that we can measure directly the intensity of the beam, using, say, a photoelectric cell, and we determine the Stokes parameters of the beam by seeing how this intensity is affected by passing the beam through various devices. As usual, we assume that the beam is travelling horizontally in the direction Oz and that, of the transverse axes, Oy is

vertical and Ox is horizontal. In general, six intensity measurements are needed to find the Stokes column
of the beam.

(1) A polaroid sheet is placed in the beam with its
pass-plane parallel to the x-axis. The polaroid
passes an intensity E_1 which is proportional to the
square of the amplitude of the vibrations parallel
to the x-axis, that is to H^2 in our previous notation. Thus $E_1 = H^2$.

(2) The polaroid is now rotated so that its pass-plane
is vertical, that is parallel to the y-axis. It
now passes an intensity E_2 proportional to the
square of the component of the vibration parallel
to the y-axis, that is $E_2 = K^2$ in our former
notation. Then, by definition of the Stokes parameters,

$$I = H^2 + K^2 = E_1 + E_2 \text{ and } Q = H^2 - K^2 = E_1 - E_2.$$

(3) The original beam is now passed through a polaroid
with its pass-plane set at $45°$ to the horizontal,
in the first and third quadrants. Using the formulae already developed for the Mueller matrix of a
polaroid, we find that the Stokes column of the beam
emerging from the polaroid must be

$$\frac{1}{2}\begin{bmatrix} 1 & 0 & 1 & 0 \\ 0 & 0 & 0 & 0 \\ 1 & 0 & 1 & 0 \\ 0 & 0 & 0 & 0 \end{bmatrix}\begin{bmatrix} I \\ Q \\ U \\ V \end{bmatrix} = \frac{1}{2}\begin{bmatrix} I+U \\ 0 \\ I+U \\ 0 \end{bmatrix}$$

Thus, the intensity E_3 of the beam emerging from
the polaroid is $\frac{1}{2}(I + U)$.

(4) We now pass the original beam through a polaroid
with its pass-planes again inclined at $45°$ to the
horizontal, but now in the second and fourth
quadrants, so that the angle is conventionally
regarded as minus $45°$. When this value for α is
substituted in the polaroid matrix, the Stokes
column of the emerging beam is seen to be

$$\frac{1}{2}\begin{bmatrix} 1 & 0 & -1 & 0 \\ 0 & 0 & 0 & 0 \\ -1 & 0 & 1 & 0 \\ 0 & 0 & 0 & 0 \end{bmatrix}\begin{bmatrix} I \\ Q \\ U \\ V \end{bmatrix} = \frac{1}{2}\begin{bmatrix} I-U \\ 0 \\ -I+U \\ 0 \end{bmatrix}$$

The intensity of the beam emerging from the polaroi
is now $E_4 = \frac{1}{2}(I - U)$. Upon subtraction, we obtain
$E_3 - E_4 = U$.

We now know three of the four Stokes parameters. To
obtain the fourth, we place in the beam a quarter-wave
plate with its fast axis horizontal. Using the formul
developed for a quarter-wave plate, we see that the
Stokes column of the beam, after passing through the
quarter-wave plate, becomes

$$\begin{bmatrix} 1 & 0 & 0 & 0 \\ 0 & 1 & 0 & 0 \\ 0 & 0 & 0 & 1 \\ 0 & 0 & -1 & 0 \end{bmatrix}\begin{bmatrix} I \\ Q \\ U \\ V \end{bmatrix} = \begin{bmatrix} I \\ Q \\ V \\ -U \end{bmatrix}$$

(5) The beam emerging from the quarter-wave plate is
now passed through a polaroid oriented as in measu
ment three. The Stokes column of the beam becomes

$$\frac{1}{2}\begin{bmatrix} 1 & 0 & 1 & 0 \\ 0 & 0 & 0 & 0 \\ 1 & 0 & 1 & 0 \\ 0 & 0 & 0 & 0 \end{bmatrix}\begin{bmatrix} I \\ Q \\ V \\ -U \end{bmatrix} = \frac{1}{2}\begin{bmatrix} I+V \\ 0 \\ I+V \\ 0 \end{bmatrix}$$

The intensity is now $E_5 = \frac{1}{2}(I + V)$

(6) The beam emerging from the quarter-wave plate is n
passed through a polaroid oriented as in measureme
four. The Stokes column of the beam now becomes

$$\tfrac{1}{2}\begin{bmatrix} 1 & 0 & -1 & 0 \\ 0 & 0 & 0 & 0 \\ -1 & 0 & 1 & 0 \\ 0 & 0 & 0 & 0 \end{bmatrix}\begin{bmatrix} I \\ Q \\ V \\ -U \end{bmatrix} = \tfrac{1}{2}\begin{bmatrix} I-V \\ 0 \\ -I+V \\ 0 \end{bmatrix}$$

The intensity is now $E_6 = \tfrac{1}{2}(I - V)$.

Subtracting the intensities, that is the first element in the Stokes columns obtained in the last two measurements, we get $V = E_5 - E_6$. All four Stokes parameters of the original beam are now determined.

We shall now see how we can use these results to determine the Mueller matrix of any optical device, by sending through it various types of light and determining the Stokes parameters of the emergent beam.

Suppose the Stokes column of the original beam is S_1, that of the emergent beam is S_2 and that the Mueller matrix of the device is T, so that

$S_2 = TS_1$

that is

$$\begin{bmatrix} I_2 \\ Q_2 \\ U_2 \\ V_2 \end{bmatrix} = \begin{bmatrix} A & B & C & D \\ E & F & G & H \\ J & K & L & M \\ N & P & R & S \end{bmatrix}\begin{bmatrix} I_1 \\ Q_1 \\ U_1 \\ V_1 \end{bmatrix}$$

We now pass beams of four types of light in turn, into the device, and measure the Stokes columns of the emerging beam, by the methods already described.

(1) We send first a beam of ordinary unpolarized light into the device. Let its intensity be α; then $I_1 = \alpha$, and Q_1, U_1 and V_1 are all zero, by the formulae already developed for unpolarized light. The equations then become

$$I_2 = A\alpha, \quad \text{that is} \quad A = I_2/\alpha$$

$$Q_2 = E\alpha, \quad \text{that is} \quad E = Q_2/\alpha$$

$$U_2 = J\alpha, \quad \text{that is} \quad J = U_2/\alpha$$

$$V_2 = N\alpha, \quad \text{that is} \quad N = V_2/\alpha$$

(2) We now put into the device plane-polarized light parallel to the x-axis. For this, $I_1 = Q_1 = \beta$, s͏a U_1 and V_1 are both zero. So, if we call S_3 the Stokes column of the emerging beam, the equations become

$$I_3 = (A + B)\beta, \quad \text{that is} \quad B = I_3/\beta - A$$

$$Q_3 = (E + F)\beta, \quad \text{that is} \quad F = Q_3/\beta - E$$

$$U_3 = (J + K)\beta, \quad \text{that is} \quad K = U_3/\beta - J$$

$$V_3 = (N + P)\beta, \quad \text{that is} \quad P = V_3/\beta - N$$

(3) We now pass into the device a beam of right-handed circularly polarized light, for which $I_1 = V_1 = \omega$ say, and Q_1 and U_1 are both zero. So, if we call S_4 the Stokes column of the emerging beam, the equations become

$$I_4 = (A + D)\omega, \quad \text{that is} \quad D = I_4/\omega - A$$

$$Q_4 = (E + H)\omega, \quad \text{that is} \quad H = Q_4/\omega - E$$

$$U_4 = (J + M)\omega, \quad \text{that is} \quad M = U_4/\omega - J$$

$$V_4 = (N + S)\omega, \quad \text{that is} \quad S = V_4/\omega - N$$

(4) We now pass into the device plane-polarized light with its vibration plane at an angle of $45°$ with the x-axis, lying in the first and third quadrant For this beam $I_1 = U_1 = \mu$, say, while Q_1 and V_1 ar both zero. So, if the Stokes column of the emerg light is called S_5, the equations become

$$I_5 = (A + C)\mu, \quad \text{that is} \quad C = I_5/\mu - A$$

$Q_5 = (E + G)\mu,$ that is $G = Q_5/\mu - E$

$U_5 = (J + L)\mu,$ that is $L = U_5/\mu - J$

$V_5 = (N + R)\mu,$ that is $R = V_5/\mu - N$

In each of the equations now developed, an unknown
element of the Mueller matrix is expressed in terms
of measured intensities and previously calculated
Mueller matrix elements. By this scheme, therefore,
all sixteen elements of the Mueller matrix of a
device can be determined.

7.5 USE OF THE JONES CALCULUS FOR TRANSFORMING A MAXWELL COLUMN

There is another calculus as well as the Mueller calculus by which problems involving fully polarized light
can be tackled. This calculus was developed by R.
Clark Jones. It is simpler in some ways than the
Mueller calculus, since it works with two-by-two matrices
instead of with four-by-fours. But the elements of
some of these matrices are complex, and it has the disadvantage that it does not work with quantities that
can be measured directly by experiment. There are,
however, no redundant elements, and every Jones matrix
that can be written corresponds to a device which, in
principle at least, can be physically realized.

For many problems in polarization optics both calculi
are equally suitable. But wherever light with only a
partial degree of polarization is involved, the Mueller
calculus should be selected; and wherever interference
effects are taking place, or the system employs coherent
light from a laser, the Jones calculus will be more
useful.

(In writing this chapter we have made some effort
to deal with the two calculi in separate sections; it
is therefore possible for the reader who is interested
only in one largely to pass over the other. By studying
both, however, the thoughtful student may gain some
insight into the differences that exist between coherent and incoherent methods in optics. Between the two
there is a borderland which still needs to be explored.)

In the Jones calculus we operate, not on the Stokes
parameters, which are proportional to the *intensity* of
the beam, but on the Maxwell column, whose elements

represent the *amplitudes* and *phases* of the components
of the transverse vibration. We shall rely on the fact
that, for devices employing fully polarized light, the
electric field components in the beam leaving the devi
are linear functions of the electric field components
in the input beam, and the matrix which connects the
components in the beam leaving the device with those
entering the device enables us to characterize that
device. We shall tabulate the Jones matrices for
various types of devices, just as we did for the Muell
calculus. (The derivation of these matrices is given
in appendix E, and, since the two calculi are closely
related, the connection between them is discussed in
appendix F.)

We have already seen in section IV.1 how the transverse vibrations corresponding to any fully polarized
disturbance can be represented by a Maxwell column. I
a plane wave which is being propagated in the directic
Oz, the electric field can be regarded as the real par
of the complex vector

$$
\begin{bmatrix} E_x \\ E_y \end{bmatrix} = \begin{bmatrix} H\exp i\left\{\omega\left(t - \frac{z}{c}\right) + \phi_x\right\} \\ K\exp i\left\{\omega\left(t - \frac{z}{c}\right) + \phi_y\right\} \end{bmatrix}
$$

In most cases the time-dependence and the z-depender
of the beam are transferred outside the column as a
scalar multiplier, so that we have

$$
\begin{bmatrix} E_x \\ E_y \end{bmatrix} = \begin{bmatrix} H\exp(i\phi_x) \\ K\exp(i\phi_y) \end{bmatrix} \exp i\left\{\omega\left(t - \frac{z}{c}\right)\right\}
$$

A useful rule for calculating the intensity of the
beam is to premultiply the Maxwell column by the
complex conjugate of its transpose, an operation
sometimes known as finding the bracket product. We
thus obtain

$$
I = \begin{bmatrix} E_x^* & E_y^* \end{bmatrix} \begin{bmatrix} E_x \\ E_y \end{bmatrix} = \begin{bmatrix} H\exp(-i\phi_x) & K\exp(-i\phi_y) \end{bmatrix} \begin{bmatrix} H\exp \\ K\exp \end{bmatrix}
$$

$$
= (H^2 + K^2)
$$

the modulus of the scalar multiplier being unity).

If only a single disturbance is involved, we can sometimes ignore the absolute phase of the vibrations and
demand arbitrarily that the phase angle, for example of
the x-component, shall be zero. In many cases, also,
the absolute amplitude is not required and can be transferred outside the column into the scalar multiplier.
For a beam of unit intensity, we must have $(H^2 + K^2)$
equal to unity, so we can write $H = \cos\theta$ and $K = \sin\theta$.
In terms of this 'normalized' Maxwell column, the
expression for the whole disturbance becomes

$$\begin{bmatrix} \cos\theta \\ \sin\theta\exp(i\Delta) \end{bmatrix} A\exp i\omega\left(t - \frac{z}{c}\right)$$

This expression provides all that is needed to represent a completely coherent plane-monochromatic wave
disturbance. As we have already mentioned, however, it
can also be used to represent less coherent radiation,
provided that we remember that the amplitude A and the
phase ϕ are themselves fluctuating functions, which
remain more or less constant only over a limited region,
measured both laterally across the beam and longitudinally in the direction of travel. The greater the coherence, the larger is the volume of this region and
the smaller, by Heisenberg's uncertainty principle, the
uncertainty in the momentum vector to be associated
with each photon - in classical terms the uncertainty
in the wavelength and in the direction of travel.

It is an experimental fact that at optical frequencies
the response of a medium to an electromagnetic wavefield is almost perfectly linear. Non-linear optical
effects do occur, of course, but to observe them it is
necessary to use a laser source and to work at very
high electric field strengths.

We can therefore predict that, if

$$E_1 = \begin{bmatrix} H_1 e^{i\phi_1} \\ K_1 e^{i\psi_1} \end{bmatrix}$$

represents the Maxwell column of the beam entering some
kind of polarizing device, then the Maxwell column for
the beam leaving that device can be represented as E_2,

where

$$E_2 = \begin{bmatrix} H_2 e^{i\phi_2} \\ K_2 e^{i\psi_2} \end{bmatrix} \quad \text{and} \quad \begin{array}{l} H_2 e^{i\phi_2} = J_{11} H_1 e^{i\phi_1} + J_{12} K_1 e^{i\psi_1} \\ K_2 e^{i\psi_2} = J_{21} H_1 e^{i\phi_1} + J_{22} K_1 e^{i\psi_1} \end{array}$$

or, in matrix form,

$$\begin{bmatrix} H_2 e^{i\phi_2} \\ K_2 e^{i\psi_2} \end{bmatrix} = \begin{bmatrix} J_{11} & J_{12} \\ J_{21} & J_{22} \end{bmatrix} \begin{bmatrix} H_1 e^{i\phi_1} \\ K_1 e^{i\psi_1} \end{bmatrix}$$

(In general, the four elements of the square matrix are themselves complex and depend only on the device.)

It will be noticed that in the equations above we have, for the sake of generality, used two separate phase angles ϕ and ψ in each of the Maxwell columns. far as the state of polarization is concerned, it is only the phase difference $(\psi - \phi) = \Delta$ which has any significance. But it is not always possible to ensure even with an input for which the phase angle ϕ_1 is zer that the corresponding output phase angle ϕ_2 for the x-component is also zero. Once the output column has been calculated, it is very easy to multiply the whole column by any desired phase factor.

Starting with the above equation, we can now develop the Jones calculus. The matrix consisting of the four Js is called the Jones matrix J of the device, so that the matrix equation can finally be written in the form

$E_2 = JE_1$

Suppose now we have two devices, whose Jones matrices are

J_a and J_b

We pass a beam of light through the two devices in series. Let the Maxwell column of the original beam of light be E_1, that of the beam between the devices E_2 and that after the second device E_3. Then, on our assumption, we can write

$E_2 = J_a E_1 \quad \text{and} \quad E_3 = J_b E_2$

Substituting from the first of these equations into

the second, we get

$$E_3 = J_b(J_a E_1) = (J_b J_a)E_1$$

by the associative property of matrices. Thus, we see that the effect of a number of devices in series on a beam of light can be obtained by multiplying together their Jones matrices. (The same rule applies here as with the refraction and translation matrices in Gaussian optics and with the Mueller matrices.)

For convenience in solving problems, we list overleaf in Table 4 the Jones matrices for various commonly used polarizing devices.

IV.6 EXPERIMENTAL DETERMINATION OF THE ELEMENTS OF A JONES MATRIX OR A MAXWELL COLUMN

We shall now describe methods for determining the Maxwell column of any beam of polarized light, whether plane or elliptically polarized, and methods by which the Jones matrix of any device can be found experimentally. These methods involve passing beams of light through the device and through various polaroids and phase plates, and then measuring the intensity of each beam that emerges.

First, we consider determination of the Maxwell column of a beam of light. Suppose the Maxwell column of the beam is

$$\begin{bmatrix} H \\ K\exp(i\Delta) \end{bmatrix}$$

so that the intensity is

$$\begin{bmatrix} H & K\exp(-i\Delta) \end{bmatrix} \begin{bmatrix} H \\ K\exp(i\Delta) \end{bmatrix} = H^2 + K^2$$

We choose our unit of intensity so that this beam is of unit intensity.

First we pass the beam through a polaroid with its pass-plane horizontal, that is parallel to the x-axis. The Maxwell column of the emerging beam is

$$\begin{bmatrix} 1 & 0 \\ 0 & 0 \end{bmatrix} \begin{bmatrix} H \\ K\exp(i\Delta) \end{bmatrix} = \begin{bmatrix} H \\ 0 \end{bmatrix}$$

Table 4. Jones matrices for ideal linear polarizers, linear retarders, rotation of axes and circular retarders

The angle θ specifies how the pass-plane of a polarizer or the fast axis of a linear retarder is oriented with respect to the x-axis.

Type of device	$\theta = 0$	$\theta = \pm\frac{\pi}{4}$	$\theta = \frac{\pi}{2}$	θ takes any value
Ideal linear polarizer at angle θ	$\begin{bmatrix} 1 & 0 \\ 0 & 0 \end{bmatrix}$	$\frac{1}{2}\begin{bmatrix} 1 & \pm 1 \\ \pm 1 & 1 \end{bmatrix}$	$\begin{bmatrix} 0 & 0 \\ 0 & 1 \end{bmatrix}$	$\begin{bmatrix} C_1^2 & C_1 S_1 \\ C_1 S_1 & S_1^2 \end{bmatrix}$ $\begin{matrix} C_1 \\ S_1 \end{matrix}$
Quarter-wave linear retarder with fast axis at angle θ	$\begin{bmatrix} 1 & 0 \\ 0 & -i \end{bmatrix}$	$\frac{1}{2}\begin{bmatrix} (1-i) & \pm(1+i) \\ \pm(1+i) & (1-i) \end{bmatrix}$	$\begin{bmatrix} -i & 0 \\ 0 & 1 \end{bmatrix}$	$\begin{bmatrix} C_1^2 - iS_1^2 & C_1 S_1(1+i) \\ C_1 S_1(1+i) & -iC_1^2 + S_1^2 \end{bmatrix}$

Matrix for $\theta = \pm\frac{\pi}{4}$ can be multiplied by $e^{i\pi/4}$ to give $\frac{1}{\sqrt{2}}\begin{bmatrix} 1 & \pm i \\ \pm i & 1 \end{bmatrix}$

Half-wave linear retarder with fast axis at angle θ	$\begin{bmatrix} 1 & 0 \\ 0 & -1 \end{bmatrix}$	$\begin{bmatrix} 0 & \pm 1 \\ \pm 1 & 0 \end{bmatrix}$	$\begin{bmatrix} -1 & 0 \\ 0 & 1 \end{bmatrix}$	$\begin{bmatrix} C_2 & S_2 \\ S_2 & -C_2 \end{bmatrix}$
Linear retarder with retardation δ and with fast axis at angle θ	$\begin{bmatrix} 1 & 0 \\ 0 & e^{-i\delta} \end{bmatrix}$	$\frac{1}{2}\begin{bmatrix} (e^{-i\delta}+1) & \pm(1-e^{-i\delta}) \\ \pm(1-e^{-i\delta}) & (e^{-i\delta}+1) \end{bmatrix}$	$\begin{bmatrix} e^{-i\delta} & 0 \\ 0 & 1 \end{bmatrix}$	$\begin{bmatrix} C_1^2 + S_1^2 e^{-i\delta} & C_1 S_1(1-e^{-i\delta}) \\ C_1 S_1(1-e^{-i\delta}) & C_1^2 e^{-i\delta} + S_1^2 \end{bmatrix}$ $\begin{matrix} C_2 \\ S_2 \end{matrix}$

or $e^{-i\delta/2}\begin{bmatrix} \cos\frac{\delta}{2} & \pm i\sin\frac{\delta}{2} \\ \pm i\sin\frac{\delta}{2} & \cos\frac{\delta}{2} \end{bmatrix}$

Rotation of x-axis to new angle θ with respect to old x-axis *or*, action of circular retarder which retards *left* circular state by 2θ radians

$\begin{bmatrix} C_1 & S_1 \\ -S_1 & C_1 \end{bmatrix}$

If we denote this 'rotator' matrix by $R(\theta)$, then we can use it to transform the Jones matrix for any polarizing device or system at a given orientation ϕ into the corresponding matrix for the same device or system at a new orientation $(\theta + \phi)$. The transformation formula is

$$J(\theta + \phi) = R(-\theta) J(\phi) R(\theta)$$

If a complete polarizing system, compounded from several devices, is to be rotated, then it may save labour to use the above unitary transformation. For a single device, however, it is usually quicker to use a matrix taken from the appropriate column in the above table. (For derivation of these matrices, and for some discussion of their normalization, see appendix E.)

Therefore the intensity $I_1 = H^2$, so that $H = \sqrt{I_1}$, where I_1 is the measured intensity transmittance.

We now pass the original beam through a polaroid with its pass-plane vertical (parallel to the Y-axis), so that the Maxwell column becomes

$$\begin{bmatrix} 0 & 0 \\ 0 & 1 \end{bmatrix} \begin{bmatrix} H \\ K\exp(i\Delta) \end{bmatrix} = \begin{bmatrix} 0 \\ K\exp(i\Delta) \end{bmatrix}$$

Therefore the intensity $I_2 = K^2$, so that $K = \sqrt{I_2}$.

We now pass the beam through a polaroid with its pass-plane at an angle of $45°$ with the X-axis in the first and third quadrants for which the Jones matrix is

$$\begin{bmatrix} 1/2 & 1/2 \\ 1/2 & 1/2 \end{bmatrix} = 1/2 \begin{bmatrix} 1 & 1 \\ 1 & 1 \end{bmatrix}$$

The new Maxwell column is

$$1/2 \begin{bmatrix} 1 & 1 \\ 1 & 1 \end{bmatrix} \begin{bmatrix} H \\ K\exp(i\Delta) \end{bmatrix} = 1/2 \begin{bmatrix} H + K\exp(i\Delta) \\ H + K\exp(i\Delta) \end{bmatrix}$$

Multiplying the Maxwell column by the transpose of its complex conjugate, we see that the intensity is

$$I_3 = 1/2\{H^2 + HK\left[\exp(i\Delta) + \exp(-i\Delta)\right] + K^2\}$$

We now pass the beam through a polaroid with its pass-plane at $45°$ to the X-axis but in the second and fourth quadrants, so that $\theta = -45°$. The Maxwell column becomes

$$1/2 \begin{bmatrix} 1 & -1 \\ -1 & 1 \end{bmatrix} \begin{bmatrix} H \\ K\exp(i\Delta) \end{bmatrix} = 1/2 \begin{bmatrix} H - K\exp(i\Delta) \\ -H + K\exp(i\Delta) \end{bmatrix}$$

The intensity is

$$I_4 = 1/2\{H^2 - HK\left[\exp(i\Delta) + \exp(-i\Delta)\right] + K^2\}$$

Therefore

$$I_3 - I_4 = 1/2 \times 2HK\left[\exp(i\Delta) + \exp(-i\Delta)\right]$$

$$= HK(\cos\Delta + i\sin\Delta + \cos\Delta - i\sin\Delta) = 2HK\cos\Delta$$

The values of H and K are already known so this equation determines $\cos\Delta$, but Δ can still be positive or negative. There is an uncertainty of sign, so we now determine $\sin\Delta$ to resolve this uncertainty.

We pass the original beam through a quarter-wave plate whose fast axis is horizontal. Using the fact that $\delta = 90°$ for a quarter-wave plate and that $\theta = 0°$, we find that the Jones matrix is

$$\begin{bmatrix} 1 & 0 \\ 0 & -i \end{bmatrix}$$

and the Maxwell column of the beam becomes

$$\begin{bmatrix} 1 & 0 \\ 0 & -i \end{bmatrix} \begin{bmatrix} H \\ K\exp(i\Delta) \end{bmatrix} = \begin{bmatrix} H \\ -iK\exp(i\Delta) \end{bmatrix}$$

We put the beam from the quarter-wave plate through a polaroid inclined at 45° to the axis. The Maxwell column becomes

$$1/2 \begin{bmatrix} 1 & 1 \\ 1 & 1 \end{bmatrix} \begin{bmatrix} H \\ -iK\exp(i\Delta) \end{bmatrix} = 1/2 \begin{bmatrix} H - iK\exp(i\Delta) \\ H - iK\exp(i\Delta) \end{bmatrix}$$

The intensity is

$$I_5 = 1/2\{H^2 - iHK\left[\exp(i\Delta) - \exp(-i\Delta)\right] + K^2\}$$

$$= 1/2(H^2 - 2HK\sin\Delta + K^2)$$

We now pass the beam from the quarter-wave plate through a polaroid with its pass-plane at $-$ 45° to the axis. The Maxwell column becomes

$$1/2 \begin{bmatrix} 1 & -1 \\ -1 & 1 \end{bmatrix} \begin{bmatrix} H \\ -iK\exp(i\Delta) \end{bmatrix} = 1/2 \begin{bmatrix} H + iK\exp(i\Delta) \\ -H - iK\exp(i\Delta) \end{bmatrix}$$

The intensity is

$$I_6 = 1/2\{H^2 - iHK\left[\exp(-i\Delta) - \exp(i\Delta)\right] + K^2\}$$

$$= 1/2(H^2 + 2HK\sin\Delta + K^2)$$

Therefore

$$I_6 - I_5 = 2HK\sin\Delta$$

This determines $\sin\Delta$. Knowing $\sin\Delta$ and $\cos\Delta$, we check that their squares sum to unity and obtain the value of Δ. We now know the whole of the Maxwell column of the original beam.

(In carrying out the above measurements, some allowance should be made for the fact that a polarizer is never perfectly transmitting even when it is fed with light vibrating in the preferred direction. If, for example, HN32 sheet polarizer is employed, the photo-cell currents obtained in measuring I_1 to I_6 will be reduced to about 64 per cent of their theoretical values for an ideal polarizer. The exact factor for a given polarizer should be determined by measurement, using an unpolarized input beam and measuring the intensity transmitted for at least two orientations of the pass-plane.)

We shall now describe the method by which the Jones matrix of any device can be determined from intensity measurements. Suppose the Jones matrix of the device is

$$J = \begin{bmatrix} J_{11} & J_{12} \\ J_{21} & J_{22} \end{bmatrix} = \begin{bmatrix} X_{11} + iY_{11} & X_{12} + iY_{12} \\ X_{21} + iY_{21} & X_{22} + iY_{22} \end{bmatrix}$$

$$= \begin{bmatrix} R_{11}\exp(i\theta_{11}) & R_{12}\exp(i\theta_{12}) \\ R_{21}\exp(i\theta_{21}) & R_{22}\exp(i\theta_{22}) \end{bmatrix}$$

(The matrix elements in a Jones matrix may, of course, be complex and we have indicated both Cartesian and polar representations.)

A. We pass into the device a beam of unit intensity plane-polarized parallel to the X-axis so that the Maxwell column is

$$\begin{bmatrix} 1 \\ 0 \end{bmatrix}$$

On emerging from the device the Maxwell column will be

$$\begin{bmatrix} J_{11} & J_{12} \\ J_{21} & J_{22} \end{bmatrix}\begin{bmatrix} 1 \\ 0 \end{bmatrix} = \begin{bmatrix} J_{11} \\ J_{21} \end{bmatrix}$$

A.1. The beam is now passed through a polaroid with its pass-plane horizontal and the Maxwell column becomes

$$\begin{bmatrix} 1 & 0 \\ 0 & 0 \end{bmatrix}\begin{bmatrix} J_{11} \\ J_{21} \end{bmatrix} = \begin{bmatrix} J_{11} \\ 0 \end{bmatrix}$$

The intensity is

$$I_2 = J_{11}^* J_{11} = (X_{11} - iY_{11})(X_{11} + iY_{11})$$

$$= X_{11}^2 + Y_{11}^2 = R_{11}^2$$

A.2. The beam from the device is now put through a polaroid with its pass-plane vertical. The Maxwell column becomes

$$\begin{bmatrix} 0 & 0 \\ 0 & 1 \end{bmatrix} \begin{bmatrix} J_{11} \\ J_{21} \end{bmatrix} = \begin{bmatrix} 0 \\ J_{21} \end{bmatrix}$$

The intensity is $I_3 = R_{21}^2$.

B. A beam of unit intensity polarized parallel to the Y-axis and having a Maxwell column

is now passed into the device. The emerging Maxwell column is

$$\begin{bmatrix} J_{11} & J_{12} \\ J_{21} & J_{22} \end{bmatrix} \begin{bmatrix} 0 \\ 1 \end{bmatrix} = \begin{bmatrix} J_{12} \\ J_{22} \end{bmatrix}$$

B.1. The beam emerging from the device is now passed through a polaroid with its pass-plane horizontal. The emerging Maxwell column is

$$\begin{bmatrix} 1 & 0 \\ 0 & 0 \end{bmatrix} \begin{bmatrix} J_{12} \\ J_{22} \end{bmatrix} = \begin{bmatrix} J_{12} \\ 0 \end{bmatrix}$$

The intensity is $I_4 = R_{12}^2$.

B.2. The beam emerging from the device is now passed through a polaroid with its pass-plane vertical. The Maxwell column is

$$\begin{bmatrix} 0 & 0 \\ 0 & 1 \end{bmatrix} \begin{bmatrix} J_{12} \\ J_{22} \end{bmatrix} = \begin{bmatrix} 0 \\ J_{22} \end{bmatrix}$$

The intensity is $I_5 = R_{22}^2$.

We have now determined the magnitude of all four Jones matrix elements. It remains only to find the angles in these matrix elements.

C. We now pass into the device a beam of right-handed circularly polarized light of unit intensity so that

$$H = K \quad \text{and } H^2 + K^2 = 1$$

whence

$$H = K = 1/\sqrt{2}$$

For right-handed circularly polarized light, $\Delta = \pi/2$, so that $\exp(i\Delta) = i$. The Maxwell column of the original beam is therefore

On emerging from the device the Maxwell column of the beam is

$$1/\sqrt{2} \begin{bmatrix} J_{11} & J_{12} \\ J_{21} & J_{22} \end{bmatrix} \begin{bmatrix} 1 \\ i \end{bmatrix} = 1/\sqrt{2} \begin{bmatrix} J_{11} + iJ_{12} \\ J_{21} + iJ_{22} \end{bmatrix}$$

C.1. The beam is now passed through a polaroid with its pass-plane horizontal so the Maxwell column becomes

$$1/\sqrt{2} \begin{bmatrix} 1 & 0 \\ 0 & 0 \end{bmatrix} \begin{bmatrix} J_{11} + iJ_{12} \\ J_{21} + iJ_{22} \end{bmatrix} = 1/\sqrt{2} \begin{bmatrix} J_{11} + iJ_{12} \\ 0 \end{bmatrix}$$

$$= 1/\sqrt{2} \begin{bmatrix} (X_{11} + iY_{11}) + i(X_{12} + iY_{12}) \\ 0 \end{bmatrix}$$

$$= 1/\sqrt{2} \begin{bmatrix} (X_{11} - Y_{12}) + i(Y_{11} + X_{12}) \\ 0 \end{bmatrix}$$

The intensity obtained by multiplying this matrix by the transpose of its complex conjugate is thus

$$I_6 = 1/2 \left[(X_{11} - Y_{12})^2 + (Y_{11} + X_{12})^2 \right]$$

Therefore

$$2I_6 = X_{11}^2 - 2X_{11}Y_{12} + Y_{12}^2 + Y_{11}^2 + 2Y_{11}X_{12} + X_{12}^2$$

Thus, substituting the known values of I_2 and I_4,

$$\frac{2I_6 - I_2 - I_4}{\sqrt{(I_2 I_4)}} = \frac{2(Y_{11}X_{12} - X_{11}Y_{12})}{R_{11}R_{12}}$$

$$= 2(\sin\theta_{11}\cos\theta_{12} - \cos\theta_{11}\sin\theta_{12})$$

(using the relationships between the Cartesian and the polar forms of the Jones matrix elements)

$$= 2\sin(\theta_{11} - \theta_{12})$$

C.2. The beam from the device is now passed through a polaroid with its pass-plane vertical. The Maxwell column becomes

$$\frac{1}{\sqrt{2}}\begin{bmatrix} 0 & 0 \\ 0 & 1 \end{bmatrix}\begin{bmatrix} J_{11} + iJ_{12} \\ J_{21} + iJ_{22} \end{bmatrix} = \frac{1}{\sqrt{2}}\begin{bmatrix} 0 \\ J_{21} + iJ_{22} \end{bmatrix}$$

so that the intensity is

$$I_7 = 1/2\left[(X_{21} - Y_{22})^2 + (Y_{21} + X_{22})^2\right]$$

Thus,

$$\frac{2I_7 - I_3 - I_5}{\sqrt{(I_3 I_5)}} = 2\sin(\theta_{21} - \theta_{22})$$

(Here, and in later parts of this section, we have left a good deal of detailed calculation to the student.)

We now know the sines of the angles $(\theta_{11} - \theta_{12})$ and $(\theta_{21} - \theta_{22})$, but this still leaves doubt as to the value of the angles because

$$\sin(\pi - \theta) = \sin\theta$$

To determine the angles completely we need to know the cosines of the angles as well.

D. We pass into the device a beam of light of unit intensity plane-polarized at $45°$ to the X-axis. For this beam $H = K = 1/\sqrt{2}$ and $\Delta = 0$, so that $\exp(i\Delta) = 1$. For the original beam the Maxwell column is

$$1/\sqrt{2} \begin{bmatrix} 1 \\ 1 \end{bmatrix}$$

so that after the device the Maxwell column is

$$1/\sqrt{2} \begin{bmatrix} \mathcal{J}_{11} & \mathcal{J}_{12} \\ \mathcal{J}_{21} & \mathcal{J}_{22} \end{bmatrix} \begin{bmatrix} 1 \\ 1 \end{bmatrix} = 1/\sqrt{2} \begin{bmatrix} \mathcal{J}_{11} + \mathcal{J}_{12} \\ \mathcal{J}_{21} + \mathcal{J}_{22} \end{bmatrix}$$

D.1. The beam is now passed through a polaroid with its pass-plane horizontal. The Maxwell column becomes

$$1/\sqrt{2} \begin{bmatrix} 1 & 0 \\ 0 & 0 \end{bmatrix} \begin{bmatrix} \mathcal{J}_{11} + \mathcal{J}_{12} \\ \mathcal{J}_{21} + \mathcal{J}_{22} \end{bmatrix} = 1/\sqrt{2} \begin{bmatrix} \mathcal{J}_{11} + \mathcal{J}_{12} \\ 0 \end{bmatrix}$$

so that the intensity is

$$I_8 = 1/2 \left[(X_{11} + X_{12})^2 + (Y_{11} + Y_{12})^2 \right]$$

By a calculation similar to the previous one, we see that

$$\frac{2I_8 - I_2 - I_4}{\sqrt{(I_2 I_4)}} = 2\cos(\theta_{11} - \theta_{12})$$

D.2. We now pass the beam from the device through a polaroid with its pass-plane vertical. The Maxwell column becomes

$$1/\sqrt{2} \begin{bmatrix} 0 & 0 \\ 0 & 1 \end{bmatrix} \begin{bmatrix} \mathcal{J}_{11} + \mathcal{J}_{12} \\ \mathcal{J}_{21} + \mathcal{J}_{22} \end{bmatrix} = 1/\sqrt{2} \begin{bmatrix} 0 \\ \mathcal{J}_{21} + \mathcal{J}_{22} \end{bmatrix}$$

The intensity is

$$I_9 = 1/2 \left[(X_{21} + X_{22})^2 + (Y_{21} + Y_{22})^2 \right]$$

and

$$\frac{2I_9 - I_3 - I_5}{\sqrt{(I_3 I_5)}} = 2\cos(\theta_{22} - \theta_{21})$$

We now know completely the angles $(\theta_{11} - \theta_{12})$ and $(\theta_{21} - \theta_{22})$. (Some check should be made that the squares of the sines and cosines obtained sum approx-

imately to unity.) To complete our task we need a
connection between these, that is we need to connect
the angle θ_{11} with either θ_{21} or θ_{22}. To do this we
again put into the device a beam plane-polarized horiz-
ontally so that, as in section A, the Maxwell column
emerging from the device is

$$\begin{bmatrix} J_{11} \\ J_{21} \end{bmatrix}$$

We pass this beam through a polaroid with its pass-plane
inclined at $45°$ to the axis. The Maxwell column becomes

$$1/2 \begin{bmatrix} 1 & 1 \\ 1 & 1 \end{bmatrix} \begin{bmatrix} J_{11} \\ J_{21} \end{bmatrix} = 1/2 \begin{bmatrix} J_{11} + J_{21} \\ J_{11} + J_{21} \end{bmatrix}$$

We now put the beam through another polaroid with its
axis horizontal. The Maxwell column becomes

$$1/2 \begin{bmatrix} 1 & 0 \\ 0 & 0 \end{bmatrix} \begin{bmatrix} J_{11} + J_{21} \\ J_{11} + J_{21} \end{bmatrix} = 1/2 \begin{bmatrix} J_{11} + J_{21} \\ 0 \end{bmatrix}$$

As in D.1 or D.2, the intensity becomes

$$I_{10} = 1/4 \left[(X_{11} + X_{21})^2 + (Y_{11} + Y_{21})^2 \right]$$

so that

$$\frac{4I_{10} - I_2 - I_3}{\sqrt{(I_2 I_3)}} = 2\cos(\theta_{11} - \theta_{21})$$

We now put the beam from the device, whose Maxwell
column is

$$\begin{bmatrix} J_{11} \\ J_{21} \end{bmatrix}$$

through a quarter-wave plate with its fast axis
vertical. The Maxwell column becomes

$$\begin{bmatrix} 1 & 0 \\ 0 & i \end{bmatrix} \begin{bmatrix} J_{11} \\ J_{21} \end{bmatrix} = \begin{bmatrix} J_{11} \\ iJ_{21} \end{bmatrix}$$

We now pass the beam through a polaroid with its optic axis at 45°. The Maxwell column becomes

$$\frac{1}{2}\begin{bmatrix} 1 & 1 \\ 1 & 1 \end{bmatrix}\begin{bmatrix} J_{11} \\ iJ_{21} \end{bmatrix} = \frac{1}{2}\begin{bmatrix} J_{11} + iJ_{21} \\ J_{11} + iJ_{21} \end{bmatrix}$$

We now pass the beam through a polaroid with its axis horizontal. The Maxwell column becomes

$$\frac{1}{2}\begin{bmatrix} 1 & 0 \\ 0 & 0 \end{bmatrix}\begin{bmatrix} J_{11} + iJ_{21} \\ J_{11} + iJ_{21} \end{bmatrix} = \frac{1}{2}\begin{bmatrix} J_{11} + iJ_{21} \\ 0 \end{bmatrix}$$

As in C.1 or C.2, the intensity is now

$$I_{11} = \frac{1}{4}\left[(X_{11} - Y_{21})^2 + (Y_{11} + X_{21})^2\right]$$

so that

$$\frac{4I_{11} - I_2 - I_3}{\sqrt{(I_2 I_3)}} = 2\sin(\theta_{11} - \theta_{21})$$

The angle $(\theta_{11} - \theta_{21})$ is thus determined and so the differences between θ_{11} and all the other three angles are now known.

As explained in the section relating the Mueller and Jones matrices, we can regard one of the θs as arbitrary, that is we can arbitrarily set θ_{11} equal to zero. This means that the angles in the other three Jones elements are now known, and we have determined the Jones matrix completely.

IV.7 ILLUSTRATIVE PROBLEMS SOLVED BY MUELLER CALCULUS AND BY JONES CALCULUS

Problem 1

A source emits plane-polarized light of unit intensity which falls on an ideal linear polarizer. Prove that the intensity of the light coming through the polaroid is $\cos^2\theta$, where θ is the angle of the polaroid measured from a position of maximum transmission. The polaroid is set to extinguish the transmitted beam and a second polaroid is then placed between the source and the first polaroid.

Show that some light can now pass through both polaroids and prove that its intensity is proportional to $\sin^2 2\phi$, where ϕ is the angle of the second polaroid measured from a position of extinction.

Solution

We shall work this problem by both the Jones and the Mueller methods.

First the Jones method. Suppose the incident beam of light is plane-polarized horizontal so that its Maxwell column is

$$E_1 = \begin{bmatrix} H \\ 0 \end{bmatrix}$$

The Jones matrix of the polaroid with its pass-plane at angle θ with the X-direction is

$$\begin{bmatrix} \cos^2\theta & \sin\theta\cos\theta \\ \sin\theta\cos\theta & \sin^2\theta \end{bmatrix}$$

so that the Maxwell column of the beam coming through the polaroid is

$$E_2 = \begin{bmatrix} \cos^2\theta & \sin\theta\cos\theta \\ \sin\theta\cos\theta & \sin^2\theta \end{bmatrix} \begin{bmatrix} H \\ 0 \end{bmatrix} = \begin{bmatrix} H\cos^2\theta \\ H\sin\theta\cos\theta \end{bmatrix}$$

We find the intensity of the emergent beam as already explained by premultiplying the emergent Maxwell column by the transpose of its complex conjugate. In order to write down the complex conjugate of a given matrix M, it is convenient to replace the double symbol $M*^T$ by the single Hermitian symbol $\mathbf{M}$. Using this Hermitian symbol, we then find for the intensity

$$\mathbf{E}_2 E_2 = \begin{bmatrix} H\cos^2\theta & H\cos\theta\sin\theta \end{bmatrix} \begin{bmatrix} H\cos^2\theta \\ H\cos\theta\sin\theta \end{bmatrix}$$

$$= H^2\cos^4\theta + H^2\cos^2\theta\sin^2\theta = H^2\cos^2\theta(\cos^2\theta + \sin^2\theta)$$

$$= H^2\cos^2\theta$$

In this particular problem, the initial intensity H^2 is known to be unity.

Suppose we now pass the beam through another polaroid with its pass-plane vertical, so that its Jones matrix is

$$\begin{bmatrix} 0 & 0 \\ 0 & 1 \end{bmatrix}$$

The Maxwell column of the beam emerging from this polaroid will thus be

$$E_3 = \begin{bmatrix} 0 & 0 \\ 0 & 1 \end{bmatrix} \begin{bmatrix} H\cos^2\theta \\ H\cos\theta\sin\theta \end{bmatrix} = \begin{bmatrix} 0 \\ H\cos\theta\sin\theta \end{bmatrix}$$

To get the intensity we multiply E_3 by the transpose of its complex conjugate, and obtain

$$\mathbf{E}_3 E_3 = \begin{bmatrix} 0 & H\cos\theta\sin\theta \end{bmatrix} \begin{bmatrix} 0 \\ H\cos\theta\sin\theta \end{bmatrix}$$

$$= H^2\cos^2\theta\sin^2\theta$$

$$= \frac{H^2\sin^2 2\theta}{4}$$

There are thus *four* orientations of the intermediate polaroid which give extinction.

We will now work the same problem using the Mueller method. We start again with a beam plane-polarized parallel to the X-axis for which the Stokes column is

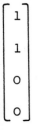

We find the Stokes column of the beam emerging through the first polaroid by multiplying this original Stokes column by the Mueller matrix of the polaroid with its pass-plane inclined at angle θ with the X-direction. The result is

$$\frac{1}{2}\begin{bmatrix} 1 & C_2 & S_2 & 0 \\ C_2 & C_2^2 & C_2S_2 & 0 \\ S_2 & S_2C_2 & S_2^2 & 0 \\ 0 & 0 & 0 & 0 \end{bmatrix} \begin{bmatrix} 1 \\ 1 \\ 0 \\ 0 \end{bmatrix} = \frac{1}{2}\begin{bmatrix} 1 + C_2 \\ C_2 + C_2^2 \\ S_2 + S_2C_2 \\ 0 \end{bmatrix}$$

The intensity of the emerging beam is thus

$$(1 + \cos 2\theta)/2 = 2\cos^2\theta/2 = \cos^2\theta$$

We now pass the beam through a polaroid with its pass-plane vertical so that its θ is $90°$. The resultant Stokes column is obtained by multiplying the Stokes column falling on it by its Mueller matrix obtaining

$$\frac{1}{2}\begin{bmatrix} 1 & -1 & 0 & 0 \\ -1 & 1 & 0 & 0 \\ 0 & 0 & 0 & 0 \\ 0 & 0 & 0 & 0 \end{bmatrix} \frac{1}{2}\begin{bmatrix} 1 + C_2 \\ C_2 + C_2^2 \\ S_2 + S_2 C_2 \\ 0 \end{bmatrix} = \frac{1}{4}\begin{bmatrix} 1 - C_2^2 \\ C_2^2 - 1 \\ 0 \\ 0 \end{bmatrix}$$

Therefore, the intensity of final beam $= (1 - C_2^2)/4$
$= S_2^2/4 = \dfrac{\sin^2 2\theta}{4}.$

Problem 2

Three polaroids are arranged in a row and a beam of light is passed through them. Find the ratio of the intensities of the transmitted and the incident light if the pass-plane of the first polaroid is vertical, that of the second makes an angle of $12°$ to the right of the vertical, as seen by an observer looking towards the light source, and that of the third makes an angle of $12°$ to the left of the vertical.

Solution

We shall do this problem using Stokes parameters and Mueller matrices, because the Jones calculus cannot cope with problems involving unpolarized beams of light We begin with an unpolarized beam of light for which the Stokes column is

$$\begin{bmatrix} 1 \\ 0 \\ 0 \\ 0 \end{bmatrix}$$

We first pass this through the polaroid with its pass-plane vertical, that is with $\theta = 90^\circ$, obtaining the Stokes column,

$$\frac{1}{2}\begin{bmatrix} 1 & -1 & 0 & 0 \\ -1 & 1 & 0 & 0 \\ 0 & 0 & 0 & 0 \\ 0 & 0 & 0 & 0 \end{bmatrix}\begin{bmatrix} 1 \\ 0 \\ 0 \\ 0 \end{bmatrix} = \frac{1}{2}\begin{bmatrix} 1 \\ -1 \\ 0 \\ 0 \end{bmatrix}$$

(Since this is now a vertically polarized beam, we could at this stage calculate the rest of the problem using the Jones calculus, allowing for the fact that 50 per cent of the original intensity has already been absorbed at the first polarizer.)

The beam now falls on the second polarizer for which the pass-plane is 12° to the right of the vertical so that $\theta = 90^\circ - 12^\circ = 78^\circ$. This gives $\cos 2\theta = -0.91$ and $\sin 2\theta = 0.41$. We obtain the Stokes column of the beam emerging from the second polaroid by multiplying the Mueller matrix of the second polaroid into the Stokes column of the beam emerging from the first polaroid. The result is

$$\frac{1}{2}\begin{bmatrix} 1 & -0.91 & ? & ? \\ -0.91 & 0.83 & ? & ? \\ 0.41 & -0.37 & ? & ? \\ 0 & 0 & 0 & 0 \end{bmatrix}\frac{1}{2}\begin{bmatrix} 1 \\ -1 \\ 0 \\ 0 \end{bmatrix} = \frac{1}{4}\begin{bmatrix} 1.91 \\ -1.74 \\ 0.77 \\ 0 \end{bmatrix}$$

It will be noted that in the Mueller matrix of the polaroid we have inserted question marks in place of some of the matrix elements. This is because in the Stokes column of the beam entering the polaroid only the first two elements are non-zero, so that the elements in the two right-hand columns of the Mueller matrix will be multiplying into zero elements from the Stokes column and producing zeros. Since the result will be uninfluenced by the values of these elements, there is no point in calculating them.

The beam now passes into the final polaroid which has its pass-plane at 12° to the left of the vertical so that $\theta = 102°$, $\cos 2\theta = -0.91$ and $\sin 2\theta = -0.41$. The Stokes column of the final emergent beam is thus given by

$$\frac{1}{2}\begin{bmatrix} 1 & -0.91 & -0.41 & ? \\ ? & ? & ? & ? \\ ? & ? & ? & ? \\ ? & ? & ? & ? \end{bmatrix} \frac{1}{4}\begin{bmatrix} 1.91 \\ -1.74 \\ 0.78 \\ 0 \end{bmatrix} = \frac{1}{8}\begin{bmatrix} 3.17 \\ ? \\ ? \\ ? \end{bmatrix}$$

Thus, the intensity of the emerging beam is $3.17/8 = 0.396$, if the intensity of the original beam is taken as unity. Here again, it will be seen that question marks have been inserted in place of all the matrix elements in the last three rows of this Mueller matrix. This is because only the intensity of the beam emerging from this polaroid is now needed, so there is no point in calculating any elements of the Stokes column except the first.

Problem 3

Consider the right-handed elliptically polarized light described by

$$X = H\cos\omega t \quad \text{and} \quad Y = K\cos(\omega t + \Delta)$$

Find the angle which the axes of this ellipse make with the X-axis and the ratio of the lengths of the axes.

Solution

If we regard the intensity I of this beam as unity, then in the usual notation $H = \cos\theta$ and $K = \sin\theta$, so that the Stokes parameters of the beam are

$$I = 1, \quad Q = \cos 2\theta, \quad U = \sin 2\theta \cos\Delta, \quad V = \sin 2\theta \sin\Delta$$

Suppose we now put the beam through a polaroid with its pass-plane at an angle α with the X-axis. The Stokes column of the emerging beam will be

$$
1/2
\begin{bmatrix}
1 & \cos2\alpha & \sin2\alpha & 0 \\
? & ? & ? & ? \\
? & ? & ? & ? \\
? & ? & ? & ?
\end{bmatrix}
\begin{bmatrix}
1 \\
\cos2\theta \\
\sin2\theta\cos\Delta \\
\sin2\theta\sin\Delta
\end{bmatrix}
$$

$$
= 1/2
\begin{bmatrix}
1 + \cos2\theta\cos2\alpha + \sin2\theta\cos\Delta\sin2\alpha \\
? \\
? \\
?
\end{bmatrix}
$$

Thus if we set $2\alpha = \beta$, the intensity of the emerging beam is proportional to

$$E = 1 + \cos2\theta\cos\beta + \sin2\theta\cos\Delta\sin\beta$$

As the polaroid is rotated, β is altered and so is the intensity. When the pass-plane of the polaroid lies along one of the axes of the ellipse the intensity will be either a maximum or a minimum, so we can find the positions of the axes of the ellipse by differentiating the intensity with respect to β and equating the result to zero. The result of differentiating is

$$dE/d\beta = -\cos2\theta\sin\beta + \sin2\theta\cos\Delta\cos\beta$$

Maxima or minima of intensity occur whenever $dE/d\beta$ is zero, that is when

$$\tan2\alpha = \tan\beta = \frac{\sin\beta}{\cos\beta} = \frac{\sin2\theta\cos\Delta}{\cos2\theta} = \tan2\theta\cos\Delta$$

If we take two solutions of this equation for 2α, $180°$ apart, the corresponding solutions for α will be $90°$ apart, and they give the orientation of the two axes of the ellipse. If we denote the two solutions by α and α^1, we have

$$\alpha^1 = \alpha + 90°$$

so that

$$\beta^1 = 2\alpha^1 = \beta + 180^\circ$$

We therefore have

$$\sin\beta^1 = -\sin\beta \quad \text{and} \quad \cos\beta^1 = -\cos\beta$$

The intensity corresponding to the second (orthogonal) setting of the polaroid is thus

$$E^1 = 1 - \cos2\theta\cos\beta - \sin2\theta\cos\Delta\sin\beta$$

Thus the ratio of the intensities, which will be the ratio of the squares of the axes of the ellipse, is given by

$$\frac{E^1}{E} = \frac{1 - (\cos2\theta\cos\beta + \sin2\theta\cos\Delta\sin\beta)}{1 + (\cos2\theta\cos\beta + \sin2\theta\cos\Delta\sin\beta)}$$

Using the known value of $\tan\beta$, we can obtain the values of $\cos\beta$ and $\sin\beta$. After substituting them into this equation and simplifying we obtain finally

$$\frac{E^1}{E} = \frac{1 - \sqrt{(1 - \sin^2 2\theta\sin^2\Delta)}}{1 + \sqrt{(1 - \sin^2 2\theta\sin^2\Delta)}}$$

The student is recommended as an exercise to solve this problem also by the Jones method. Incidentally, we have proved a result in the coordinate geometry of the ellipse which is quite difficult to obtain by geometrical methods.

Problem 4

Elliptically polarized light is passed through a quarter-wave plate and then through a polaroid. Extinction occurs when the fast axis of the plate and the pass-plane of the polaroid are inclined to the horizontal at angles of 30° and 60° respectively, the angles being measured in the same direction. Find the orientation of the ellipse and the ratio of its axes.

Solution

If we substitute the given angles into the standard formulae for the Jones matrix of the polaroid and of the quarter-wave plate, we get as the matrix of the polaroid followed by the quarter-wave plate

$$\frac{1}{4}\begin{bmatrix} 1 & \sqrt{3} \\ \sqrt{3} & 3 \end{bmatrix} \frac{1}{4}\begin{bmatrix} 3 - i & \sqrt{3}(1 + i) \\ \sqrt{3}(1 + i) & 1 - 3i \end{bmatrix}$$

(polaroid)　　(quarter-wave plate)

$$= \frac{1}{8}\begin{bmatrix} 3 + i & \sqrt{3}(1 - i) \\ \sqrt{3}(3 + i) & 3 - 3i \end{bmatrix}$$

Suppose the Maxwell column of the entering beam is

$$\begin{bmatrix} H \\ K\exp(i\Delta) \end{bmatrix}$$

If we now assign unit intensity to this beam, then in the usual notation

$$H = \cos\theta \quad \text{and} \quad K = \sin\theta$$

so that the Maxwell column is

$$\begin{bmatrix} \cos\theta \\ \sin\theta\exp(i\Delta) \end{bmatrix}$$

Since extinction is produced, the Maxwell column of the beam emerging from the combination must have both its elements zero so that

$$\begin{bmatrix} 3 + i & \sqrt{3}(1 - i) \\ \sqrt{3}(3 + i) & 3 - 3i \end{bmatrix} \begin{bmatrix} \cos\theta \\ \sin\theta\exp(i\Delta) \end{bmatrix} = \begin{bmatrix} 0 \\ 0 \end{bmatrix}$$

Now suppose we have a 2 × 2 square matrix multiplying a column matrix to produce a zero column matrix thus:

$$\begin{bmatrix} A & B \\ C & D \end{bmatrix} \begin{bmatrix} X \\ Y \end{bmatrix} = \begin{bmatrix} 0 \\ 0 \end{bmatrix}$$

That is $AX + BY = 0$　and　$CX + DY = 0$

Given that neither X nor Y vanishes for an elliptically polarized beam, we know at once from these equations that

$$\frac{X}{Y} = -\frac{B}{A} = -\frac{D}{C}$$

In our problem

$$B = \sqrt{3}(1 - i) \quad \text{and} \quad A = 3 + i$$

so that

$$\frac{B}{A} = \frac{\sqrt{3}(1 - i)}{3 + i} = \frac{\sqrt{3}(1 - i)(1 + i)}{(3 + i)(1 + i)}$$

$$= \frac{\sqrt{3} \times 2}{2 + 4i} = \frac{\sqrt{3}}{1 + 2i}$$

(The student should verify here that D/C has the same value.) Therefore

$$\frac{\cos\theta}{\sin\theta\exp(i\Delta)} = \frac{X}{Y} = -\frac{B}{A}$$

$$= \frac{\sqrt{3}}{-2i - 1}$$

Thus, inverting,

$$\frac{\sin\theta\cos\Delta + i\sin\theta\sin\Delta}{\cos\theta} = \frac{-2i - 1}{\sqrt{3}}$$

Equating real parts,

$$\tan\theta\cos\Delta = -1/\sqrt{3}$$

Equating imaginary parts,

$$\tan\theta\sin\Delta = -2/\sqrt{3}$$

Dividing the second equation by the first and cancelling $\tan\theta$, we get

$$\tan\Delta = \sin\Delta/\cos\Delta = 2$$

so that

$$\Delta = 63^\circ \, 26', \quad \cos\Delta = 0 \cdot 447$$

The equation between the real parts above now gives

$$\tan\theta = \frac{-1}{1 \cdot 732 \times 0 \cdot 446} = -1 \cdot 292$$

so that

$$\theta = -52^\circ \, 16'$$

The formulae developed in the last problem now give

$$\tan 2\alpha = \tan 2\theta \cos\Delta$$

and

$$(\text{Ratio of axes})^2 = \frac{1 + \sqrt{(1 - \sin^2 2\theta \sin^2 \Delta)}}{1 - \sqrt{(1 - \sin^2 2\theta \sin^2 \Delta)}}$$

giving α' (the angle between the minor axis of the ellipse and the x-axis) as 30° and the ratio of the lengths of the axes as $\sqrt{3} : 1$.

Problem 5

A beam of right-handed circularly polarized light is passed normally through (a) a quarter-wave plate and (b) an eighth-wave plate. Both plates are to be regarded as having their fast axes vertical. Describe the state of polarization of the light as it emerges from each plate.

Solution

The circularly polarized beam has its components H and K parallel to the two axes equal. For a right-handed state, the phase difference Δ between them is 90° so that the normalized Maxwell column is

$$\frac{1}{2}\begin{bmatrix} 1 \\ 1 \times (\cos \pi/2 + i \sin \pi/2) \end{bmatrix} = \frac{1}{2}\begin{bmatrix} 1 \\ i \end{bmatrix}$$

The Jones matrix of the quarter-wave plate with its fast axis vertical can be written as

$$\begin{bmatrix} 1 & 0 \\ 0 & i \end{bmatrix}$$

so that the Maxwell column of the beam emerging from the quarter-wave plate is

$$\frac{1}{2}\begin{bmatrix} 1 & 0 \\ 0 & i \end{bmatrix}\begin{bmatrix} 1 \\ i \end{bmatrix} = \frac{1}{2}\begin{bmatrix} 1 \\ -1 \end{bmatrix}$$

There is no imaginary component in the second element of the Maxwell column of the emerging beam so that for this beam $\sin\Delta$ must be zero, that is $\Delta = 0$ or π. Since the real component is negative, Δ must equal π so that $K = 1$ and $H = 1$. The beam is thus linearly polarized, the vibration making an angle of -45° with the X-axis and lying in the second and fourth quadrants.

For the eighth-wave plate the phase shift produced between the ordinary and extraordinary waves is 45°, so that $\cos\delta$ and $\sin\delta$ both equal $1/\sqrt{2}$. In this case $\theta = 90^\circ$, and the Jones matrix can be written in the form

$$\begin{bmatrix} 1 & 0 \\ 0 & \frac{1}{\sqrt{2}}(1 + i) \end{bmatrix}$$

so that the emerging Maxwell column is

$$\frac{1}{2}\begin{bmatrix} 1 & 0 \\ 0 & \frac{1}{\sqrt{2}}(1 + i) \end{bmatrix}\begin{bmatrix} 1 \\ i \end{bmatrix} = \frac{1}{2}\begin{bmatrix} 1 \\ \frac{1}{\sqrt{2}}(i - 1) \end{bmatrix}$$

From the ratios of the real and imaginary parts of the second component we see that $\tan\Delta = -1$ so that Δ is 135°, $\sin\Delta = 1/\sqrt{2}$ and $\cos\Delta = -1/\sqrt{2}$. Inserting these values in the matrix, we see that both H and $K = 1$, so that using the standard form for the equation of the ellipse

$$\frac{x^2}{H^2} - \frac{2xy}{HK}\cos\Delta + \frac{y^2}{K^2} = \sin^2\Delta$$

we obtain

$$\frac{x^2}{1^2} - \frac{2xy}{1 \times 1 \times \sqrt{2}} + \frac{y^2}{1^2} = 1$$

that is

$$x^2 - \sqrt{2}(xy) + y^2 = 1$$

Because Δ lies between 0° and 180°, the elliptical state is right-handed.

Problem 6
  A wave described by

$$x = A\cos(\omega t + \pi/4) \quad \text{and} \quad y = A\cos\omega t$$

is incident on a polaroid which is rotated in its own plane until the transmitted intensity is a maximum.

(a) In what direction does the pass-plane of the
polaroid now lie?
(b) Compute the ratio of the transmitted intensities
observed with the polaroid so oriented and
oriented with its pass-plane in the Y-direction.

Solution
We can find the intensity for any orientation of the
polaroid using either the Mueller or the Jones matrices.
In the Mueller formulation we have

$$H = A, \quad K = A, \quad \Delta = - \pi/4$$

Therefore

$$I = 2A^2, \quad Q = 0, \quad U = 2A^2 \cos\pi/4, \quad v = - 2A^2 \sin\pi/4$$

We now pass it through a polaroid at angle θ and find
the resulting intensity. It will be noticed that we
only need to use the first row of the Mueller matrix
of the polaroid. The resulting intensity is

$$I_1 = 1/2 \begin{bmatrix} 1 & \cos2\theta & \sin2\theta & 0 \end{bmatrix} \begin{bmatrix} 2A^2 \\ 0 \\ \sqrt{2}A^2 \\ -\sqrt{2}A^2 \end{bmatrix}$$

$$= A^2 (1 + \sin2\theta/\sqrt{2})$$

This has clearly a maximum value when $\sin2\theta = 1$, that
is when $\theta = 45^\circ$. Its value is then

$$A^2 (1 + 1/\sqrt{2})$$

When the pass-plane is in the Y-direction $\theta = 90^\circ$, so
that $\sin2\theta$ is zero and the intensity is

$$A^2$$

Thus the ratio of the intensities is

1·707 : 1

If we wish to use the Jones method to find the intensity we first set up the Maxwell column of the original
beam. This is

$$\begin{bmatrix} A \\ \overset{\diamond}{A}\exp(-i\pi/4) \end{bmatrix} = A\begin{bmatrix} 1 \\ \exp(-i\pi/4) \end{bmatrix}$$

After passing through the polarizer at an angle θ, the Maxwell column becomes

$$A\begin{bmatrix} \cos^2\theta & \sin\theta\cos\theta \\ \cos\theta\sin\theta & \sin^2\theta \end{bmatrix}\begin{bmatrix} 1 \\ \exp(-i\pi/4) \end{bmatrix}$$

$$= \begin{bmatrix} \cos^2\theta + \sin\theta\cos\theta\exp(-i\pi/4) \\ \sin\theta\cos\theta + \sin^2\theta\exp(-i\pi/4) \end{bmatrix}$$

As usual we find the intensity by multiplying this Maxwell column by the transpose of its complex conjugate, obtaining

Intensity =

$$A^2\begin{bmatrix} \cos^2\theta + \sin\theta\cos\theta\exp(+i\pi/4) & \sin\theta\cos\theta + \sin^2\theta\exp(+i\pi/4) \end{bmatrix}$$

$$\times \begin{bmatrix} \cos^2\theta + \sin\theta\cos\theta\exp(-i\pi/4) \\ \sin\theta\cos\theta + \sin^2\theta\exp(-i\pi/4) \end{bmatrix}$$

which multiplies out to the same result as before, namely

$$A^2(1 + \sin2\theta/\sqrt{2})$$

Problem 7

A beam of right-handed elliptically polarized light is represented by an ellipse with major axis H and minor axis K. The beam of light falls on a polaroid with its pass-plane at an angle α with the major axis of the ellipse

Prove that the intensity of the beam passing through the polaroid is

$$I = H^2\cos^2\alpha + K^2\sin^2\alpha$$

Solution

If we take the major axis of the ellipse to be horizontal, the Maxwell column of the incident beam is

$$\begin{bmatrix} H \\ K\exp(i\pi/2) \end{bmatrix} = \begin{bmatrix} H \\ iK \end{bmatrix}$$

(See type (4) light in appendix C.) The Maxwell column of the beam after passing through the polaroid is obtained by multiplying this by the Jones matrix of the polaroid. The result is

$$\begin{bmatrix} \cos^2\alpha & \sin\alpha\cos\alpha \\ \sin\alpha\cos\alpha & \sin^2\alpha \end{bmatrix} \begin{bmatrix} H \\ iK \end{bmatrix} = \begin{bmatrix} H\cos^2\alpha + iK\sin\alpha\cos\alpha \\ H\sin\alpha\cos\alpha + iK\sin^2\alpha \end{bmatrix}$$

We obtain the intensity by multiplying this Maxwell column by the transpose of its complex conjugate, obtaining

$$\begin{bmatrix} H\cos^2\alpha - iK\sin\alpha\cos\alpha & H\sin\alpha\cos\alpha - iK\sin^2\alpha \end{bmatrix}$$

$$\times \begin{bmatrix} H\cos^2\alpha + iK\sin\alpha\cos\alpha \\ H\sin\alpha\cos\alpha + iK\sin^2\alpha \end{bmatrix}$$

which multiplies out and simplifies to the required result.

It is interesting to notice that this is a result which is easily tested experimentally using a photo-cell to measure the intensities; we thus obtain a direct confirmation of the theories of elliptically polarized light. The results can also be proved, though nothing like so easily, from the coordinate geometry of the ellipse. (For a left-handed beam, the calculation is very similar, except that wherever i appears it is changed in sign. Once the bracket product has been formed, however, these imaginary terms are eliminated.)

Problem 8
Develop the theory of photoelasticity using Maxwell columns and the Jones matrices.

Solution
It will be convenient to consider polaroids making angles of 45° and of - 45° (in the conventional sense) with the X-axis. We suppose that we work with an

original beam in which $H = K =$ unity (so that the intensity is 2 units), and of plane-polarized light, so that the Maxwell column for the original beam is

$$\begin{bmatrix} 1 \\ 1 \end{bmatrix}$$

For the first polaroid $\theta = 45^{\circ}$, so that $\cos\theta$ and $\sin\theta$ are both $1/\sqrt{2}$, and the Jones matrix is

$$\begin{bmatrix} 1/2 & 1/2 \\ 1/2 & 1/2 \end{bmatrix} = 1/2 \begin{bmatrix} 1 & 1 \\ 1 & 1 \end{bmatrix}$$

For the second polaroid $\theta = -45^{\circ}$, so that $\cos\theta$ is $1/\sqrt{2}$ but $\sin\theta = -1/\sqrt{2}$ and the Jones matrix is

$$\begin{bmatrix} 1/2 & -1/2 \\ -1/2 & 1/2 \end{bmatrix} = 1/2 \begin{bmatrix} 1 & -1 \\ -1 & 1 \end{bmatrix}$$

(This matrix is among those listed in Table 4.) Using the expression for the general linear retarder to represent the effect due to the specimen at any point, we see that the Maxwell column of the emerging beam is

$$E = \frac{1}{2} \underbrace{\begin{bmatrix} 1 & -1 \\ -1 & 1 \end{bmatrix}}_{\substack{\text{(second} \\ \text{polaroid)}}} \underbrace{\begin{bmatrix} \cos^2\alpha + \sin^2\alpha\exp(-i\delta) & \cos\alpha\sin\alpha(1 - \exp(-i\delta) \\ \cos\alpha\sin\alpha(1 - \exp(-i\delta)) & \sin^2\alpha + \cos^2\alpha\exp(-i\delta) \end{bmatrix}}_{\text{(phase plate)}}$$

$$\times \frac{1}{2} \underbrace{\begin{bmatrix} 1 & 1 \\ 1 & 1 \end{bmatrix}}_{\substack{\text{(first} \\ \text{polaroid)}}} \underbrace{\begin{bmatrix} 1 \\ 1 \end{bmatrix}}_{\substack{\text{(original} \\ \text{beam)}}}$$

This multiplies out eventually to

$$(1/2)\cos 2\alpha (1 - \exp(-i\delta)) \begin{bmatrix} 1 \\ -1 \end{bmatrix} = G \begin{bmatrix} 1 \\ -1 \end{bmatrix}, \text{ say}$$

To get the intensity of the emerging beam, as already explained, we multiply the Maxwell column by the transpose of its complex conjugate, obtaining

$$\text{Intensity} = \text{E}\bar{E} = G\bar{G}\begin{bmatrix} 1 & -1 \end{bmatrix}\begin{bmatrix} 1 \\ -1 \end{bmatrix} = 2G\bar{G}$$

which works out to

Intensity $= 2\cos^2 2\alpha \sin^2 \delta/2 = I$, say

Since the original intensity I_0 was 2 units, this leads to

$$I = I_0 \cos^2 2\alpha \sin^2 (\delta/2)$$

It will be noticed that this is different from the formula which we found using the Mueller matrices; $\cos^2 2\alpha$ appears instead of $\sin^2 2\alpha$. This is because we have chosen to have the pass-planes of the polaroids not horizontal and vertical, but at angles of 45° and −45° with the axes. Vanishing of $\cos^2 2\alpha$ now implies that the optic axis of the phase plate is parallel to one or other of the pass-planes of the polaroids.

We will now see the effect of putting in the quarter-wave plates. Since these are to be perpendicular to one another but at 45° with the pass-planes of the polaroids, one of them must be horizontal and the other vertical. If we take the first quarter-wave plate to have its fast axis vertical, then $\theta = 90°$, $\cos\theta = 0$, $\sin\theta = 1$ and $\delta = \pi/2$, so that $\exp(i\delta) = i$. Using Table 4 for these values, we find that the matrix for the quarter-wave plate becomes

$$\begin{bmatrix} 1 & 0 \\ 0 & i \end{bmatrix} \quad \text{or, alternatively,} \quad \begin{bmatrix} -i & 0 \\ 0 & 1 \end{bmatrix}$$

For the second quarter-wave plate $\theta = 0°$, so that $\cos\theta$ and $\sin\theta$ interchange their values and the matrix is

$$\begin{bmatrix} i & 0 \\ 0 & 1 \end{bmatrix} \quad \text{or, alternatively,} \quad \begin{bmatrix} 1 & 0 \\ 0 & -i \end{bmatrix}$$

Using the same initial beam as before, we find that the Maxwell column of the emerging beam becomes

$$E = \frac{1}{2}\begin{bmatrix} 1 & -1 \\ -1 & 1 \end{bmatrix}\begin{bmatrix} i & 0 \\ 0 & 1 \end{bmatrix}\begin{bmatrix} C_1^2+S_1^2e^{-i\delta} & C_1S_1(1-e^{-i\delta}) \\ C_1S_1(1-e^{-i\delta}) & S_1^2+C_1^2e^{-i\delta} \end{bmatrix}$$

(second   (second   (retardation δ at angle θ)
polaroid) $\lambda/4$)

$$\times \begin{bmatrix} 1 & 0 \\ 0 & i \end{bmatrix} \tfrac{1}{2}\begin{bmatrix} 1 & 1 \\ 1 & 1 \end{bmatrix} \begin{bmatrix} 1 \\ 1 \end{bmatrix}$$

(first   (first   (input polarized
$\lambda/4$)   polaroid)   at 45°)

In the above matrix for the retarder, the symbols C_1 and S_1 are used to denote $\cos\theta$ and $\sin\theta$ respectively. In the next stage of the calculation, we simplify the two ends of the matrix chain and substitute $e^{-i\delta} = \beta - i\mu$, where $\beta = \cos\delta$ and $\mu = \sin\delta$. The result is

$$E = \frac{1}{2}\begin{bmatrix} i & -1 \\ -i & 1 \end{bmatrix}\begin{bmatrix} C_1^2+S_1^2\beta-iS_1^2\mu & C_1S_1(1-\beta)+iC_1S_1\mu \\ C_1S_1(1-\beta)+iC_1S_1\mu & S_1^2+C_1^2\beta-iC_1^2\mu \end{bmatrix}$$

$$= \frac{1}{2}\begin{bmatrix} i & -1 \\ -i & 1 \end{bmatrix}\begin{bmatrix} (C_1^2+S_1^2\beta-C_1S_1\mu) + i(C_1S_1-C_1S_1\beta-S_1^2\mu) \\ (C_1S_1-C_1S_1\beta+C_1^2\mu) + i(S_1^2+C_1^2\beta+C_1S_1\mu) \end{bmatrix}$$

Multiplying out the last product, we obtain for our output the Maxwell column

$$E = \frac{1}{2}\begin{bmatrix} (-S_2(1-\beta)-C_2\mu) + i(C_2(1-\beta)-S_2\mu) \\ (S_2(1-\beta)+C_2\mu) + i(-C_2(1-\beta)+S_2\mu) \end{bmatrix}$$

(where C_2 denotes $\cos2\theta$ and S_2 denotes $\sin2\theta$).

It will be noticed that, apart from a change in sig the E_x and E_y components of this column are the same an inevitable consequence of the fact that the beam i question has just emerged from a polarizer set at – On multiplying either element by its complex conjugat we find eventually that the intensity is given by

$$I = \tfrac{1}{2}(S_2(1-\beta)+C_2\mu)^2 + \tfrac{1}{2}(S_2\mu-C_2(1-\beta))^2 = \tfrac{1}{2}(1-\beta)^2 + \tfrac{1}{2}\mu^2$$

$$= (1-\beta)$$

his result for the intensity can be re-expressed as
$$I = 2\sin^2(\delta/2)$$

independent of the value of θ).
Since the original intensity I was 2 units, this leads
)

$I_{output} = I_{input} \times \sin^2(\delta/2)$, the same result as was

otained by the Mueller calculus.

In our opinion, unless coherent combination of more han one beam is involved, it is often preferable to roceed using wholly real numbers according to the heller method.

One of the authors, A. Gerrard, has prepared a set programs for dealing with many of the polarized ght calculations, both in ALGOL and FORTRAN. Readers no would like copies of these programs are invited write to him at the University of Bath when he 11 supply copies of the programs and be willing to scuss supplying cards or magnetic tape versions of e programs for use in the computer.

e authors gratefully acknowledge permission to use e following copyright material:

o Messrs. Longmans, Green and Co. and to the Oxford elegacy of Local Examinations, for problem 4 on age 546-7 of *Geometrical and Physical Optics*, (1957) y R.S. Longhurst.

o Messrs. Addison-Wesley Publishing Co. Ltd. for roblem 6.4 on page 307, and problem 6.7 on page 308 f *Optics* (1957) by Bruno Rossi.

V

Propagation of Light through Crystals

V.1 INTRODUCTORY DISCUSSION

When a beam of light passes through certain types of crystals it is well known that the phenomenon of double refraction takes place, two beams differently polarized being produced from the initial beam. This phenomenon can be described by using Fresnel's extension of Huygen's construction, that is to say by using two Huygens wavefronts, a spheroid as well as a sphere, the spheroid representing the extraordinary wave and the sphere the ordinary wave. The existence of the sphere and the spheroid can be explained in terms of the electromagnetic theory of light by assuming that the dielectric constant or permittivity of the crystal is not a scalar but a tensor, whose components can be represented by a matrix.

In order to see how this arises, without using tensor analysis, we shall first express the ordinary operations of vector analysis in matrix form, then obtain the matrix form of Maxwell's equations, assuming that the medium through which the waves are passing is a crystal, and finally solve Maxwell's equations, showing that they lead to the known behaviour of light in crystals. In this treatment we shall restrict ourselves to the case of a uniaxial crystal.

V.2 EXPRESSION OF VECTOR OPERATIONS IN MATRIX FORM

V.2.1

A vector $\mathbf{V}$ of components V_1, V_2 and V_3 parallel to the x-, y- and z-axes is written

$$\mathbf{V} = \mathbf{i}V_1 + \mathbf{j}V_2 + \mathbf{k}V_3$$

here **i**, **j** and **k** are unit vectors in the direction of
ne positive x-, y- and z-axes respectively.

We now define four matrices, two row matrices and
wo column matrices, as follows:

$$V = \begin{bmatrix} V_1 \\ V_2 \\ V_3 \end{bmatrix} \qquad V^T = \text{its transpose} \begin{bmatrix} V_1 & V_2 & V_3 \end{bmatrix}$$

$$A = \begin{bmatrix} \mathbf{i} \\ \mathbf{j} \\ \mathbf{k} \end{bmatrix} \qquad A^T = \begin{bmatrix} \mathbf{i} & \mathbf{j} & \mathbf{k} \end{bmatrix}$$

he defining equation for the vector **V** can now be
ritten in the form

$$\mathbf{V} = \begin{bmatrix} \mathbf{i} & \mathbf{j} & \mathbf{k} \end{bmatrix} \begin{bmatrix} V_1 \\ V_2 \\ V_3 \end{bmatrix} = A^T V$$

r in the form

$$\mathbf{V} = \begin{bmatrix} V_1 & V_2 & V_3 \end{bmatrix} \begin{bmatrix} \mathbf{i} \\ \mathbf{j} \\ \mathbf{k} \end{bmatrix} = V^T A$$

2.2

The scalar product can be expressed as follows. If
have another vector **U**, say, defined by

$$\mathbf{U} = \mathbf{i}U_1 + \mathbf{j}U_2 + \mathbf{k}U_3$$

$$= A^T U \quad \text{where} \quad U = \begin{bmatrix} U_1 \\ U_2 \\ U_3 \end{bmatrix}$$

$$= U^T A \quad \text{where} \quad U^T = \begin{bmatrix} U_1 & U_2 & U_3 \end{bmatrix}$$

then the scalar product of **U** and **V** is equal to

$$\mathbf{U} \cdot \mathbf{V} = \mathbf{V} \cdot \mathbf{U} = U_1 V_1 + U_2 V_2 + U_3 V_3$$

This we can write in either of the forms

$$\begin{bmatrix} U_1 & U_2 & U_3 \end{bmatrix} \begin{bmatrix} V_1 \\ V_2 \\ V_3 \end{bmatrix} = U^T V$$

or

$$\begin{bmatrix} V_1 & V_2 & V_3 \end{bmatrix} \begin{bmatrix} U_1 \\ U_2 \\ U_3 \end{bmatrix} = V^T U$$

V.2.3
   The vector product of two vectors can be written in the form

$$\mathbf{V} \times \mathbf{U} = i(V_2 U_3 - V_3 U_2) + j(V_3 U_1 - V_1 U_3) + k(V_1 U_2 - $$

$$= V_1(-jU_3 + kU_2) + V_2(iU_3 - kU_1) + V_3(-iU_2 + jU_1)$$

We can write this as

$$\begin{bmatrix} V_1 & V_2 & V_3 \end{bmatrix} \begin{bmatrix} -jU_3 + kU_2 \\ iU_3 - kU_1 \\ -iU_2 + jU_1 \end{bmatrix} = \begin{bmatrix} V_1 & V_2 & V_3 \end{bmatrix} \begin{bmatrix} 0 & +k & -j \\ -k & 0 & +i \\ +j & -i & 0 \end{bmatrix} \begin{bmatrix} \\ \\ \end{bmatrix}$$

$$= V^T \alpha U$$

where

$$\alpha = \begin{bmatrix} 0 & +k & -j \\ -k & 0 & +i \\ +j & -i & 0 \end{bmatrix}$$

(similarly $\mathbf{V} \times \mathbf{U} = U^T \alpha^T V$, where α^T = the transpose of

$$\alpha^T = \begin{bmatrix} 0 & -k & +j \\ +k & 0 & -i \\ -j & +i & 0 \end{bmatrix} = -\alpha$$

At this stage we need to use matrices for which each individual element represents not a scalar quantity but a differential operator, for example the partial derivative $\partial/\partial x$. In using such operators we must remember that,'while they obey the distributive law

$$\frac{\partial (A)}{\partial x} + \frac{\partial (B)}{\partial x} = \frac{\partial (A+B)}{\partial x}$$

the commutative law must not be assumed. Thus, for example, $\dfrac{U \,\partial\, (V)}{\partial x}$ is *not* the same as $\dfrac{\partial\, U(V)}{\partial x}$ – unless it happens that U is independent of x.

We now express the gradient of a scalar ϕ, say, and the divergence and curl of the vector **V** in matrix notation.

V.2.4

The gradient of a scalar ϕ, say, is given by

$$\mathrm{grad}\phi = i\partial\phi/\partial x + j\partial\phi/\partial y + k\partial\phi/\partial z$$

$$= \begin{bmatrix} i & j & k \end{bmatrix} \begin{bmatrix} \partial\phi/\partial x \\ \partial\phi/\partial y \\ \partial\phi/\partial z \end{bmatrix}$$

$$= \begin{bmatrix} i & j & k \end{bmatrix} \begin{bmatrix} \partial/\partial x \\ \partial/\partial y \\ \partial/\partial z \end{bmatrix} \phi = A^T G \phi$$

where

$$G = \begin{bmatrix} \partial/\partial x \\ \partial/\partial y \\ \partial/\partial z \end{bmatrix} \qquad \text{and} \qquad G^T = \begin{bmatrix} \partial/\partial x & \partial/\partial y & \partial/\partial z \end{bmatrix}$$

V.2.5

The divergence of a vector **V** is

$$\mathrm{div}\mathbf{V} = \partial V_1/\partial x + \partial V_2/\partial y + \partial V_3/\partial z$$

$$= \begin{bmatrix} \partial/\partial x & \partial/\partial y & \partial/\partial z \end{bmatrix} \begin{bmatrix} V_1 \\ V_2 \\ V_3 \end{bmatrix} = G^T V$$

V.2.6

The curl of a vector **V** is

$$\mathrm{curl}\mathbf{V} = \mathbf{i}(\partial V_3/\partial y - \partial V_2/\partial z) + \mathbf{j}(\partial V_1/\partial z - \partial V_3/\partial x)$$
$$+ \mathbf{k}(\partial V_2/\partial x - \partial V_1/\partial y)$$

$$= \begin{bmatrix} \mathbf{i} & \mathbf{j} & \mathbf{k} \end{bmatrix} \begin{bmatrix} \partial V_3/\partial y - \partial V_2/\partial x \\ \partial V_1/\partial z - \partial V_3/\partial x \\ \partial V_2/\partial x - \partial V_1/\partial y \end{bmatrix}$$

$$= \begin{bmatrix} \mathbf{i} & \mathbf{j} & \mathbf{k} \end{bmatrix} \begin{bmatrix} 0 & -\partial/\partial z & +\partial/\partial y \\ +\partial/\partial z & 0 & -\partial/\partial x \\ -\partial/\partial y & +\partial/\partial x & 0 \end{bmatrix} \begin{bmatrix} V_1 \\ V_2 \\ V_3 \end{bmatrix}$$

$$= A^T \Delta V$$

where

$$\Delta = \begin{bmatrix} 0 & -\partial/\partial z & +\partial/\partial y \\ +\partial/\partial z & 0 & -\partial/\partial x \\ -\partial/\partial y & +\partial/\partial x & 0 \end{bmatrix}$$

We shall now use the matrix method to prove some well-known identities.

V.2.7

If $\mathrm{div}\,\mathbf{V} = \theta$ (a scalar), then $\theta = G^T V$ (see V.2.5) grad div$\mathbf{V}$ = gradθ = $A^T G \theta$ (see V.2.4)

$\qquad\qquad\qquad$ = $A^T G G^T V$

$$= A^T \begin{bmatrix} \partial/\partial x \\ \partial/\partial y \\ \partial/\partial z \end{bmatrix} \begin{bmatrix} \partial/\partial x & \partial/\partial y & \partial/\partial z \end{bmatrix} V$$

$$= A^T \begin{bmatrix} \partial^2/\partial x^2 & \partial^2/\partial x \partial y & \partial^2/\partial x \partial z \\ \partial^2/\partial y \partial x & \partial^2/\partial y^2 & \partial^2/\partial y \partial z \\ \partial^2/\partial z \partial x & \partial^2/\partial z \partial y & \partial^2/\partial z^2 \end{bmatrix} V$$

$$= A^T M V, \text{ say}$$

(By inspection, M is its own transpose, $M^T = M$.)

V.2.8

Vector $\mathbf{V}$ in matrix form is $A^T V$ (see V.2.1). Its curl $\mathbf{C}$ is $A^T \Delta V$, that is $A^T \Delta$ multiplying the matrix representation of $\mathbf{V}$ with the A^T factor removed from the front of the representation of $\mathbf{V}$. Thus

$\mathbf{T}$ = curl curl$\mathbf{V}$ = curl$\mathbf{C}$ = $A^T \Delta (\Delta V)$

$$= A^T \begin{bmatrix} 0 & -\partial/\partial z & +\partial/\partial y \\ +\partial/\partial z & 0 & -\partial/\partial x \\ -\partial/\partial y & +\partial/\partial x & 0 \end{bmatrix} \begin{bmatrix} 0 & -\partial/\partial z & +\partial/\partial y \\ +\partial/\partial z & 0 & -\partial/\partial x \\ -\partial/\partial y & +\partial/\partial x & 0 \end{bmatrix} V$$

$$= A^T \begin{bmatrix} -\partial^2/\partial z^2 - \partial^2/\partial y^2 & +\partial^2/\partial x \partial y & +\partial^2/\partial x \partial z \\ +\partial^2/\partial x \partial y & -\partial^2/\partial z^2 - \partial^2/\partial x^2 & +\partial^2/\partial z \partial y \\ +\partial^2/\partial x \partial z & +\partial^2/\partial y \partial z & -\partial^2/\partial y^2 - \partial^2/\partial x^2 \end{bmatrix} V$$

$$= A^T (M - L) V$$

where

$$L = \begin{bmatrix} \partial^2/\partial x^2 + \partial^2/\partial y^2 + \partial^2/\partial z^2 & & O & & \\ & O & & \partial^2/\partial x^2 + \partial^2/\partial y^2 + \partial^2/\partial z^2 & \cdots \\ & O & & O & \\ & & & & O \\ & & \cdots & & O \\ & & & & \partial^2/\partial x^2 + \partial^2/\partial y^2 + \partial^2/\partial z^2 \end{bmatrix}$$

V.2.9

curl grad θ (with θ any scalar point function)

$= \mathrm{curl}(A^T G\theta)$ (see V.2.4)

$= A^T \Delta G\theta$ (see V.2.6)

$$= A^T \begin{bmatrix} O & -\partial/\partial z & +\partial/\partial y \\ +\partial/\partial z & O & -\partial/\partial x \\ -\partial/\partial y & +\partial/\partial x & O \end{bmatrix} \begin{bmatrix} \partial/\partial x \\ \partial/\partial y \\ \partial/\partial z \end{bmatrix} \theta$$

$$= A^T \begin{bmatrix} -\partial^2/\partial z\partial y + \partial^2/\partial y\partial z \\ +\partial^2/\partial z\partial x - \partial^2/\partial x\partial z \\ -\partial^2/\partial y\partial x + \partial^2/\partial x\partial y \end{bmatrix} \theta$$

$$= A^T \begin{bmatrix} O \\ O \\ O \end{bmatrix} \theta = O$$

V.2.10

div curl**V** $= \mathrm{div}(A^T \Delta V)$ (see V.2.6)

$= G^T \Delta V$ (see V.2.5)

$$= \begin{bmatrix} \partial/\partial x & \partial/\partial y & \partial/\partial z \end{bmatrix} \begin{bmatrix} 0 & -\partial/\partial z & +\partial/\partial y \\ +\partial/\partial z & 0 & -\partial/\partial x \\ -\partial/\partial y & +\partial/\partial x & 0 \end{bmatrix} V$$

$$= \left[(+\partial^2/\partial y \partial z - \partial^2/\partial z \partial y) \quad (-\partial^2/\partial x \partial z + \partial^2/\partial z \partial x) \dots \right.$$
$$\left. \dots (-\partial^2/\partial x \partial y + \partial^2/\partial y \partial x) \right] V$$

$$= \begin{bmatrix} 0 & 0 & 0 \end{bmatrix} V = 0$$

V.3 DIELECTRIC PROPERTIES OF AN ANISOTROPIC MEDIUM
V.3.1 *Statement of Maxwell's equations in matrix form and derivation of the general wave equation for the electric field*

It is assumed that the reader is already familiar with the physical ideas leading to Maxwell's equations and has already seen these equations stated in vector form. Using the notation developed in the last section, we see that these four equations can be written in matrix form, thus:

$$\mathrm{div}\,D = 0 \quad \text{becomes} \quad G^T D = 0 \quad \text{(see V.2.5)} \tag{V.1}$$

$$\mathrm{div}\,B = 0 \quad \text{becomes} \quad G^T B = 0$$
$$\tag{V.2}$$
$$\mathrm{div}\,H = 0 \quad \text{becomes} \quad G^T H = 0$$

(We assume that we are in a magnetically isotropic region, so that the vanishing of divB implies the vanishing of divH.)

$$\mathrm{curl}\,E = -\partial B/\partial t \quad \text{becomes} \quad A^T \Delta E = -\partial B/\partial t \quad \text{(see V.2.6)} \tag{V.3}$$

$$\mathrm{curl}\,H = \partial D/\partial t \quad \text{becomes} \quad A^T \Delta H = \partial D/\partial t \tag{V.4}$$

We shall now derive the general wave equation for the electric field. We assume a sinusoidal variation of all the field vectors with time - that, for instance, the electric field is given by

$$\mathscr{E} = E\sin(\omega t + \theta)$$

with ω the angular frequency and θ a phase angle. Then

$$d\mathscr{E}/dt = \omega\mathbf{E}\cos(\omega t + \theta)$$

and

$$d^2\mathscr{E}/dt^2 = -\omega^2\mathbf{E}\sin(\omega t + \theta)$$
$$= -\omega^2\mathscr{E}$$

Thus a double differentiation with respect to time is equivalent to a multiplication by $-\omega^2$.

V.3.2 *Derivation of the wave equations in matrix form*
From equation (V.3)

$$A^T\Delta E = -\partial\mathbf{B}/\partial t = -A^T\partial B/\partial t, \text{ where } B = \begin{bmatrix} B_1 \\ B_2 \\ B_3 \end{bmatrix}$$

Taking curl of both sides, and noting that the order of differentiation is immaterial,

$$A^T\Delta(\Delta E) = -A^T\Delta\partial B/\partial t = -\mu A^T\Delta\partial H/\partial t \quad \text{(see V.2.6)}$$

$$= -\mu\partial/\partial t(A^T\Delta H)$$

$$= -\mu\partial/\partial t(\partial\mathbf{D}/\partial t) \quad \text{(using equation V.4)}$$

$$= -\mu\partial^2\mathbf{D}/\partial t^2 = \mu\omega^2\mathbf{D}$$

Therefore,

$$A^T(M - L)E = \mu\omega^2 A^T D \quad \text{(see V.2.8)}$$

So far the treatment has been quite general and will apply in any type of medium. We shall now assume that we are considering waves passing through an anisotropic medium. The anisotropy means that the permittivity K is not an ordinary scalar, so that the equation $D = KE$ does not apply. Instead, each component of D depends on all three components of E, so that the equation $D = KE$ becomes

$$D_1 = K_{11}E_1 + K_{12}E_2 + K_{13}E_3$$

$$D_2 = K_{21}E_1 + K_{22}E_2 + K_{23}E_3$$

$$D_3 = K_{31}E_1 + K_{32}E_2 + K_{33}E_3$$

That is

$$\begin{bmatrix} D_1 \\ D_2 \\ D_3 \end{bmatrix} = \begin{bmatrix} K_{11} & K_{12} & K_{13} \\ K_{21} & K_{22} & K_{23} \\ K_{31} & K_{32} & K_{33} \end{bmatrix} \begin{bmatrix} E_1 \\ E_2 \\ E_3 \end{bmatrix}$$

or

$$D = KE$$

where K is now a square matrix. Furthermore, by using conservation of energy considerations it can be shown that the matrix K is symmetric, and that by suitable choice of axes it can be made diagonal; that is to say, all of the non-diagonal elements can be made sero. We shall now assume that this has been done, so that the connection between D and E is now

$$\begin{bmatrix} D_1 \\ D_2 \\ D_3 \end{bmatrix} = \begin{bmatrix} K_1 & 0 & 0 \\ 0 & K_2 & 0 \\ 0 & 0 & K_3 \end{bmatrix} \begin{bmatrix} E_1 \\ E_2 \\ E_3 \end{bmatrix}$$

Note that we now need only one subscript on the elements of the square matrix.

We shall now make a further simplifying assumption that two of the Ks are equal. This, we shall see, corresponds to the crystal being uniaxial. Let us suppose that K_2 and K_3 are equal, both equal to K_2, so that the final form of the equation is

$$\begin{bmatrix} D_1 \\ D_2 \\ D_3 \end{bmatrix} = \begin{bmatrix} K_1 & 0 & 0 \\ 0 & K_2 & 0 \\ 0 & 0 & K_2 \end{bmatrix} \begin{bmatrix} E_1 \\ E_2 \\ E_3 \end{bmatrix}$$

Our wave equation now becomes

$$A^T(M - L)E = \mu\omega^2 A^T KE$$

μ and ω being constants, and it is now convenient to incorporate them in the matrix K. We now define a new matrix S by

$$S = \mu\omega^2 K$$

that is

$$
\begin{bmatrix}
S_1 & 0 & 0 \\
0 & S_2 & 0 \\
0 & 0 & S_3
\end{bmatrix}
=
\begin{bmatrix}
\mu\omega^2 K_1 & 0 & 0 \\
0 & \mu\omega^2 K_2 & 0 \\
0 & 0 & \mu\omega^2 K_3
\end{bmatrix}
$$

We shall also write $M - L = P$, a new matrix, where

$$P = M - L$$

$$
=
\begin{bmatrix}
\left(\dfrac{-\partial^2}{\partial y^2} - \dfrac{\partial^2}{\partial z^2}\right) & \dfrac{\partial^2}{\partial x \partial y} & \dfrac{\partial^2}{\partial z \partial x} \\[2ex]
\dfrac{\partial^2}{\partial x \partial y} & \left(\dfrac{-\partial^2}{\partial x^2} - \dfrac{\partial^2}{\partial z^2}\right) & \dfrac{\partial^2}{\partial y \partial z} \\[2ex]
\dfrac{\partial^2}{\partial x \partial z} & \dfrac{\partial^2}{\partial y \partial z} & \left(\dfrac{-\partial^2}{\partial x^2} - \dfrac{\partial^2}{\partial y^2}\right)
\end{bmatrix}
$$

so that the wave equation becomes

$$A^{T}(S - P)E = 0 \tag{V.6}$$

This is the general form of the wave equation for E. For use later we shall now see how H can be obtained, knowing the value of E. One of the Maxwell equations is

$$\text{curl}\,\mathbf{E} = \frac{-\partial \mathbf{B}}{\partial t} = -\mu\,\frac{\partial \mathbf{H}}{\partial t}$$

$$H = -\left(\frac{1}{\mu}\right)\int \text{curl}\,\mathbf{E}\;dt$$

Since differentiation with respect to time multiplies by ω and advances the phase by $\pi/2$, so integration with respect to time must divide by ω and retard the phase by $\pi/2$. Thus, if we omit the phase factor,

$$H = \left(\frac{1}{\mu\omega}\right)\text{curl}\,\mathbf{E}$$

$$= \left(\frac{1}{\mu\omega} \right) A^{\mathrm{T}} \Delta E \qquad\qquad (V.7)$$

We shall now use this formalism to investigate what plane waves can travel through the crystal.

V.4 PROPAGATION OF PLANE WAVES IN A UNIAXIAL CRYSTAL

The reader will know that, in this crystal, in almost all directions there are two possible plane waves travelling through the crystal at different speeds, but that in the direction of the optic axis there is only one possible wave. Where there are two waves, they are polarized always at right angles to each other. The ordinary wave, which corresponds to the spherical Huygens wavelet, is always polarized so that its vibrations are perpendicular to the principal section. The extraordinary wave, which corresponds to the ellipsoidal Huygens wavelet, is polarized so that its vibrations are in the principal section. We shall now derive these results using the newly developed matrix form of the electromagnetic theory. We consider first two special cases corresponding to waves travelling along and perpendicular to the optic axis of the crystal.

V.4.1 First special case of plane wavefronts travelling parallel to the optic axis of the crystal

We consider wavefronts in which the values of E and H are uniform over planes perpendicular to the x-axis, that is to the axis associated with the special value K_1 of the diagonal element of the permittivity matrix. For these waves, by definition, all the y- and z-derivatives are zero, so that the matrix P becomes

$$\begin{bmatrix} 0 & 0 & 0 \\ 0 & -\partial^2/\partial x^2 & 0 \\ 0 & 0 & -\partial^2/\partial x^2 \end{bmatrix}$$

and the wave equation for E becomes

$$A^{\mathrm{T}} \begin{bmatrix} S_1 & 0 & 0 \\ 0 & S_2 + \partial^2/\partial x^2 & 0 \\ 0 & 0 & S_2 + \partial^2/\partial x^2 \end{bmatrix} E = 0$$

Multiplying out, this gives

$$iS_1E_1 + j(S_2E_2 + \partial^2E_2/\partial x^2) + k(S_2E_3 + \partial^2E_3/\partial x^2) = 0$$

In this equation there is only one term in i, so that E_1 must be zero. That is to say, there is no component of **E** in the x-direction, that is in the direction normal to the wavefront, so that here **E** is purely transverse.

Inserting the value of S_2 in terms of K_2, and remembering that multiplication by $-\omega^2$ is equivalent to double differentiation with respect to time, equating of the y- and z-components on the left-hand side to zero gives

$$\frac{\partial^2E_2}{\partial x^2} = \mu K_2 \frac{\partial^2E_2}{\partial t^2} \quad \text{and} \quad \frac{\partial^2E_3}{\partial x^2} = \mu K_2 \frac{\partial^2E_3}{\partial t^2}$$

Comparing this with the standard wave equation,

$$\frac{\partial^2\phi}{\partial x^2} = \frac{1}{V^2}\frac{\partial^2\phi}{\partial t^2}$$

we see that waves of both y- and x-components of **E** travel along the x-direction with speed $1/\sqrt{(\mu K_2)}$.

Using the general formula which we have established for H and the fact that E and all the y- and z-derivatives are now zero, we find

$$H = (1/\mu\omega)A^T\begin{bmatrix} 0 & 0 & 0 \\ 0 & 0 & -\partial/dx \\ 0 & +\partial/dx & 0 \end{bmatrix}\begin{bmatrix} 0 \\ E_2 \\ E_3 \end{bmatrix}$$

$$= (1/\mu\omega)\left[-j\,\partial E_3/\partial x + k\,\partial E_2/\partial x\right]$$

From this equation for H we see that (a) there is no x-component of H, that is to say, H, like E, is purely transverse;   (b) if E_2 equals 0, H is purely in the y-direction, while if E_3 equals 0, H is purely in the z-direction. This suggests that E and H are mutually perpendicular;   (c) for the waves to be propagated without change of waveform, the waves of $\partial E_2/\partial x$ must travel at the same speed as those of E_2, and similarly for E_3, therefore both the y- and the z-components of H travel at the same speed as those of E, that is at a speed of $1/(\mu K_2)$. Thus, the complete ordinary and extraordinary wavefronts, both the E and the H parts, all travel at the same speed in this direction. This corresponds with the known fact that there is no double refraction for wavefronts travelling along the direction of the optic axis of a uniaxial crystal.

V.4.2 A second special case of plane wavefronts travelling through uniaxial crystals

Let us now consider plane wavefronts perpendicular to the z-axis, so that all the x- and y-derivatives are zero. By symmetry, since S_2 and S_3 are equal, a similar discussion, leading to results in which y and z are simply interchanged, could be given for wavefronts perpendicular to the y-axis. Since div$H$ = 0 and the x- and y-derivatives are assumed zero, this means that $\partial H_3/\partial z$ = 0, that is that H_3 is zero so far as varying fields are concerned; in other words, H is here purely transverse. The matrix P becomes

$$\begin{bmatrix} -\partial^2/\partial z^2 & 0 & 0 \\ 0 & -\partial^2/\partial z^2 & 0 \\ 0 & 0 & 0 \end{bmatrix}$$

and equation (V.6) becomes

$$A^{\mathrm{T}} \begin{bmatrix} S_1+\partial^2/\partial z^2 & 0 & 0 \\ 0 & S_2+\partial^2/\partial z^2 & 0 \\ 0 & 0 & S_2 \end{bmatrix} E = 0$$

that is

$$i(S_1 + \partial^2/\partial z^2)E_1 + j(S_2 + \partial^2/\partial z^2)E_2 + kS_2E_3 = 0$$

Again, there is only one term in $\mathbf{k}$, so that E_3 is zero. There is no z-component of E, so that E, like H, is again purely transverse. Equating the other two terms to zero, inserting the values of S_1 and S_2 and remembering that multiplication by minus ω^2 is equivalent to double differentiation, we get the two equations

$$\partial^2 E_1/\partial z^2 = \mu K_1 \ \partial^2 E_1/\partial t^2 \quad \text{and} \quad \partial^2 E_2/\partial z^2 = \mu K_2 \ \partial^2 E_2/\partial t^2$$

Again comparing with the general equation for wave motion, we see that E_1 and E_2 travel as plane waves along the z-direction, but at different speeds, E_1 at speed $1/\sqrt{(\mu K_1)}$, E_2 at speed $1/\sqrt{(\mu K_2)}$. Thus, E_2 travels at the same speed as did both E_2 and E_3 along the x-direction. We have therefore two waves polarized perpendicular to each other travelling at different speeds along the z-direction. E_2 will prove to be associated with the ordinary wavefront, E_1 with the extraordinary wavefront. We now again use the general equation for H, and remember that all x- and y-derivatives and E_3 are now zero. The equation for H thus becomes

$$H = (1/\mu\omega) \begin{bmatrix} \mathbf{i} & \mathbf{j} & \mathbf{k} \end{bmatrix} \begin{bmatrix} 0 & -\partial/\partial z & 0 \\ +\partial/\partial z & 0 & 0 \\ 0 & 0 & 0 \end{bmatrix} \begin{bmatrix} E_1 \\ E_2 \\ 0 \end{bmatrix}$$

that is

$$H = -\ \mathbf{i}\partial E_2/\partial z + \mathbf{j}\partial E_1/\partial z$$

Again, for no variation in the wave shape, $\partial E_2/\partial z$ must travel at the same speed as E_2 and $\partial E_1/\partial z$ at the same speed as E_1. Thus, we can see that (a) there is no z-component of H, so that it is purely transverse; (b) the x-component of H travels at the same speed as the y-component of E. These constitute the ordinary waves; (c) the y-component of H travels at the same speed as the x-component of E. These constitute the extraordinary wave. Thus, we see that the ordinary and extraordinary waves are polarized at right angles to each other.

V.4.3 A more general case of plane waves travelling through a uniaxial crystal

Suppose E is constant over planes like α in Figure V.1 parallel to the y-axis, so that all y-derivatives are zero. We let these planes make an angle θ with the positive direction of the x-axis, so that the normal makes an angle $(\theta - \pi/2) = \beta$ with that direction. All derivatives of E and H along directions

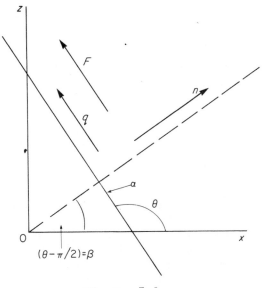

Figure 5.1

lying in the plane α will be zero. That is $\partial E/\partial q = 0$
in the diagram. By the formula for directional deriv-
atives which can be found, for instance, in R. Courant,
Differential and Integral Calculus, Volume 2, page 63,
Blackie and Son, London and Glasgow, 1936,

$$\partial/\partial x = \cos(-\beta)\partial/\partial n + \sin(-\beta)\partial/\partial q = \cos\beta\ \partial/\partial n$$

$$\partial/\partial z = -\cos\theta\ \partial/\partial n + \sin\theta\ \partial/\partial q = \sin\beta\ \partial/\partial n$$

Therefore, the matrix P becomes

$$\begin{bmatrix} -\partial^2/\partial z^2 & 0 & \partial^2/\partial z\partial x \\ 0 & (-\partial^2/\partial x^2 - \partial^2/\partial z^2) & 0 \\ \partial^2/\partial x\partial z & 0 & -\partial^2/\partial x^2 \end{bmatrix}$$

$$= \begin{bmatrix} -\sin^2\beta\ \partial^2/\partial n^2 & 0 & \sin\beta\cos\beta\ \partial^2/\partial n^2 \\ 0 & -\partial^2/\partial n^2 & 0 \\ \sin\beta\cos\beta\ \partial^2/\partial n^2 & 0 & -\cos^2\beta\ \partial^2/\partial n^2 \end{bmatrix}$$

and the general equation for E becomes

$$A^T \begin{bmatrix} (S_1 + \sin^2\beta \; \partial^2/\partial n^2) & 0 & (-\sin\beta\cos\beta \; \partial^2/\partial n^2) \\ 0 & \partial^2/\partial n^2 + S_2 & 0 \\ (-\sin\beta\cos\beta \; \partial^2/\partial n^2) & 0 & (S_2 + \cos^2\beta \; \partial^2/\partial n^2) \end{bmatrix} E =$$

that is

$$i\left[\sin^2\beta \; \partial^2 E_1/\partial n^2 + S_1 E_1 - \sin\beta\cos\beta \; \partial^2 E_3/\partial n^2\right]$$

$$+ \; j\left[S_2 E_2 + \partial^2 E_2/\partial n^2\right]$$

$$+ \; k\left[-\sin\beta\cos\beta \; \partial^2 E_1/\partial n^2 + S_2 E_3 + \cos^2\beta \; \partial^2 E_3/\partial n^2\right] = 0$$

Equating the coefficient of j to zero gives a wave equation for E_2 identical with that in section V.4.1, except that the double differentiation is now with respect to n, not with respect to x. This means that the y-component of E travels along the normal at a speed $1/\sqrt{(\mu K_2)}$, irrespective of the direction of the normal. This component of E is part of the ordinary wavefront and its speed found here is consistent with the values of the speed found in sections V.4.1 and V.4.2. We cannot now assume that E is purely transverse, that is that the direction of E lies in the plane α. (The y-component of E certainly lies in this plane, by choice of the y-axis.) If the magnitude of the **E** vector is $|E|$, then

$$|E|^2 = E_1^2 + E_2^2 + E_3^2$$

$$= E_2^2 + F^2 \quad \text{where } F^2 = E_1^2 + E_3^2$$

Here, F is the magnitude of the component of E in the xz-plane. It is convenient to work with the transverse and longitudinal components of F, A and C, say, respectively. By the usual formulae for the rotation of coordinate axes (see, for instance, Courant, *Differential and Integral Calculus*, Volume 2, page 6), we have

$$E_1 = A\cos\beta - C\sin\beta$$

$$E_3 = A\sin\beta + C\cos\beta$$

Equating the coefficient of i to zero, in the general
equation for E, and substituting in these values of E_1
and E_3, we get, after a little simplification

$$S_1 A \cos\beta - S_1 C \sin\beta - \sin\beta \, \partial^2 C/\partial n^2 = 0$$

Equating the coefficient of k to zero gives

$$\cos\beta \, \partial^2 C/\partial n^2 + S_2 A \sin\beta + S_2 C \cos\beta = 0$$

Elimination of A between these two last equations (by
multiplying the first by $S_2 \sin\beta$, multiplying the second
by $S_1 \cos\beta$ and subtracting), gives

$$\partial^2 C/\partial n^2 \, (S_1 \cos^2\beta + S_2 \sin^2\beta) + S_1 S_2 C = 0$$

Putting in the values of S_1 and S_2 and using again the
fact that multiplication by $-\omega^2$ is equivalent to double
differentiation with respect to time, we obtain

$$\partial^2 C/\partial n^2 = \frac{1}{\dfrac{\cos^2\beta}{\mu K_2} + \dfrac{\sin^2\beta}{\mu K_1}} \, \partial^2 C/\partial t^2 = \frac{1}{N^2} \, \partial^2 C/\partial t^2, \text{ say}$$

Comparing with the standard equation of wave motion,
we see that plane waves of **E** are propagated in the
direction defined by angle β at speed N. (If $\beta = 0$,
N simplifies to $1/\sqrt{(\mu K_2)}$, while if $\beta = \pi/2$, N simplifies to $1/\sqrt{(\mu K_1)}$. These simple results agree with
the conclusions of sections V.4.1 and V.4.2.)

We shall now find the magnetic field using the
formula

$$H = (1/\mu\omega) A^T \Delta E.$$

Since all y-derivatives are now zero, this formula
becomes

$$H = (1/\mu\omega) A^T \begin{bmatrix} 0 & -\partial/\partial z & 0 \\ +\partial/\partial z & 0 & -\partial/\partial x \\ 0 & +\partial/\partial x & 0 \end{bmatrix} \begin{bmatrix} E_1 \\ E_2 \\ E_3 \end{bmatrix}$$

that is

$$H = \left(\frac{1}{\mu\omega}\right) \left[-i \frac{\partial E_2}{\partial z} + j \left(\frac{\partial E_1}{\partial z} - \frac{\partial E_3}{\partial x}\right) + k \frac{\partial E_2}{\partial x} \right]$$

Using the formula for directional differentiation again, this becomes

$$H = (1/\mu\omega)\left[\mathbf{i}(-\sin\beta\ \partial E_2/\partial n) + \mathbf{j}(\sin\beta_1\partial E\ /\partial n\right.$$
$$\left. - \cos\beta\ \partial E_3/\partial n) + \mathbf{k}(+\cos\beta\ \partial E_2/\partial n)\right]$$

The component of H in the xz-plane is $\partial E_2/\partial n\left[-\mathbf{i}\sin\beta + \mathbf{k}\cos\beta\right]$. This travels at the same speed as $\partial E_2/\partial n$, that is at the same speed as E, namely $1/\sqrt{(\mu K_2)}$, this speed being the same for all directions of propagation. This is the H-part of the ordinary wave and is perpendicular to the E-part of the ordinary wave. We can see that it is purely transverse, for the tangent of the angle ϕ which it makes with the x-axis is given by

$$\tan\phi = \frac{z\text{-component of }H}{x\text{-component of }H} = \frac{+\cos\beta}{-\sin\beta} = -\cot\beta$$

Therefore,

$$\tan\phi\tan\beta = -1$$

so that the directions defined by the angles ϕ and β are perpendicular to one another. The y-component of H is $\sin\beta\ \partial E_1/\partial n - \cos\beta\ \partial E_3/\partial n$. Substituting for E_1 and E_3 in terms of A and B, this simplifies to $-\partial B/\partial n$. It thus travels at the same speed as B, and so is the H-part of the extraordinary wave.

Since it is in the y-direction, it is perpendicular both to the electric field and to the direction of propagation, that is to say, H is transverse in both the extraordinary and the ordinary waves.

We shall now find the value of D. Since we are in an anistropic crystal, we cannot now assume that D and E are in the same direction. From an equation that we derived earlier,

$$\partial D/\partial t = \text{curl}H = A^T\Delta H$$

$$= A^T\begin{bmatrix} 0 & -\partial/\partial z & 0 \\ +\partial/\partial z & 0 & -\partial/\partial x \\ 0 & +\partial/\partial x & 0 \end{bmatrix}\begin{bmatrix} H_1 \\ H_2 \\ H_3 \end{bmatrix}$$

$$= -\mathbf{i}\partial H_2/\partial z + \mathbf{j}(\partial H_1/\partial z - \partial H_3/\partial x) + \mathbf{k}\partial H_2/\partial x$$

Since, as we have seen, differentiation with respect to time is equivalent to multiplying by ω and by a phase factor, the components of D are proportional to those of $\partial D/\partial t$. The y-component of D is certainly in the wavefront. The angle Σ which the component of D in the xz-plane makes with the positive x-direction is given by

$$\tan\Sigma = \frac{z\text{-component of } D}{x\text{-component of } D} = \frac{\partial H_2/\partial x}{-\partial H_2/\partial z} = \frac{\cos\beta \; \partial H_2/\partial n}{-\sin\beta \; \partial H_2/\partial n} = -\cot\beta$$

(using the formulae for directional differentiation). Thus,

$$\tan\Sigma\tan\beta = -1$$

so that the component of D in the xz-plane lies in the wavefront. This means that D, like H, is pure transverse. (So also is B, which is simply H multiplied by the scalar μ.) This is obviously true both in the ordinary and in the extraordinary wave. Thus, in both the ordinary and the extraordinary waves, D and H are both transverse, but E may not be so.

V.5 HUYGENS WAVELETS IN A UNIAXIAL CRYSTAL

We have shown that the speed N of the plane wavefront, whose normal makes an angle β with the x-axis, is given by

$$N^2 = V_1^2\cos^2\beta + V_2^2\sin^2\beta$$

where $V_1^2 = 1/\mu K_2$ is the square of the speed of wavefronts travelling along the x-direction, and $V_2^2 = 1/\mu K_1$ is the square of the speed of wavefronts travelling along the z-direction.

In an anisotropic medium, the energy of the disturbance travels in the wavefront. It does not travel along the wavefront normal, but along the Poynting vector. Thus, if we consider a point source (origin of Huygens wavelets) in the medium, at each instant the energy travelling along the Poynting vector in any direction has reached the junction of the Poynting vector with the corresponding wavefront. Let us consider the ray surface corresponding to a short time delay such as one picosecond, which gives the locus after one picosecond of the energy in disturbances leaving the origin simultaneously and in different directions. This surface

260

will be the envelope of all the plane wavefronts corre
ponding to different directions of propagation (see
Figure V.2.)

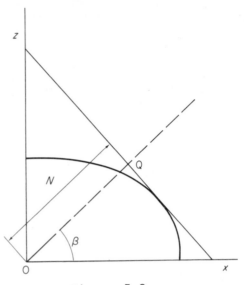

Figure 5.2

If Q is the point at which a particular plane wavefront (one picosecond after passing through the origin
intersects its normal (passing through the origin and
making an angle β with the x-axis) then (if N is
measured in metres per picosecond) the coordinates of
Q are $(N\cos\beta, N\sin\beta)$. The slope of the wavefront is
$\tan\theta = -\cot\beta$, so that the equation of the wavefront i

$$(z - N\sin\beta) = -\cot\beta (x - N\cos\beta)$$

that is

$$x\cos\beta + z\sin\beta = N \tag{V.8}$$

The equations to wavefronts corresponding to different directions are all of this form, and differ from
each other only in the different values of β (and henc
of N). To find their envelope, we eliminate β from
this equation and the one obtained by differentiating
it with respect to β. (See, for example, Courant,
Differential and Integral Calculus Volume 2, pp.171 -
174.)

Differentiating equation (V.8) with respect to β:

$$-x\sin\beta + z\cos\beta = dN/d\beta = \cos\beta\sin\beta(V_2^2 - V_1^2)/N \tag{V.9}$$

(using the definition of N). Multiplying equation (V.8) by $\sin\beta$, equation (V.9) by $\cos\beta$ and adding:

$$z = N\sin\beta + \left(\frac{dN}{d\beta}\right)\cos\beta$$

Multiplying equation (V.8) by $\cos\beta$, equation (V.9) by $\sin\beta$ and subtracting:

$$x = N\cos\beta - \left(\frac{dN}{d\beta}\right)\sin\beta$$

Substituting for $dN/d\beta$ in terms of N and β and then for N^2 in terms of β, we obtain

$$z = \frac{V_2^2\sin\beta}{N} \quad \text{and} \quad x = \frac{V_1^2\cos\beta}{N}$$

From these equations we have

$$\cos^2\beta = N^2x^2/V_1^4 \quad \text{and} \quad \sin^2\beta = N^2z^2/V_2^4$$

Substituting for N and simplifying, these equations become

$$\tan^2\beta = \frac{(1 - x^2/V_1^2)V_1^4}{x^2V_2^2} \quad \text{and} \quad \tan^2\beta = \frac{z^2V_1^2}{V_2^4(1 - z^2/V_2^2)}$$

Equating for $\tan^2\beta$ and simplifying, these equations lead to

$$\frac{x^2}{V_1^2} + \frac{z^2}{V_2^2} = 1$$

This is the equation of the intersection of the extraordinary ray surface (Huygens wavelet) with the Oxz-plane.

Since we have seen that speed of the ordinary wave is independent of direction, the ordinary Huygens wavelet is a sphere of radius $V_1 = 1/\sqrt{(\mu K_2)}$. The semi-axis parallel to the x-direction of the ellipse representing the extraordinary wavelet is of length V_1. This corresponds to the fact that the ordinary and extraordinary wavelets touch on the optic axis of the crystal. Since we are working with the uniaxial crystal in which K_2 equals K_3 by definition, an exactly similar analysis would have followed if we had chosen to work in the Oxy-plane so that the intersection of the extraordinary wavelet with the Oxy-plane is given by

$$\frac{x^2}{V_1^2} + \frac{y^2}{V_2^2} = 1$$

This is an ellipse identical with that obtained for the xz-plane.

We now consider propagation in a general plane passing through the x-axis. Suppose this plane makes an angle ϕ with the z-direction; and let the component of E and of D in this direction be E_ϕ and D_ϕ (see Figure V.3). Now $D_2 = K_2 E_2$ and $D_3 = K_2 E_3$. So

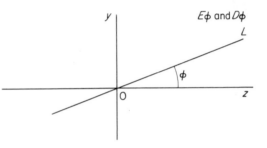

Figure 5.3

$$E_\phi = jE_2 + kE_3$$

$$D_\phi = jD_2 + kD_3$$

$$= j(K_2 E_2) + k(K_2 E_3)$$

$$= K_2(jE_2 + kE_3) = K_2 E_\phi$$

Since the same relation holds between D_ϕ and E_ϕ as held between D_2 and E_2 or between D_3 and E_3, the same ellipse is obtained. Thus the extraordinary wavelet is a spheroid which touches the ordinary wavelet at points on the optic axis of the crystal and has revolution symmetry about that axis.

Appendix A
Aperture Properties of Centred Lens Systems

The matrix methods discussed in chapter II enable us to calculate the position and magnification of the images produced by an optical system, but they do not predict how sharp they will be. Unless the system is a very rudimentary one, it produces images over an appreciable field and an appreciable aperture, and there will be defects of image sharpness which need to be considered in terms of third-order and even higher-order aberration theory. Nevertheless, there is still one job for which first-order methods are still adequate, and that is to calculate the way in which the *stops* of the system, its finite apertures, actually set a limit to the images that can be formed.

Every lens in a system has a finite diameter, but in many cases stops are actual diaphragms with an opaque periphery that have been deliberately introduced; their purpose is to let through all the useful light, but to exclude those regions of the aperture or field for which image aberrations would be unacceptably bad. Stops can also help to ensure that those images which are seen are uniformly illuminated, and they may act as baffles for any stray light that has been scattered, for example from the inside walls of a tube or from a previous stop in a coronagraph. Small stops or pinholes are also sometimes used as spatial filters.

It is probable that the student has already met such terms as the iris, the pupils and the windows of an optical system. For completeness, however, definitions of the technical terms used will now be given.

264

A.1 STOPS THAT LIMIT THE APERTURE

Consider a system that images a plane containing an axial object point O onto an image plane containing the axial image point I (see Figure A.1). Out of the pencil of rays that leave O, only the central portion will be able to reach the image I; the remainder will be cut off either by the boundaries of the lenses or by the opaque peripheries of any apertures that are included in the system. There will always be one of these stops whose radius primarily determines what cone of rays will get right through the system. It is this physical stop or aperture that we call the *apertur* *stop*, or *iris*.

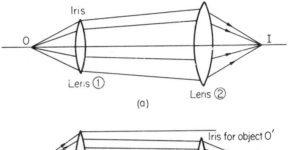

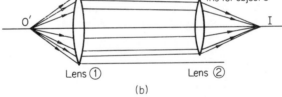

Figure A.1

In Figure A.1a it is the first lens which acts as th iris, but in Figure A.1b the second lens cuts off some of the light transmitted by the first, and therefore replaces it as the iris. In Figure A.2, two examples are shown where the object is at infinity; in both cases, contrary to what we might expect, the raypaths are such that it is the *larger* of the two lenses which acts as the effective iris.

Associated with the physical iris are the real or virtual images of it that can be seen from either side of the system. The *entrance pupil* is the image of the iris formed by any lenses of the system which the ligh reaches *before* reaching the the iris, that is the imag

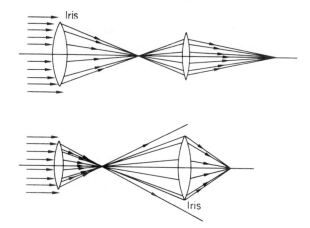

Figure A.2

of the iris as seen when looking in from the left (see
Figure A.3). The *exit pupil* is the image of the iris
formed by the lenses which the light reaches *after* the
iris, that is looking from the right in Figure A.3.

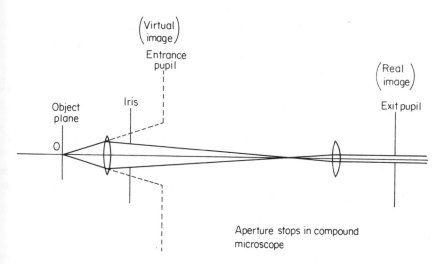

Figure A.3

From the point of view of excluding stray light, it
is sometimes advantageous to locate the iris in front
of the lens system, in which case the iris and the
entrance pupil are identical. But for an adjustable
iris, for example in a photographic objective, there

may be mechanical as well as optical reasons for burying the iris somewhere inside the system. Particularly in projectors, where the size of an object is being measured or compared, or in image processing systems, it is often advantageous for either the entrance pupil or the exit pupil to be located at infinity. Such a system is referred to as being *telecentric* on the object side or the image side. If the system is afocal, it can be telecentric on both sides.

Since the entrance pupil is imaged by the first part of the system onto the iris and by the second part of the system onto the exit pupil, it follows that any ray from the object space that is inclined to the axis but passes through the centre of the entrance pupil will emerge through the centre of the exit pupil into the image space. Such a ray will be referred to as a *principal ray*, or *chief ray*. (Since the entrance pupil does not necessarily coincide with the first principal plane, the principal ray defined above does not necessarily pass through the principal points of the system.

A.2 STOPS THAT LIMIT THE FIELD

Having found for a given object plane the position and size of the iris, and its corresponding entrance and exit pupils, we now seek to determine how large a field in the object space will be imaged through the system. To do this, we consider first all the possible principal rays that pass from different parts of the object plane through the centre of the entrance pupil.

The *field stop* is the lens or aperture which determines what pencil of principal rays emerges through the centre of the exit pupil into the image space (see Figure A.4). The *entrance window* (sometimes called the

(Entrance and exit windows are at infinity)

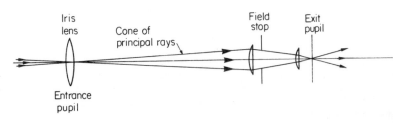

Figure A.4

entrance port) is the image of the field stop seen from the entrance (object) side of the system. The *exit window*, or *exit port*, is the corresponding image seen from the exit side. (See Figure A.5.) In a well-designed system, these entrance and exit windows are located in the object plane and in the image plane respectively.

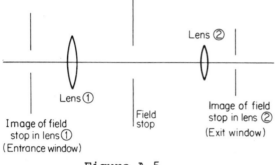

Figure A.5

Provided that an object point lies inside the entrance window, it can send its rays to any point of the entrance pupil, and hence to the exit pupil; for an extended object of uniform brightness, the result is an extended image whose brightness also remains constant until a sharp cut-off occurs at the edge of the exit window.

If the field stop is incorrectly located, however, or if more than one stop is used in addition to the iris, the field illumination in the image plane exhibits a shaded border. There is a *field of full illumination* (see Figure A.6a) surrounded by a penumbral region, and the sum of the two is called the *total field*. As the image point which we are considering moves through the penumbral region, the percentage of the exit pupil that is filled with light falls from 100 per cent at the edge of the field of full illumination to 0 per cent at the edge of the total field (see Figure A.6b). When images are formed with a partially obscured exit pupil, they are said to suffer from *vignetting*.

It frequently happens that a telescope or camera is designed to operate with an object at infinity. We then use a linear measure for the aperture and an angular measure for the field diameters. For a micro-

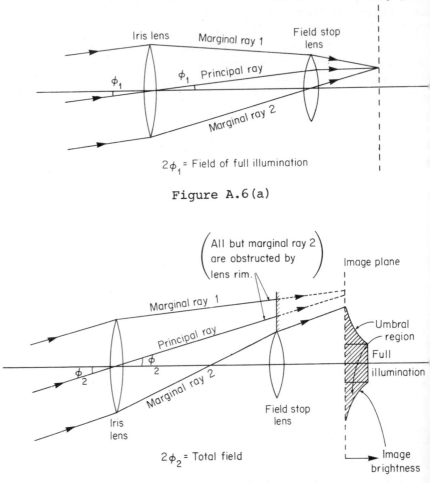

Figure A.6(a)

Figure A.6(b)

scope objective, however, it is the entrance pupil
that is located at infinity and is measured as an
angle or the sine of an angle (numerical aperture).

A.3 SYSTEMATIC CALCULATION OF APERTURE AND FIELD STOPS

A.3.1 Aperture stop

In section II.5 the overall matrix M for a complete
system was calculated as the product of a chain of
component $\mathcal{R}$- and $\mathcal{T}$-matrices. It will now be necessary
however, to calculate the effect of each stop of radius
J_i located at each intermediate reference plane RP_i.
To achieve this, we break the matrix chain and use

only the portion that links RP_i to RP_1. The relation between an intermediate ray vector K_i and an input ray vector K_1 then takes the form

$$K_i = M_{i-1}M_{i-2} \cdots M_3M_2M_1K_1 = L_iK_1$$

Here the matrix symbol L_i is used to denote the product of the matrix sequence $(M_{i-1}M_{i-2} \cdots M_3M_2M_1)$. We shall need to calculate an L-matrix for each of the various stops in the system, but their evaluation involves little extra work, as they can emerge as subproducts at the same time as the overall matrix M is determined. We shall denote the components of the matrix L_i by $(L_{11})_i$, $(L_{12})_i$, etc.

Consider now a ray which leaves an axial object point O located at a distance R to the left of RP_1. If V is the angle at which this ray is travelling, it will reach RP_1 with the ray vector description

$$K_1 = \begin{bmatrix} 1 & R \\ 0 & 1 \end{bmatrix} \begin{bmatrix} 0 \\ V \end{bmatrix} = \begin{bmatrix} RV \\ V \end{bmatrix}$$

When this ray reaches a given intermediate reference plane RP_i, its ray height will be given by

$$y_i = \begin{bmatrix} 1 & 0 \end{bmatrix} K_i = \begin{bmatrix} 1 & 0 \end{bmatrix} L_i \begin{bmatrix} RV \\ V \end{bmatrix} = V \begin{bmatrix} 1 & 0 \end{bmatrix} L_i \begin{bmatrix} R \\ 1 \end{bmatrix}$$

taking V outside as a scalar multiplier. On matrix multiplication, this becomes

$$y_i = V(R(L_{11})_i + (L_{12})_i)$$

If the ray from the axial object point is to pass through the stop in RP_i, the maximum angle V that we can choose for it is that which makes y_i equal to $\pm J_i$ - the stop radius of the stop in RP_i.

(*Footnote*. In this appendix we shall frequently refer to intermediate reference planes, but shall have little occasion to speak of the output reference plane. Where necessary, the output reference plane will be referred to as RP_2 with a bold $_2$ suffix, to distinguish it from the second intermediate reference plane RP_2.)

If we call this maximum ray angle $V_{\max(i)}$, we can write

$$V_{\max(i)} = \frac{\pm J_i}{R(L_{11})_i + (L_{12})_i}$$

If this calculation is repeated for all stops and lens rims in the system, there will be a particular stop, at RP_s for instance, which generates the *smallest* value for $V_{\max}$, that is such that $V_{\max(s)}$ is smaller than any other $V_{\max(i)}$. The stop in RP_s is then the iris of the system and J_s is its radius.

In cases where the distance R to the object plane is very great (or even infinite), the above calculation needs to be re-expressed in terms of the ray height y at the input plane, the ray angle being taken as y/R, which vanishes if R tends to infinity. (On the other hand, if RP_1 is itself the object plane, so that R vanishes, it is necessary to use the earlier method, in which V is used as the trial variable.) We then find

$$y_i = \begin{bmatrix} 1 & 0 \end{bmatrix} L_i \begin{bmatrix} y \\ y/R \end{bmatrix} = y\left[(L_{11})_i + (L_{12})_i/R\right]$$

Hence $y_{\max(i)}$, the greatest value of y in the input plane which allows the ray to pass through the stop in RP_i, is given by

$$y_{\max(i)} = \pm \frac{RJ_i}{R(L_{11})_i + (L_{12})_i}$$

If R tends to infinity, this becomes

$$y_{\max(i)} = \pm \frac{J_i}{(L_{11})_i}$$

Once again, the iris belongs to the plane which generates the *smallest* value for $y_{\max}$.

The position and size of the entrance and exit pupil are now determined using the same methods as in Problems 1 and 2 of chapter II. If the entrance pupil is located a distance E_1 to the left of RP_1, the transfer matrix from the iris plane RP_s back to the entrance pupil will be $L_s \begin{bmatrix} 1 & E_1 \\ 0 & 1 \end{bmatrix}$. For the two planes to be conjugate, the distance E_1 must be equal to $-(L_{12}/L_{11})_s$, and the ratio (entrance pupil radius)/(iris radius J_s) has the value $|1/(L_{11})_s|$.

In similar fashion, if the exit pupil is located at a distance E_2 to the right of the output reference plane RP_2, the transfer matrix from the exit pupil back to the iris will be of the form $\begin{bmatrix} 1 & E_2 \\ 0 & 1 \end{bmatrix} Q_s$, where Q_s represents the portion of the matrix chain between RP_s and the output plane RP_2.

The condition for conjugacy of the exit pupil with the iris then requires that $E_2 = -(Q_{12}/Q_{22})_s$, and the ratio (exit pupil radius)/(iris radius J_s) has the value $|1/(Q_{22})_s|$.

(In general, only two of these Q-matrices will need to be evaluated - one for the exit pupil and one for the exit window. Since the L-matrices will already be known, there is no need to multiply out the second half of the matrix chain. For each ith reference plane, we have $Q_i L_i = M$, so that $Q_i = M L_i^{-1}$, and only one additional multiplication is needed.)

A.3.2 Field stop

Having determined the aperture limitations imposed by the iris in the sth reference plane, we now use as our trial input a principal ray travelling at an angle V through the centre of the entrance pupil. Since the entrance pupil is at a distance E_1 to the left of RP_1, the input ray vector will be of the form $\begin{bmatrix} E_1 V \\ V \end{bmatrix}$ and the ray height y_i in each ith reference plane will be given by

$$y_i = \begin{bmatrix} 1 & 0 \end{bmatrix} L_i \begin{bmatrix} E_1 V \\ V \end{bmatrix} = V \begin{bmatrix} 1 & 0 \end{bmatrix} L_i \begin{bmatrix} E_1 \\ 1 \end{bmatrix}$$

If the trial input principal ray is to pass through the stop in the ith plane, the maximum angle at which we can launch it will be given by

$$V_{max(i)} = \pm \frac{J_i}{[1 \quad 0]L_i\begin{bmatrix} E_i \\ 1 \end{bmatrix}} = \pm \frac{J_i}{E_1(L_{11})_i + (L_{12})_i}$$

Notice that, so far as our original iris plane RP_s is concerned, the denominator of this expression vanishes, since E_1 has already been determined as $-(L_{12}/L_{11})_s$.

By inspection of all the other values calculated for V_{max} from this formula, we now choose the smallest and identify the corresponding reference plane (say RP_f) as the field stop. (In some systems, the field stop comes before the iris, in some systems it occurs afterwards, so the fact that alphabetically f precedes s is of no significance.)

Determination of the size and position of the entrance and exit windows is now performed in the same way as for the pupils, using matrices referring to RP_f rather than to RP_s. The entrance window is located a distance F_1 to the left of RP_1, where $F_1 = -(L_{12}/L_{11})_f$. The ratio (radius of entrance window)/(radius of field stop J_f) has the value $|1/(L_{11})_f|$. The exit window is located a distance F_2 to the right of RP_2, where $F_2 = -(Q_{12}/Q_{22})_f$ and $Q_f = ML_f^{-1}$. The ratio (radius of exit window)/(radius of field stop J_f) has the value $|1/(Q_{22})_f|$.

Finally, measured from the entrance (object) side, the maximum permissible angle for a principal ray is given by

$$V_{max(field)} = \frac{\pm J_f}{E_1(L_{11})_f + (L_{12})_f}$$

$$= \pm \frac{(L_{11})_s J_f}{(L_{12})_f(L_{11})_s - (L_{11})_f(L_{12})_s}$$

(because $E_1 = -(L_{12}/L_{11})_s$).

A.4 DETERMINATION OF THE FIELDS OF ILLUMINATION

Although it is in the exit (image) side of the system that the effects of stops become apparent, we shall refer all these effects back to the conjugate object plane on the entrance side of the system.

A.4.1

In a well-designed system, the entrance window is arranged to coincide with the object plane, and we have

$$F_1 = - (L_{12}/L_{11})_f = R$$

The field then has a sharp boundary whose radius is the product of $V_{max(field)}$ and the distance $(E_1 - F_1)$ $= (E_1 - R)$, which separates the entrance pupil from the object plane. By inserting the values already given for $V_{max(field)}$, E_1 and F_1, it will be found that

$$V_{max(field)} \times (E_1 - F_1) = \frac{\pm J_f}{(L_{11})_f}$$

(agreeing with the expression already given for the radius of the entrance window).

In the case where the entrance window is located in front of or behind the object plane, the radius of the illuminated field in the object plane (as observed through the system) becomes well defined only if the iris is an adjustable one, and is stopped down to a small central region. Projecting the entrance window onto the object plane, we use a scaling factor $(E_1 - R)/(E_1 - F_1)$ and obtain a field radius

$$\left| \frac{(E_1 - R)}{(E_1 - F_1)} \times \frac{J_f}{(L_{11})_f} \right|$$

beyond which the central region of the exit pupil will be obscured. If R tends to infinity, this becomes an angular field radius

$$\left| \frac{J_f}{(E_1 - F_1)(L_{11})_f} \right|$$

A.4.2

Unless a small iris is used, however, the penumbral shading or vignetting may be considerable, and it is necessary to determine separately the total field and the field of full illumination.

The simplest case arises when there are only two stops in the system, one functioning as the iris and the other as a field stop (but not conjugate to the object plane). The required field radii are then obtained by tracing a pair of marginal rays that pass respectively through the top and bottom of the iris and through the top of the field stop. It is usually convenient to make this calculation in the object space, using the calculated size and location of the entrance pupil and the entrance window.

If many stops are present in addition to the iris, this procedure may not suffice, and the usual approach is to construct some sort of diagram. One possibility is to image all of the stops into the image space; each kth stop will form its image a distance $- (L_{12}/L_{11})_k$ to the left of RP_1, and its radius in the object space will be $\left| J_k/(L_{11})_k \right|$.

Once this information has been obtained, we can construct for any chosen field point a vignetting diagram in which all the stop images are projected onto the entrance pupil so as to indicate what portion of the latter remains unobscured. Several such diagrams may be needed, however.

A.4.3

An alternative method is to use a (y, η) diagram, in which the coordinate y denotes the height at which the ray crosses the object plane and the coordinate η denotes the ratio (ray height)/(pupil radius) for the same ray crossing the entrance pupil (or, indeed, any conjugate pupil). (See Figure A.7.)

In such a diagram, every ray is represented not by a line but by a single (y, η) 'ray point'. We shall find that the action of each kth stop can be represented on the diagram by means of a parallel strip, centred on the origin and bounded by two parallel lines $a_k y + b_k \eta = \pm c_k$; a ray escapes obscuration by the kth stop only if its representative ray point lies inside this parallel strip. In terms of the coefficients just given, the width of the kth strip is $2c_k/\sqrt{(a_k^2 + b_k^2)}$, and the tangent of the angle that it makes with the η-axis has the value $(- b_k/a_k)$, in other words, the ratio (negative intercept on the y-axis)/(intercept on η-axis).

The aperture restriction imposed by the presence of an iris - the sth stop - is represented by the boundary lines $\eta = \pm 1$; obviously no ray can be transmitted through the iris at more than 100 per cent of its

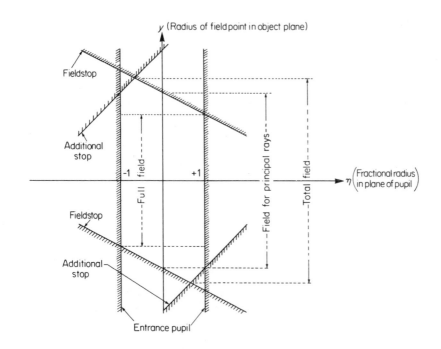

Figure A.7

radius. For all stops other than the iris, the region on the η-axis between $η = + 1$ and $η = - 1$ must by definition remain unobscured.

To determine the obscuration conditions introduced by each of the other stops, let us take a general ray represented by the two parameters y and $η$ and calculate its successive K-values. Such a ray starts from a field point R to the left of RP_1 and y above the axis. It then passes through a second point in the entrance pupil, $E_1 = - (L_{12}/L_{11})_S$ to the left of RP_1 and $ηJ_S/(L_{11})_S$ above the axis, because $J_S/(L_{11})_S$ is the radius of the entrance pupil. The slope of such a ray is given by

$$V = \frac{ηJ_S/(L_{11})_S - y}{(L_{12})_S/(L_{11})_S + R} = \frac{ηJ_S - y(L_{11})_S}{(L_{12})_S + R(L_{11})_S}$$

On reaching RP_1, the ray vector is therefore

$$K_1 = \begin{bmatrix} 1 & R \\ 0 & 1 \end{bmatrix} \begin{bmatrix} y \\ V \end{bmatrix} = \begin{bmatrix} y+RV \\ V \end{bmatrix} = \begin{bmatrix} y_1 \\ V_1 \end{bmatrix}$$

At any kth reference plane, this input ray vector K_1 will generate a ray height

$$y_k = \begin{bmatrix} 1 & 0 \end{bmatrix} L_k K_1 = (L_{11})_k (y + RV) + (L_{12})_k V$$

and the condition for just avoiding obscuration by the kth stop of radius J_k is $|y_k| = J_k$. On substituting the values obtained above for y_k and for V, we obtain, after some rearrangement, the following basic equation for the two boundary lines in the (y, η) diagram:

$$y \left[(L_{11})_k (L_{12})_s - (L_{12})_k (L_{11})_s \right] + \eta J_s \left[(L_{12})_k + R(L_{11})_k \right]$$

$$= \pm J_k \left[(L_{12})_s + R(L_{11})_s \right]$$

This equation has the same form as that already mentioned:

$$a_k y + b_k \eta = \pm c_k$$

The a_k, b_k and c_k coefficients can thus be calculated directly in terms of the L and J data. Construction of the boundary lines on the (y, η) diagram can then proceed either by joining the intercept points on the two axes or (if the a or b coefficients are awkwardly small) by using the tangent relationships mentioned above.

For the simplest case of a single field stop not conjugate with the object plane, the (y, η) diagram is a parallelogram. As more stops are added, however, the boundary lines produce further truncations of the central unobscured region, and much more complicated figures may result. As shown in Figure A.7, it will be a simple matter to determine from the total extent of the diagram in the y-direction the diameter of the total field. If more accuracy is needed, then the common y-values of the two relevant intersecting boundary lines can be calculated (but in that case the third-order aberrations also may need to be considered) The diameter of the field of full illumination corresponds to the y-range over which the diagram retains its full width of 2 measured in the η-direction. It is always the field stop which determines the height of the diagram measured along the central y-axis - the field for principal rays.

For systems such as cameras or telescopes where the object is at a great distance it becomes necessary to replace the field radius y by a corresponding field angle $\phi = y/R$. The equation for the two boundary lines in a (ϕ, η) diagram then becomes

$$R\phi \left[(L_{11})_k (L_{12})_s - (L_{12})_k (L_{11})_s \right] + \eta J_s \left[(L_{12})_k + R(L_{11})_k \right]$$
$$= \pm J_k \left[(L_{12})_s + R(L_{11})_s \right]$$

In the limit where R becomes infinite, we obtain the slightly simpler equation

$$\phi \left[(L_{11})_k (L_{12})_s - (L_{12})_k (L_{11})_s \right] + \eta J_s (L_{11})_k = \pm J_k (L_{11})_s$$

It is worth noticing that, if it is decided to construct each boundary line by the intercept method, most of the calculations needed will already have been done during the initial determinations of the aperture and field stops.

In order to determine the iris, for example, we calculated for each stop, using the axial object point $y = 0$, the maximum angle $V_{\max(i)}$ or the maximum ray height $y_{\max(i)}$ that was permitted. If each of these values is divided by $V_{\max(s)}$ or $y_{\max(s)}$, the value for the iris, the resulting ratio can be used as the required intercept on the η-axis. Similarly, in order to determine the field stop, we used principal rays for which $\eta = 0$, and calculated for each stop the maximum angle permitted. If the object plane is at infinity, each of these angles can be used directly as the required intercept on the ϕ-axis. For a (y, η) diagram, however, we need to multiply each angle $V_{\max(i)}$ by $(E_1 - R) = \left[-R - (L_{12}/L_{11})_s \right]$ to convert it into a ray height in the object plane.

These points will be illustrated further in the numerical example with which we conclude this appendix.

A.5 ILLUSTRATIVE PROBLEM

A simple astronomical finder telescope consists of an objective of diameter 30 mm and focal length 100 mm, arranged to be confocal with a single lens eyepiece of diameter 5 mm and focal length 10 mm. An aperture of diameter 6·5 mm is located at the common focus.

(a) Find the aperture and field stops, and the angular fields of view, of this instrument.

(b) If the observer's eye has a pupil 4 mm in diameter, where should he position it and what happens to the fields of view if he moves his eye 5 mm too close to the eye-lens or 5 mm too far away from it?

Solution

Since the object plane is at infinity, we do not attempt to use it as a reference plane. As shown in Figure A.8, we take RP_1 and RP_2 on either side of the objective lens, RP_3 at the common focal plane, and RP_4 and RP_5 on either side of the eye-lens. For the second half of the problem, we shall need RP_6 in a variable position (which will range 5 mm on either side of the exit pupil).

A.5.1

The first step is to calculate an overall matrix M from RP_6 back to RP_1, listing L-matrices on the right as the calculation proceeds. Since each L_i-matrix is the product $(M_{i-1}M_{i-2} \cdot \cdot \cdot M_2M_1)$, it can be considered as the product of M_{i-1} and the corresponding L-matrix L_{i-1}. (See page 280 for this calculation.) Before proceeding further, we check that the determinants of all these matrices are unity. Inspection of the overall matrix M confirms that it represents an afocal system of (-10) angular magnification. It is also clear that if $d = 0 \cdot 011$ (11 mm), the planes RP_1 and RP_6 will be conjugate.

Having evaluated the L-matrices, we now construct a table which lists for each relevant reference plane the three quantities that we shall need - namely, the stop radius J_i and the matrix elements $(L_{11})_i$ and $(L_{12})_i$. Working in metres and dioptres, we have:

279

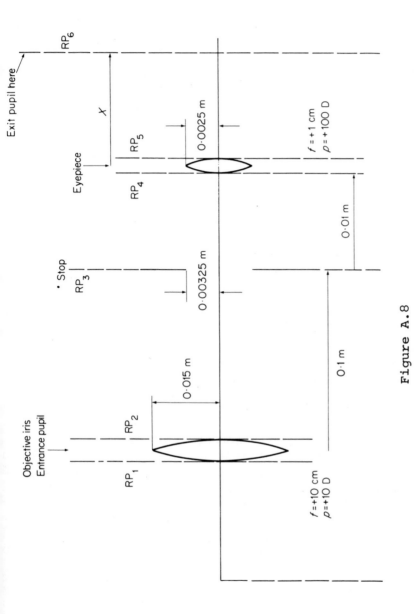

Figure A.8

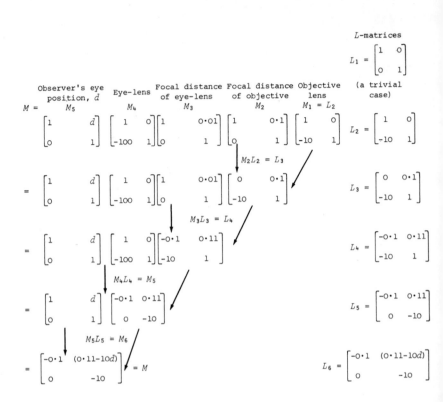

Table of data required for calculating aperture effects in a lens system

Reference plane	RP_1	RP_3	RP_4	RP_6
Stop radius J_i	0·015	0·00325	0·0025	0·002
$(L_{11})_i$	1	0	- 0·1	- 0·1
$(L_{12})_i$	0	0·1	0·11	(0·11-10d)

A.5.2 Determination of aperture stop

Since the object is at infinity, we identify the iris by seeking a minimum value for

$$y_{max(i)} = \left| J_i / (L_{11})_i \right|$$

The values obtained can be added as an additional row to the above table:

Reference plane	RP$_1$	RP$_3$	RP$_4$	RP$_6$
$y_{max(i)} = \left\| \dfrac{J_i}{(L_{11})_i} \right\|$	0·015	∞	0·025	0·02

The smallest value of $y_{max(i)}$ is shown boxed, and this indicates that the iris is provided at RP$_1$ by the lens rim of the objective. Since RP$_1$ is at one end of the system, the iris is itself the entrance pupil. For location of the exit pupil, it is clear that the Q-matrix Q_1 is the same as M, since L_1 is just a unit matrix. We thus obtain $E_2 = - (M_{12}/M_{22}) = (- d + 0·011)$. But, since RP$_6$ is itself located a distance d to the right of the eye-lens, this means that the exit pupil or Ramsden disc is 11 mm to the right of the eye-lens. Its radius is $|J_1/M_{22}| = 0·0015$. We can see already that if the observer positions himself 'correctly', so that $d = 0·011$, his own 4 mm pupil diameter will be sufficient to transmit all the light from the 3 mm diameter exit pupil. But the lateral tolerance on his head position is only ± 0·5 mm.

A.5.3 Determination of field stops

To determine which reference plane provides the field stop, we now consider the maximum available field angle V for a principal ray through the centre of the entrance pupil. Tabulating the data in the form of an additional row, we find:

Reference plane	RP$_1$	RP$_3$	RP$_4$	RP$_6$
$V_{max(i)} = \dfrac{J_i}{(L_{12})_i}$	∞ (iris)	0·0325	0·0227	$\dfrac{0·002}{0·11-10d}$ (∞ if $d = 0·011$)

Ignoring for the moment the effects of the observer's eye position at RP$_6$, we see that it is the rim of the eye-lens at RP$_4$ rather than the focal plane diaphragm at RP$_3$ that must be considered the field stop.

Location of the entrance window is at a distance $F_1 = - (L_{12}/L_{11})_4 = +$ 1·1 metres to the left of the objective at RP_1. Its radius is $|J_4/(L_{11})_4| =$ 0·025 metres (that is 5 cm diameter). Since the field stop is the rim of the last lens in the system, it is itself the exit window, and no calculation is needed.

(Notice that, when a lens rim provides a stop, it makes no difference which of two reference planes we use in calculating its obscuration effect. The intervening lens matrix is such that it leaves the L_{11} and L_{12} elements unchanged.)

A.5.4 *Determination of fields of the instrument*

Since the entrance window unfortunately does not coincide with the object plane at infinity, we shall encounter some effects of vignetting. The angular field radius corresponding to the cone of principal rays is obtained, as in section A.4.1, by projection of the entrance window onto the object plane, and its value is

$$\left| \frac{J_f}{(E_1 - F_1)(L_{11})_f} \right| = \frac{0\cdot0025}{1\cdot1 \times 0\cdot1} = 0\cdot0227 \text{ radians}$$

(the angle-value already boxed under RP_4 in the calculation of V_{max}). (Here, $f = 4$.)

To determine the effects of vignetting, we shall construct a (ϕ, η) diagram. For each kth stop, the boundary lines for an object field at infinity are given by the equations

$$\phi\left[(L_{11})_k(L_{12})_s - (L_{12})_k(L_{11})_s\right] + \eta J_s(L_{11})_k = \pm J_k(L_{11}$$

Since $s = 1$ for this instrument - the iris is at the front - we can immediately put $(L_{11})_s = 1$ and $(L_{12})_s = 0$, and obtain the simple equation

$$\phi\left[- (L_{12})_k\right] + \eta J_s(L_{11})_k = \pm J_k$$

One of the boundary lines is that joining the point $\phi = 0$, $\eta = J_k/J_s(L_{11})_k$ to the point $\phi = - J_k/(L_{12})_k$, $\eta = 0$. The other line runs parallel to it at the same distance on the other side of the origin.

It will be noticed that, apart from division by a normalising factor, $y_{max}(s) = J_s$, we have already cal-

culated for each stop all the data necessary to plot
the intercepts. The ϕ-intercepts are listed in section
A.5.3, and, on division by $J_s = 0\cdot015$ and reinserting
signs, the data in section A.5.2 yield the η-intercepts:

Reference plane	RP_1	RP_3	RP_4	RP_6
η-intercepts	1	∞	$-1\cdot667$	$-1\cdot333$

Using these calculated values for the intercepts, we
now construct (ϕ,η) diagrams as shown in Figure A.9.
In the central diagram, for which the observer's eye
is assumed to be in its correct position at the exit
pupil, the main field-limiting action of the eyepiece
is represented by a parallelogram. Only the extreme
tips of this parallelogram are truncated by the lines
$\phi = \pm 0\cdot0325$ which represent the focal plane diaphragm.
For visual observation, this Keplerian arrangement is
extremely unsatisfactory, since only a small central
portion of the field remains free from vignetting. It
is for this reason that nearly all astronomical eyepieces incorporate a field lens which helps to image
the iris onto a near-central region of the eye-lens.
(For a telescope to be used backwards as a beam-
expander for a low-power laser, the situation is quite
different; the above arrangement might well be used,
supplemented by an extremely small focal plane diaphragm to act as a spatial filter.)

Either from the diagram, or by calculating ϕ-values
for the relevant intersection points, we obtain the
following angular radii for the object field at infinity:

	Radius (in radians)	*Diameter* (in degrees)
Field of full illumination	$0\cdot0091$	$1\cdot04$
Field for principal rays	$0\cdot0227$	$2\cdot60$
Total field	$0\cdot0325$	$3\cdot72$

Since the system functions as a ×10 telescope, the
angles subtended by fields observed in the image space
will be ten times larger. But for rays travelling at
0·2 radians to the axis, the paraxial approximation
will become imperfect: unless the eyepiece has been
designed to be orthoscopic, we may expect discrepancies
of 1 or 2 per cent between actual and nominal values.

A.5.5 *Effect of observer's eye position*

In the left-hand diagram of Figure A.9, an additional
pair of boundary lines has been added to indicate the
effect of moving the observer's eye 5 mm too close to
the eyepiece. (For d = 0·006, the first ϕ-intercept,
0·002/(0·11 - 10d) becomes 0·04.) In this case, no
additional obscuration is caused, since the situation
is still dominated by the inadequate diameter of the
eye-lens

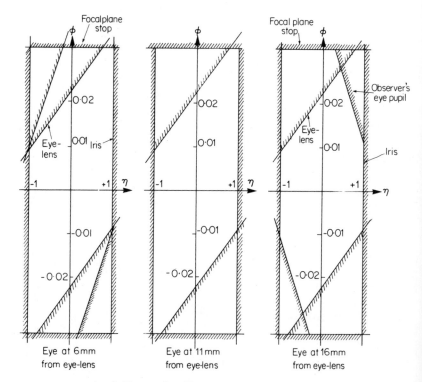

(ϕ, η) Diagrams for ×10 telescope with single eye-lens

Figure A.9

In the right-hand diagram, however, we see that moving the head 5 mm away from the instrument (so that $d = 0 \cdot 016$ and the first ϕ-intercept is $- 0 \cdot 04$) begins to introduce additional obscuration on the right-hand side of the diagram, the shape of which is now slightly more symmetrical. It is clear, however, that the focal plane diaphragm will cease to be visible from this eye position, the total field now being reduced to $0 \cdot 0281$ radians radius, or $3 \cdot 22°$ diameter.

In a well-designed hand-held telescope or binocular, the entrance window is at infinity, and the resultant (ϕ, η) diagram has an ideal rectangular shape. But, unless the head is centred carefully with respect to the exit pupil, both laterally and longitudinally, the performance of the instrument can still be spoiled by vignetting, and also by unnecessary eye aberrations. Caveat observator!

Appendix B
Matrix Representation of Centring and Squaring Errors

B.1 THE USE OF AUGMENTED 3×3 MATRICES

In an ideal centred optical system, we are able to use a 2×2 ray-transfer matrix to represent the homogeneous equations

$$y_2 = Ay_1 + BV_1$$

$$V_2 = Cy_1 + DV_1$$

in the form

$$\begin{bmatrix} y_2 \\ V_2 \end{bmatrix} = \begin{bmatrix} A & B \\ C & D \end{bmatrix} \begin{bmatrix} y_1 \\ V_1 \end{bmatrix}$$

In an actual system, however, the position of the optical axis in RP₂ may be displaced slightly from its assumed position by a small distance Δy_1, for example because a plane-parallel window has been tilted slightly so that it is no longer square to the z-axis. In many cases, also, the exact direction in which the optical axis is pointing deviates slightly from its assumed direction by a small angle ΔV_1. This effect can occur if a 'plane-parallel' plate is slightly wedge-shaped or if one of the lenses has not been properly centred.

Under these circumstances, constant terms Δy_1 and ΔV_1 have to be added to the above two equations, with the result that they are no longer homogeneous. We can, if we wish, represent the transformation by the equation:

$$\begin{bmatrix} y_2 \\ V_2 \end{bmatrix} = \begin{bmatrix} A & B \\ C & D \end{bmatrix} \begin{bmatrix} y_1 \\ V_1 \end{bmatrix} + \begin{bmatrix} \Delta y_1 \\ \Delta V_1 \end{bmatrix}$$

But the output ray vector is no longer obtained by simple matrix multiplication of the input ray vector. There is, however, a trick which enables us to retain this feature of matrix multiplication: this is the use of a 3×3 'augmented matrix'. To see how such a matrix is formed, we write down our two ray-transfer equations and then set down a third 'dummy' equation, which at first sight seems completely trivial:

$$y_2 = Ay_1 + BV_1 + \Delta y_1$$

$$V_2 = Cy_1 + DV_1 + \Delta V_1$$

$$1 = O(1) + O(1) + 1$$

This can be rewritten

$$\begin{bmatrix} y_2 \\ V_2 \\ 1 \end{bmatrix} = \begin{bmatrix} A & B & \Delta y_1 \\ C & D & \Delta V_1 \\ O & O & 1 \end{bmatrix} \begin{bmatrix} y_1 \\ V_1 \\ 1 \end{bmatrix}$$

It is easily verified that, if $\begin{bmatrix} A & B \\ C & D \end{bmatrix}$ is unimodular, then so is the above 3×3 'augmented matrix'. Because two of the elements in the added row are zero, and the third is $+1$, their determinants must be the same.

B.2 MULTIPLICATION OF AUGMENTED MATRICES

Consider now an optical system for which the multiplication of the normal 2×2 matrix chain gives the overall matrix:

$$M = M_n M_{n-1} \cdots M_r \cdots M_3 M_2 M_1$$

where $M_r = \begin{bmatrix} A_r & B_r \\ C_r & D_r \end{bmatrix}$ is the unimodular matrix representing transfer from the rth intermediate reference plane to the $(r+1)$th. Let us suppose that for each matrix M_r there are additional small misalignment terms $\Delta y_r, \Delta V_r$ that need to be considered.

We shall therefore need to consider a corresponding chain of augmented 3×3 matrices, which we shall represent by the curly symbols $\mathcal{M}_1, \mathcal{M}_2$, etc. We can write

$$\mathcal{M} = \mathcal{M}_n \mathcal{M}_{n-1} \cdots \mathcal{M}_r \cdots \mathcal{M}_3 \mathcal{M}_2 \mathcal{M}_1$$

where

$$\mathcal{M}_r = \begin{bmatrix} A_r & B_r & \Delta y_r \\ C_r & D_r & \Delta V_r \\ 0 & 0 & 1 \end{bmatrix}$$

Consider now the multiplication of two such matrices:

$$\mathcal{M}_2 \mathcal{M}_1 = \begin{bmatrix} A_2 & B_2 & \Delta y_2 \\ C_2 & D_2 & \Delta V_2 \\ 0 & 0 & 1 \end{bmatrix} \begin{bmatrix} A_1 & B_1 & \Delta y_1 \\ C_1 & D_1 & \Delta V_1 \\ 0 & 0 & 1 \end{bmatrix}$$

$$= \begin{bmatrix} A_2 A_1 + B_2 C_1 & A_2 B_1 + B_2 D_1 & A_2 \Delta y_1 + B_2 \Delta V_1 + \Delta y_2 \\ C_2 A_1 + D_2 C_1 & C_2 B_1 + D_2 D_1 & C_2 \Delta y_1 + D_2 \Delta V_1 + \Delta V_2 \\ 0 & 0 & 1 \end{bmatrix}$$

If we now examine the elements of this product matrix, we find that there is a square of four elements which are exactly the same as would be obtained by calculating the 2×2 product

$$M_2 M_1 = \begin{bmatrix} A_2 & B_2 \\ C_2 & D_2 \end{bmatrix} \begin{bmatrix} A_1 & B_1 \\ C_1 & D_1 \end{bmatrix}$$

Of the other elements, those in the third row are of little interest; but the top two elements of the third column can be regarded as new effective values of Δy and ΔV for the two matrices taken together. Writing them as Δy_{21} and ΔV_{21} and extracting them from the $\mathcal{M}_2\mathcal{M}_1$ product matrix that we have just calculated, we find by inspection that

$$\begin{bmatrix} \Delta y_{21} \\ \Delta V_{21} \end{bmatrix} = M_2 \begin{bmatrix} \Delta y_1 \\ \Delta V_1 \end{bmatrix} + \begin{bmatrix} \Delta y_2 \\ \Delta V_2 \end{bmatrix}$$

Proceeding in the same way to form the product $\mathcal{M}_3\,\mathcal{M}_2\,\mathcal{M}_1 = \mathcal{M}_3\,(\,\mathcal{M}_2\,\mathcal{M}_1)$, we find that

$$\begin{bmatrix} \Delta y_{321} \\ \Delta V_{321} \end{bmatrix} = M_3 \begin{bmatrix} \Delta y_{21} \\ \Delta V_{21} \end{bmatrix} + \begin{bmatrix} \Delta y_3 \\ \Delta V_3 \end{bmatrix}$$

$$= M_3M_2 \begin{bmatrix} \Delta y_1 \\ \Delta V_1 \end{bmatrix} + M_3 \begin{bmatrix} \Delta y_2 \\ \Delta V_2 \end{bmatrix} + \begin{bmatrix} \Delta y_3 \\ \Delta V_3 \end{bmatrix}$$

and, similarly,

$$\begin{bmatrix} \Delta y_{4321} \\ \Delta V_{4321} \end{bmatrix} = M_4 \begin{bmatrix} \Delta y_{321} \\ \Delta V_{321} \end{bmatrix} + \begin{bmatrix} \Delta y_4 \\ \Delta V_4 \end{bmatrix}$$

$$= M_4M_3M_2 \begin{bmatrix} \Delta y_1 \\ \Delta V_1 \end{bmatrix} + M_4M_3 \begin{bmatrix} \Delta y_2 \\ \Delta V_2 \end{bmatrix} + M_4 \begin{bmatrix} \Delta y_3 \\ \Delta V_3 \end{bmatrix} + \begin{bmatrix} \Delta y_4 \\ \Delta V_4 \end{bmatrix}$$

If we use the symbol Q_r, as in appendix A, to denote the product of a matrix chain beginning with M_n and finishing with M_r, then the effective values of Δy and ΔV for the complete matrix chain $M = M_nM_{n-1} \cdot \cdot \cdot M_2M_1$ can be represented by the following expression:

$$\begin{bmatrix} \Delta y \\ \Delta V \end{bmatrix}_{\text{effective}} = \sum_{r=1}^{n} Q_{r+1} \begin{bmatrix} \Delta y_r \\ \Delta V_r \end{bmatrix}$$

Alternatively, since $Q_{r+1}L_{r+1} = M$ and $Q_{r+1} = ML_{r+1}^{-1}$, we can express this result in terms of L-matrices, and

$$\begin{bmatrix} \Delta y \\ \Delta V \end{bmatrix}_{\text{effective}} = M \sum_{r=1}^{n} L_{r+1}^{-1} \begin{bmatrix} \Delta y_r \\ \Delta V_r \end{bmatrix}$$

An example of the calculation of the various L-matric has been discussed in appendix A.

The overall matrix of the system can now be written down as

$$\mathcal{M} = \begin{bmatrix} A & B & \Delta y \\ C & D & \Delta V \\ O & O & 1 \end{bmatrix}$$

where A, B, C and D are the elements of the usual over-all ray-transfer matrix M, and Δy and ΔV are the overal misalignment terms, calculated as above.

B.3 EFFECT OF MISALIGNMENT OF AN OPTICAL RESONATOR

If we assume that the overall augmented matrix $\mathcal{M}$ tha we have calculated represents the effect of a single round trip in an optical resonator, it is evident that

a ray $\begin{bmatrix} O \\ O \\ 1 \end{bmatrix}$ travelling down the axis in RP$_1$ generates in

RP$_2$ a ray $\begin{bmatrix} \Delta y \\ \Delta V \\ 1 \end{bmatrix}$ which has a different path. To find a

new effective optical axis in this situation, we seek

an input ray $\begin{bmatrix} y_0 \\ V_0 \\ 1 \end{bmatrix}$ that repeats itself exactly - in

effect an eigenvector of the matrix $\mathcal{M}$. Obviously, we must have

$$
\begin{bmatrix} A & B & \Delta y \\ C & D & \Delta V \\ O & O & 1 \end{bmatrix} \begin{bmatrix} y_0 \\ V_0 \\ 1 \end{bmatrix} = \begin{bmatrix} y_0 \\ V_0 \\ 1 \end{bmatrix}
$$

for this to be achieved.

Multiplying out, we obtain

$$
Ay_0 + BV_0 + \Delta y = y_0
$$

$$
Cy_0 + DV_0 + \Delta V = V_0
$$

and

$$
1 = 1
$$

a trivial result again. Rearranging the first two equations, we find

$$
(1 - A)y_0 - BV_0 = \Delta y
$$

$$
- Cy_0 + (1 - D)V_0 = \Delta V
$$

Hence

$$
y_0 = \frac{\begin{vmatrix} \Delta y & -B \\ \Delta V & (1 - D) \end{vmatrix}}{\begin{vmatrix} (1 - A) & -B \\ -C & (1 - D) \end{vmatrix}} \quad \text{and} \quad V_0 = \frac{\begin{vmatrix} (1 - A) & \Delta y \\ -C & \Delta V \end{vmatrix}}{\begin{vmatrix} (1 - A) & -B \\ -C & (1 - D) \end{vmatrix}}
$$

The determinant in the denominator has the value $(1 - A)(1 - D) - BC = (2 - A - D)$, since $(AD - BC) = 1$. Provided that $(2 - A - D)$ does not vanish, we then obtain the solution

$$
y_0 = \frac{(1 - D)\Delta y + B\Delta V}{(2 - A - D)} \quad \text{and} \quad V_0 = \frac{C\Delta y + (1 - A)\Delta V}{(2 - A - D)}
$$

Alternatively, this solution can be expressed in the matrix form

$$
\begin{bmatrix} y_0 \\ V_0 \end{bmatrix} = \frac{1}{(2 - A - D)} (I - M^{-1}) \begin{bmatrix} \Delta y \\ \Delta V \end{bmatrix}
$$

See section III.4 for a discussion of some of the implications of this result for a simple two-mirror resonator.

Appendix C

Statistical Derivation of the Stokes Parameters

In this appendix we consider the problem of specifying the state of polarization for a light beam which contains a large number of independent oscillating components.

For a given single *fully polarized* disturbance, the Stokes parameters were defined in section I V.3 in terms of the amplitudes and phases of the transverse electric field components

$$E_x = He^{i\phi}$$

and

$$E_y = Ke^{i\psi}$$

where H and K are real and positive amplitudes and the phase difference $(\psi - \phi)$ is given by Δ. In terms of H, K and Δ, the defining equations were

$$I = H^2 + K^2, \quad Q = H^2 - K^2, \quad U = 2HK\cos\Delta$$

$$\text{and} \quad V = 2HK\sin\Delta$$

The first parameter I represents the intensity, or irradiance; it will be recalled that the usual method of calculating this from the electric field vector is to form the bracket product

$$\begin{bmatrix} E_x^* & E_y^* \end{bmatrix} \begin{bmatrix} E_x \\ E_y \end{bmatrix} = \begin{bmatrix} He^{-i\phi} & Ke^{-i\psi} \end{bmatrix} \begin{bmatrix} He^{i\phi} \\ Ke^{i\psi} \end{bmatrix} = (H^2 + K^2)$$

If we replace each of the above matrices by its complex conjugate transpose, we obtain a new product which is not a scalar but a 2×2 Hermitian matrix, sometimes referred to as a 'coherency matrix'.

$$\begin{bmatrix} He^{i\phi} \\ Ke^{i\psi} \end{bmatrix} \begin{bmatrix} He^{-i\phi} & Ke^{-i\psi} \end{bmatrix} = \begin{bmatrix} H^2 & HKe^{-i\Delta} \\ HKe^{i\Delta} & K^2 \end{bmatrix}$$

since $(\psi - \phi) = \Delta$. It will be seen immediately that each of the elements in the above matrix can be re-expressed in terms of the four real Stokes parameters. If we denote the coherency matrix by Z, we can write

$$Z = \tfrac{1}{2} \begin{bmatrix} I+Q & U-iV \\ U+iV & I-Q \end{bmatrix}$$

For any 2×2 matrix to be Hermitian, the two diagonal elements must necessarily be real, and one off-diagonal element must be the complex conjugate of the other. In our choice of elements for a Z-matrix, therefore, we have just four degrees of freedom, and there is one and only one Z-matrix which corresponds to the four real elements of a given Stokes column.

Let us now proceed to calculate a Z-matrix

$$Z = \begin{bmatrix} E_x \\ E_y \end{bmatrix} \begin{bmatrix} E_x^\star & E_y^\star \end{bmatrix}$$

for the case where the transverse electric field components arise from the superposition of a large number N of individual disturbances. We shall assume that each rth disturbance can be represented by a Maxwell column $\begin{bmatrix} H_r e^{i\phi_r} \\ K_r e^{i\psi_r} \end{bmatrix}$ multiplied by the time-dependence factor $\exp(i\omega t)$, which is taken to be the same for all N disturbances.

(There are perhaps two points to be noted here. First, the fact that each rth disturbance will tend to radiate in a slightly different direction, and at a

slightly different optical frequency, can be accommo-
dated by taking both ϕ_r and ψ_r to be rapidly varying
functions of time and of position in the xy-plane.
Second, the student should note that in many textbooks
the time-dependence in a complex exponential expression
for an electromagnetic wave is written with the *opposite* sign $\exp(-iwt)$.)

Using the principle that for any electromagnetic
field the contributions can be added *linearly* for each
component we obtain for the resultant electric field
vector the Maxwell column.

$$
\begin{bmatrix} E_x \\ E_y \end{bmatrix} = \begin{bmatrix} \sum_r^N H_r e^{i\phi_r} \\ \sum_r^N K_r e^{i\psi_r} \end{bmatrix}
$$

The corresponding Z-matrix is then given by

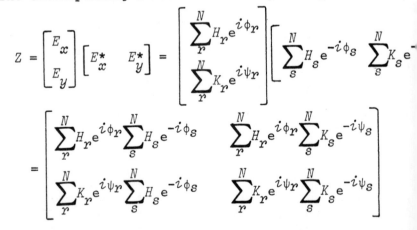

$$
Z = \begin{bmatrix} E_x \\ E_y \end{bmatrix} \begin{bmatrix} E_x^* & E_y^* \end{bmatrix} = \begin{bmatrix} \sum_r^N H_r e^{i\phi_r} \\ \sum_r^N K_r e^{i\psi_r} \end{bmatrix} \begin{bmatrix} \sum_s^N H_s e^{-i\phi_s} & \sum_s^N K_s e^{-} \end{bmatrix}
$$

$$
= \begin{bmatrix} \sum_r^N H_r e^{i\phi_r} \sum_s^N H_s e^{-i\phi_s} & \sum_r^N H_r e^{i\phi_r} \sum_s^N K_s e^{-i\psi_s} \\ \sum_r^N K_r e^{i\psi_r} \sum_s^N H_s e^{-i\phi_s} & \sum_r^N K_r e^{i\psi_r} \sum_s^N K_s e^{-i\psi_s} \end{bmatrix}
$$

In order to proceed further, we now need to make three
assumptions.

(a) We assume that our main purpose in calculating the
    Z-matrix is to obtain values for four Stokes para-
    meters - real numbers which can be compared with
    those obtained by direct photoelectric measurement
    In making such measurements, however, we shall need
    to collect the light energy carried by a *finite*
    area of the beam over a *finite* period of observa-

tion, and the values so obtained will correspond to a process of time-averaging and space-averaging applied to the elements of the above Z-matrix. The mean value of Z obtained by such averaging will be denoted by $<Z>$.

(b) We shall also assume that for both x-vibrations and y-vibrations all N oscillating disturbances are 'mutually incoherent'. What this statement means is that, for any pair of oscillators r and s, the averaged intensity observed when both are operating simultaneously is the same as the sum of their individual intensities. As far as the x-vibrations are concerned, when both oscillators operate together they generate an intensity

$$I_{rs} = (H_r e^{-i\phi_r} + H_s e^{-i\phi_s})(H_r e^{i\phi_r} + H_s e^{i\phi_s})$$

$$= H_r^2 + H_s^2 + 2H_r H_s \cos(\phi_r - \phi_s)$$

$$= I_r + I_s + 2H_r H_s \cos(\phi_r - \phi_s)$$

If the two disturbances are mutually incoherent, this indicates that the *averaged* value of the 'interference term' or 'mutual intensity' must be vanishingly small, that is $<2H_r H_s \cos(\phi_r - \phi_s)> = 0$. Since both H-values are positive, and in some cases can be considered constant over the beam area and the observation time that is being sampled, the vanishing of this term implies that the phase difference $(\phi_r - \phi_s)$ is passing through many complete cycles of variation, and is equally likely, at any given instant, to have any value between 0 and 2π. Under these circumstances, we can say that all averaged terms such as $<\cos(\phi_r - \phi_s)>$, $<\exp[i(\phi_r - \phi_s)]>$ or $<\exp[-i(\phi_r - \phi_s)]>$ are vanishingly small.

By similar reasoning applied to the y-vibrations, we find that $<\exp[i(\psi_r - \psi_s)]> = 0$ and $<\exp[-i(\psi_r - \psi_s)]> = 0$. There is, of course, an important exception to this rule, in that, when $r = s$, all of these expressions, instead of vanishing, are equal to unity.

(c) Finally, we shall assume that for each individual rth oscillator the phase difference $(\psi_r - \phi_r) = \Delta_r$ is constant. In other words, although ϕ_r itself varies through many cycles during an observation, exactly the same sequence of variations is followed by ψ_r. The rth disturbance can therefore be regarded as the product of a Maxwell column $\begin{bmatrix} H_r \\ K_r e^{i\Delta_r} \end{bmatrix}$ and a rapidly varying phase factor $\exp(i\phi_r)$.

It follows that the averaged expressions $\langle \exp[\pm i(\psi_r - \phi_s)] \rangle$ can be regarded as vanishingly small if $r \neq s$, and as equal to $\exp(\pm i\Delta)$ for the special case where $r = s$.

Armed with these three assumptions, we can now apply the averaging process to the Z-matrix, and obtain

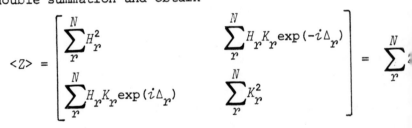

$$\langle Z \rangle = \begin{bmatrix} \sum_r^N \sum_s^N H_r H_s \langle \exp[i(\phi_r - \phi_s)] \rangle & \sum_r^N \sum_s^N H_r K_s \langle \exp[i(\phi_r - \psi_s)] \\ \sum_r^N \sum_s^N K_r H_s \langle \exp[-i(\phi_s - \psi_r)] \rangle & \sum_r^N \sum_s^N K_r K_s \langle \exp[i(\psi_r - \psi_s)] \end{bmatrix}$$

Because, as we have seen, each of the averaged phase factors vanishes except when $r = s$, we can eliminate the double summation and obtain

$$\langle Z \rangle = \begin{bmatrix} \sum_r^N H_r^2 & \sum_r^N H_r K_r \exp(-i\Delta_r) \\ \sum_r^N H_r K_r \exp(i\Delta_r) & \sum_r^N K_r^2 \end{bmatrix} = \sum_r^N$$

We thus obtain the important result that, when the disturbances being added are mutually incoherent, the averaged Z-matrix for the compound disturbance is just the sum of the Z-matrices for the individual disturbances. And, similarly, each of the four resultant Stokes parameters is equal to the sum of the individual parameters

$$I = \sum_r^N I_r, \quad Q = \sum_r^N Q_r, \quad U = \sum_r^N U_r \quad \text{and} \quad V = \sum_r^N V_r$$

It follows that, if we know the Stokes parameters

for each individual disturbance, or if we can predict by statistical reasoning their most probable values, then the parameters for the complete beam can be obtained by straightforward summation. Alternatively, we can start by determining the (averaged) Z-matrix elements and then use the formulae

$$I = (z_{11} + z_{22}), \quad Q = (z_{11} - z_{22}), \quad U = (z_{12} + z_{21})$$

and $V = i(z_{12} - z_{21})$

We shall conclude this appendix by listing, for several types of light, the Z-matrix, the Stokes column and, in most cases, the Maxwell column by which each type may be represented. In listing the columns, however, we shall normalize them so that they each represent a beam of unit intensity.

Type (1) Unpolarized light
In this case, the H_r-values and K_r-values will all be positive, and they will have the same mean-square value $\overline{H^2} = \overline{K^2}$. On the other hand the phase differences Δ_r will have no preferred value, so that, on summation over many disturbances, $\sum_{r}^{N} \exp(i\Delta_r)$ and $\sum_{r}^{N} \exp(-i\Delta_r)$ will be very small in comparison with N.
The Z-matrix thus becomes

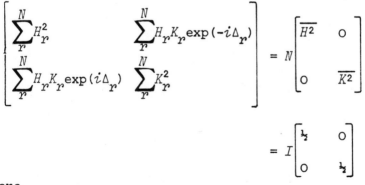

$$\begin{bmatrix} \sum_r^N H_r^2 & \sum_r^N H_r K_r \exp(-i\Delta_r) \\ \sum_r^N H_r K_r \exp(i\Delta_r) & \sum_r^N K_r^2 \end{bmatrix} = N \begin{bmatrix} \overline{H^2} & 0 \\ 0 & \overline{K^2} \end{bmatrix}$$

$$= I \begin{bmatrix} \tfrac{1}{2} & 0 \\ 0 & \tfrac{1}{2} \end{bmatrix}$$

where

$$I = N(\overline{H^2} + \overline{K^2}) = 2N\overline{H^2} = 2N\overline{K^2}$$

When normalized to unit intensity, the corresponding

Stokes column is evidently $\begin{bmatrix} 1 \\ 0 \\ 0 \\ 0 \end{bmatrix}$ There is, of course,

no Maxwell column corresponding to this type of light.

Type (2a) Light polarised parallel to the x-axis

For this case, every K is zero, so the Z-matrix becomes

$$Z = \begin{bmatrix} \overline{NH^2} & 0 \\ 0 & 0 \end{bmatrix} = \overline{NH^2} \begin{bmatrix} 1 & 0 \\ 0 & 0 \end{bmatrix}$$

If type (2a) light is produced by passing type (1) through an (ideal) linear polarizer, the intensity remaining is only one-half of the original - namely $\overline{NH^2}$ instead of $(\overline{NH^2} + \overline{NK^2})$.

When normalized to unit intensity, however, the

Stokes column for this light is $\begin{bmatrix} 1 \\ 1 \\ 0 \\ 0 \end{bmatrix}$ and the corres-

ponding Maxwell column is $\begin{bmatrix} 1 \\ 0 \end{bmatrix}$

Type (2b) Light plane-polarized parallel to the y-axis

For this case, every H is zero, so the Z-matrix becomes

$$Z = \begin{bmatrix} 0 & 0 \\ 0 & \overline{NK^2} \end{bmatrix} = \overline{NK^2} \begin{bmatrix} 0 & 0 \\ 0 & 1 \end{bmatrix}$$

The corresponding normalized Stokes column and Maxwell column are evidently $\begin{bmatrix} 1 \\ -1 \\ 0 \\ 0 \end{bmatrix}$ and $\begin{bmatrix} 0 \\ 1 \end{bmatrix}$

Type (2c) Light plane-polarized with the vibration-plane at an angle θ to the x-axis

This light can be obtained by passing type (1) light through a polarizer with its pass-plane at angle θ to the x-axis. If we describe the transverse vibration components emerging from the polarizer in terms of (x',y') coordinates, such that the x'-axis coincides with the pass-plane, we shall have all the K_r'-values equal to zero. If we denote each surviving H_r'-value by A_r, then in terms of the original (x,y) coordinates we shall have transverse components

$$H_r = A_r\cos\theta \quad \text{and} \quad K_r = A_r\sin\theta$$

with Δ_r equal to zero. Our Z-matrix then becomes

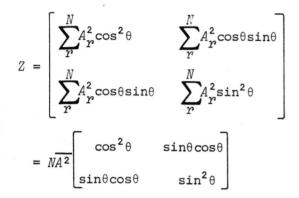

$$Z = \begin{bmatrix} \sum_r^N A_r^2\cos^2\theta & \sum_r^N A_r^2\cos\theta\sin\theta \\ \sum_r^N A_r^2\cos\theta\sin\theta & \sum_r^N A_r^2\sin^2\theta \end{bmatrix}$$

$$= \overline{NA^2}\begin{bmatrix} \cos^2\theta & \sin\theta\cos\theta \\ \sin\theta\cos\theta & \sin^2\theta \end{bmatrix}$$

The corresponding normalized Stokes column and Maxwell column are then $\begin{bmatrix} 1 \\ \cos2\theta \\ \sin2\theta \\ 0 \end{bmatrix}$ and $\begin{bmatrix} \cos\theta \\ \sin\theta \end{bmatrix}$

Type (2d) Light polarized at angle $-\theta$ with respect to the x-axis

The results for this case are obtained by substituting $-\theta$ for θ throughout section (2c) above. Only the odd sine functions are changed in sign, and we obtain $\begin{bmatrix} 1 \\ \cos2\theta \\ -\sin2\theta \\ 0 \end{bmatrix}$ and $\begin{bmatrix} \cos\theta \\ -\sin\theta \end{bmatrix}$

Type (3) Light in circular state of polarization

We start with light of type (2c), with $\theta = \pi/4$, or 45°, so that $\cos2\theta = 0$ and $\sin2\theta = 1$. If we now insert a quarter-wave plate with its fast axis along the x-direction, the y-vibrations of the light will be *retarded* in phase by $\pi/2$ with respect to the x-

vibrations. For each rth disturbance, therefore, we shall have *left-handed* circularly polarized light for which $\Delta_r = -\pi/2$. Because both $\cos\theta$ and $\sin\theta$ are $1/\sqrt{2}$, the Z-matrix becomes

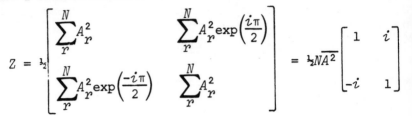

$$Z = \tfrac{1}{2}\begin{bmatrix} \sum_r^N A_r^2 & \sum_r^N A_r^2 \exp\left(\frac{i\pi}{2}\right) \\ \sum_r^N A_r^2 \exp\left(\frac{-i\pi}{2}\right) & \sum_r^N A_r^2 \end{bmatrix} = \tfrac{1}{2} N\overline{A^2} \begin{bmatrix} 1 & i \\ -i & 1 \end{bmatrix}$$

The corresponding Stokes column and Maxwell column are therefore $\begin{bmatrix} 1 \\ 0 \\ 0 \\ -1 \end{bmatrix}$ and $\frac{1}{\sqrt{2}} \begin{bmatrix} 1 \\ -i \end{bmatrix}$

   If we rotate the quarter-wave plate through $90°$, so that it *advances* the y-vibrations by $\pi/2$, the phase difference Δ_r will become $+\pi/2$ and we shall generate *right-handed* circularly polarized light, for which the normalized Stokes column and Maxwell column are $\begin{bmatrix} 1 \\ 0 \\ 0 \\ 1 \end{bmatrix}$ and $\frac{1}{\sqrt{2}} \begin{bmatrix} 1 \\ i \end{bmatrix}$

Type (4) Elliptically polarized light with axes of the ellipse in the x- and y-directions

   As with type (3) light, we assume that type (2c) light is sent through a quarter-wave plate with its fast axis parallel to the x-axis. In this case, however, the initial plane of the polarizer is set at some general angle θ which is not equal to $45°$.

   The Z-matrix is then

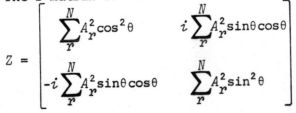

$$Z = \begin{bmatrix} \sum_r^N A_r^2 \cos^2\theta & i \sum_r^N A_r^2 \sin\theta\cos\theta \\ -i \sum_r^N A_r^2 \sin\theta\cos\theta & \sum_r^N A_r^2 \sin^2\theta \end{bmatrix}$$

$$= \overline{NA^2} \begin{bmatrix} \cos^2\theta & i\sin\theta\cos\theta \\ -i\sin\theta\cos\theta & \sin^2\theta \end{bmatrix}$$

For this *left-handed* elliptical state of polarization, therefore, the normalized Stokes column and Maxwell column will be $\begin{bmatrix} 1 \\ \cos2\theta \\ 0 \\ -\sin2\theta \end{bmatrix}$ and $\begin{bmatrix} \cos\theta \\ -i\sin\theta \end{bmatrix}$

As with type (3) light, if the quarter-wave plate is rotated through 90° then it is *right-handed* elliptically polarized light that will be generated, and the corresponding normalized Stokes column and Maxwell column will be $\begin{bmatrix} 1 \\ \cos2\theta \\ 0 \\ \sin2\theta \end{bmatrix}$ and $\begin{bmatrix} \cos\theta \\ i\sin\theta \end{bmatrix}$

Appendix D
Derivation of
Mueller Matrices

D.1 THE POLARIZER

Let the pass-plane of a linear polarizer (for example a sheet of polaroid or a Nicol prism) make an angle θ with the x-axis of an arbitrary rectangular coordinate system. Let $C_1 = \cos\theta$, $C_2 = \cos2\theta$, $S_1 = \sin\theta$ and $S_2 = \sin2\theta$. Let the matrix be

$$Z = \begin{bmatrix} X & B & T & D \\ E & F & G & H \\ J & K & L & M \\ N & P & R & S \end{bmatrix}$$

(This 4×4 matrix Z is not to be confused with the 2×2 complex 'coherency matrix' Z used in appendix C.)

(a) The device polarized unpolarized light at an angle θ with the x-axis, that is it turns type (1) light into type (2c) light (see the list of Stokes parameters in appendix C). (We shall use the symbol $W = I/2$, where I represents the intensity of the unpolarized input beam.) Therefore,

W(Column matrix of Stokes $= Z \times 2W$(Column matrix of
    parameters of type 2c)   Stokes parameters
                                              of type 1)

$$\begin{bmatrix} W \\ WC_2 \\ WS_2 \\ 0 \end{bmatrix} = \begin{bmatrix} X & B & T & D \\ E & F & G & H \\ J & K & L & M \\ N & P & R & S \end{bmatrix} \begin{bmatrix} 2W \\ 0 \\ 0 \\ 0 \end{bmatrix}$$

that is $W = 2WX + B.0 + T.0 + D.0$, $WC_2 = 2WE + \ldots$
$WS_2 = 2WJ + \ldots$, $0 = 2WN + \ldots$ (A string of dots
means some zero terms in the sum.) On cancelling
W, we obtain $2X = 1$, $X = 1/2$. Similarly, $E = C_2/2$,
$J = S_2/2$, $N = 0$.

(b) The device leaves light already polarized at angle
θ with the x-axis unaffected, that is it turns type
(2c) light into type (2c) light.

$$
\begin{bmatrix} W \\ WC_2 \\ WS_2 \\ 0 \end{bmatrix}
=
\begin{bmatrix} 1/2 & B & T & D \\ C_2/2 & F & G & H \\ S_2/2 & K & L & M \\ 0 & P & R & S \end{bmatrix}
\begin{bmatrix} W \\ WC_2 \\ WS_2 \\ 0 \end{bmatrix}
$$

that is

$$W = W/2 + BWC_2 + TWS_2 + D.0 \tag{D.1}$$

$$WC_2 = WC_2/2 + FWC_2 + GWS_2 + H.0 \tag{D.2}$$

$$WS_2 = WS_2/2 + KWC_2 + LWS_2 + M.0 \tag{D.3}$$

$$0 = 0.0 + PWC_2 + RWS_2 + S.0 \tag{D.4}$$

From equation (D.4) $PC_2 = - RS_2$ for *all* values of
C_2 and of S_2. Put $C_2 = 0$; then $R = 0$. Put $S_2 = 0$;
then $P = 0$. Equation (D.1) gives $BC_2 + TS_2 = 1/2$,
which is used later.

(c) The device turns light of amplitude A polarized
along Ox into light of amplitude $A\cos\theta = AC_1$, polarized along angle θ (by resolution). For the
original beam $H^2 = A^2$, $K = 0$, so that $I = Q = A^2$,
$U = V = 0$. The component of emergent beam along
Ox is AC_1^2; along Oy it is AC_1S_1. Therefore,
$H^2 = A^2C_1^4$, $K^2 = A^2C_1^2S_1^2$, $\Delta = 0$. These lead to
$I = A^2C_1^2$, $Q = A^2C_1^2C_2$, $U = A^2C_1^2S_2$, $V = 0$ (as for
type 2c light). Thus, cancelling A^2 throughout,

$$
\begin{bmatrix} C_1^2 \\ C_1^2C_2 \\ C_1^2S_2 \\ 0 \end{bmatrix}
=
\begin{bmatrix} 1/2 & B & T & D \\ C_2/2 & F & G & H \\ S_2/2 & K & L & M \\ 0 & 0 & 0 & S \end{bmatrix}
\begin{bmatrix} 1 \\ 1 \\ 0 \\ 0 \end{bmatrix}
$$

that is

$$C_1^2 = 1/2 + B, \qquad B = C_1^2 - 1/2 = C_2/2$$

$$C_1^2 C_2 = C_2/2 + F, \qquad F = C_2(C_1^2 - 1/2) = C_2 \times C_2/2$$

$$= C_2^2/2$$

$$C_1^2 S_2 = S_2/2 + K, \qquad K = 2C_1^3 S_1 - S_2/2 = 2C_1^3 S_1 - C_1 S_1$$

$$= C_1 S_1(2C_1^2 - 1) = S_2 C_2/2$$

Using this B, equation (D.1) now gives

$$C_2^2/2 + TS_2 = 1/2$$

Therefore, $T = S_2/2$.

(d) Consider circular light of amplitude A (= radius).
The component amplitudes parallel to Ox and to Oy
are both A, so (see the Stokes column for light of
type 3) $I = 2A^2$, $Q = U = 0$ and $V = 2A^2$. The device
turns this into a beam which is linearly polarized
at angle θ with Ox and of amplitude A, so that
$H = AC_1$, $K = AS_1$, $\Delta = 0$, whence $I = A^2$,
$Q = A^2(C_1^2 - S_1^2) = A^2 C_2$, $U = 2AC_1 \cdot AS_1 \cdot 1 = A^2 S_2$,
$V = 0$ (light of type 2c).
On cancelling A^2 throughout, we obtain

$$
\begin{bmatrix} 1 \\ C_2 \\ S_2 \\ 0 \end{bmatrix}
=
\begin{bmatrix}
1/2 & C_2/2 & S_2/2 & D \\
C_2/2 & C_2^2/2 & G & H \\
S_2/2 & C_2 S_2/2 & L & M \\
0 & 0 & 0 & S
\end{bmatrix}
\begin{bmatrix} 2 \\ 0 \\ 0 \\ 2 \end{bmatrix}
$$

that is

$$1 = 1 + 2D, \qquad D = 0$$

$$C_2 = C_2 + 2H, \qquad H = 0$$

$$S_2 = S_2 + 2M, \qquad M = 0$$

$$0 = 2S, \qquad S = 0$$

Using known F, equation (D.2) becomes

$$C_2 = C_2/2 + C_2^3/2 + GS_2$$

$$GS_2 = C_2/2 - C_2^3/2 = (C_2/2)(1 - C_2^2) = C_2 S_2^2/2$$

$$G = C_2 S_2/2$$

Using known K and M, equation (D.3) becomes

$$S_2 = S_2/2 + S_2 C_2^2/2 + LS_2$$

$$LS_2 = S_2/2 - S_2 C_2^2/2 = (1 - C_2^2)S_2/2 = S_2^3/2$$

$$L = S_2^2/2$$

Thus, the Mueller matrix of the polarizer is

$$1/2 \begin{bmatrix} 1 & C_2 & S_2 & 0 \\ C_2 & C_2^2 & C_2 S_2 & 0 \\ S_2 & C_2 S_2 & S_2^2 & 0 \\ 0 & 0 & 0 & 0 \end{bmatrix}$$

D.2 THE HALF-WAVE PLATE (optic axis at angle θ with the x-direction)

This device introduces a phase shift of π or $180°$ between the O- and E-vibrations passing through it. We shall assume that it advances the extraordinary vibration by π with respect to the ordinary vibration. The geometrical effect of the plate is to turn the plane of vibration of plane-polarized light through an angle 2θ, where θ is the angle between the pass-plane and the optic axis. No energy is absorbed, and the vibration-plane is shifted as if by reflection, to its own mirror image in the optic axis.

Let the matrix of the device be

$$Z = \begin{bmatrix} W & B & Y & D \\ E & F & G & H \\ K & L & M & N \\ P & R & X & T \end{bmatrix}$$

We shall use the abbreviated code: $C_1 = \cos\theta$, $C_2 = \cos2\theta$, $C_4 = \cos4\theta$, $S_1 = \sin\theta$, $S_2 = \sin2\theta$, $S_4 = \sin4\theta$.

(a) It converts light of unit intensity whose vibration plane is in the x-direction into a vibration-plane at angle 2θ with x-direction without loss of intensity, that is it turns type (2a) light into light of type (2c), except that 2θ replaces θ in the latter Stokes column to give $I = 1$, $Q = (C_2^2 - S_2^2)$, $U = 2S_2C_2$, and $V = 0$.

$$
\begin{bmatrix} 1 \\ C_2^2 - S_2^2 \\ 2S_2C_2 \\ 0 \end{bmatrix}
=
\begin{bmatrix} W & B & Y & D \\ E & F & G & H \\ K & L & M & N \\ P & R & X & T \end{bmatrix}
\begin{bmatrix} 1 \\ 1 \\ 0 \\ 0 \end{bmatrix}
$$

that is

$$W + B = 1 \tag{D.5}$$

$$E + F = C_2^2 - S_2^2 \tag{D.6}$$

$$K + L = 2S_2C_2 \tag{D.7}$$

$$P + R = 0 \tag{D.8}$$

(b) The device does not affect unpolarized light, that is type (1) light is turned into type (1) light. Therefore,

$$
\begin{bmatrix} 1 \\ 0 \\ 0 \\ 0 \end{bmatrix}
=
\begin{bmatrix} W & B & Y & D \\ E & F & G & H \\ K & L & M & N \\ P & R & X & T \end{bmatrix}
\begin{bmatrix} 1 \\ 0 \\ 0 \\ 0 \end{bmatrix}
\qquad
\begin{aligned}
&\text{that is } W = 1 \\
&\quad\; E = 0 \\
&\quad\; K = 0 \\
&\quad\; P = 0
\end{aligned}
$$

Combined with equations (D.5) to (D.8) above, these give

$B = 0$

$F = C_2^2 - S_2^2 = C_4$

$L = 2S_2 C_2 \quad = S_4$

$R = 0$

(c) The device leaves light vibrating at angle θ with x-direction unchanged, that is it turns type (2c) light into type (2c) light. So,

$$\begin{bmatrix} 1 \\ C_2 \\ S_2 \\ 0 \end{bmatrix} = \begin{bmatrix} 1 & 0 & Y & D \\ 0 & C_2^2-S_2^2 & G & H \\ 0 & 2C_2 S_2 & M & N \\ 0 & 0 & X & T \end{bmatrix} \begin{bmatrix} 1 \\ C_2 \\ S_2 \\ 0 \end{bmatrix}$$

that is

$1 = 1 + YS_2, \quad Y = 0$

$C_2 = (C_2^2 - S_2^2)C_2 + GS_2, \quad G = \left[C_2 - (1 - 2S_2^2)C_2^2 \right]/S_2$

$\quad = 2S_2 C_2 = S_4$

$S_2 = 2C_2^2 S_2 + MS_2, \quad M = 1 - 2C_2^2$

$\quad = S_2^2 + C_2^2 - 2C_2^2 = S_2^2 - C_2^2 = -C_4$

$0 = XS_2, \quad X = 0$

(d) Consider right-handed circularly polarized light (type 3). The intensity is unaffected, but the phase difference Δ, originally $\pi/2$, becomes $3\pi/2$, or equivalently $-\pi/2$. This corresponds merely to a circle described in the opposite sense, so H^2 and K^2 are unaffected. $\cos\Delta = 0$, still, but $\sin\Delta$ is now -1. The Stokes parameters are thus $I = 2A^2$, $Q = 0$, $U = 0_1$, $V = -2A^2$, in accordance with the left-handed Stokes column for type (3) light. On cancelling $2A^2$, therefore, we obtain

$$
\begin{bmatrix} 1 \\ 0 \\ 0 \\ -1 \end{bmatrix} = \begin{bmatrix} 1 & 0 & 0 & D \\ 0 & C_2^2-S_2^2 & 2C_2S_2 & H \\ 0 & 2C_2S_2 & S_2^2-C_2^2 & N \\ 0 & 0 & 0 & T \end{bmatrix} \begin{bmatrix} 1 \\ 0 \\ 0 \\ 1 \end{bmatrix}
$$

that is

$$1 = 1 + D, \qquad D = 0$$

$$0 = H, \qquad H = 0$$

$$0 = N, \qquad N = 0$$

$$-1 = T, \qquad T = -1$$

The matrix of the half-wave plate is thus

$$
\begin{bmatrix}
1 & 0 & 0 & 0 \\
0 & C_4 & S_4 & 0 \\
0 & S_4 & -C_4 & 0 \\
0 & 0 & 0 & -1
\end{bmatrix}
$$

D.3 QUARTER-WAVE PLATE: GENERAL ORIENTATION (optic axis at angle θ with x-direction)

For a negative uniaxial crystal, the E-vibration in the plane of the optic axis travels faster than the O-vibration. We shall assume that the plate has a thickness such that the O-wave is retarded by $\pi/2$ relative to the E-wave. Let $C_1 = \cos\theta$, $S_1 = \sin\theta$, $C_2 = \cos2\theta$, $S_2 = \sin2\theta$. Let the Mueller matrix be

$$
Z = \begin{bmatrix}
W & B & Y & D \\
E & F & G & H \\
K & L & M & N \\
P & R & X & T
\end{bmatrix}
$$

(a) The device leaves unpolarized light unchanged, that is it turns type (1) light into type (1) light. Therefore,

$$
\begin{bmatrix} 1 \\ O \\ O \\ O \end{bmatrix} = Z \begin{bmatrix} 1 \\ O \\ O \\ O \end{bmatrix} = \begin{bmatrix} W & B & Y & D \\ E & F & G & H \\ K & L & M & N \\ P & R & X & T \end{bmatrix} \begin{bmatrix} 1 \\ O \\ O \\ O \end{bmatrix} = \begin{bmatrix} W \\ E \\ K \\ P \end{bmatrix}
$$

that is

$W = 1$

$E = O$

$K = O$

$P = O$

(b) The device leaves alone vibrations along its own optic axis, at an angle θ with the x-direction, that is it turns type (2c) light into type (2c) light. (For convenience we assume unit intensity.) Therefore,

$$
\begin{bmatrix} 1 \\ C_2 \\ S_2 \\ O \end{bmatrix} = \begin{bmatrix} 1 & B & Y & D \\ O & F & G & H \\ O & L & M & N \\ O & R & X & T \end{bmatrix} \begin{bmatrix} 1 \\ C_2 \\ S_2 \\ O \end{bmatrix} = \begin{bmatrix} 1 + BC_2 + YS_2 \\ FC_2 + GS_2 \\ LC_2 + MS_2 \\ RC_2 + XS_2 \end{bmatrix}
$$

that is

$1 = 1 + BC_2 + YS_2$ \hfill (D.9)

$C_2 = FC_2 + GS_2$ \hfill (D.10)

$S_2 = LC_2 + MS_2$ \hfill (D.11)

$O = RC_2 + XS_2$ \hfill (D.12)

(c) To investigate the action of the device on a vibration parallel to the x-direction, it is very useful to use the rotation matrices, which convert the coordinates of a point, referred to one set of rectangular axes, into the coordinates referred to a pair of axes inclined at an angle θ with the original pair (see Figure D.1). We need to be able to go both ways. In the diagram:

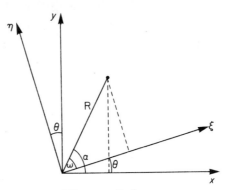

Figure D.1

$\eta = R\sin(\alpha - \theta)$ $x = R\cos(\theta + \omega)$

$\eta = R\sin\alpha\cos\theta - R\cos\alpha\sin\theta$ $x = R\cos\theta\cos\omega - R\sin\theta\sin\omega$

$\eta = y\cos\theta - x\sin\theta$ $x = \xi\cos\theta - \eta\sin\theta$

$\xi = R\cos(\alpha - \theta)$ $y = R\sin(\theta + \omega)$

$\xi = R\cos\alpha\cos\theta + R\sin\alpha\sin\theta$ $y = R\sin\theta\cos\omega + R\cos\theta\sin\omega$

$\xi = x\cos\theta + y\sin\theta$ $y = \xi\sin\theta + \eta\cos\theta$

In matrix form, we write

$$\begin{bmatrix} \xi \\ \eta \end{bmatrix} = \begin{bmatrix} \cos\theta & \sin\theta \\ -\sin\theta & \cos\theta \end{bmatrix} \begin{bmatrix} x \\ y \end{bmatrix} \quad \text{and} \quad \begin{bmatrix} x \\ y \end{bmatrix} = \begin{bmatrix} \cos\theta & -\sin\theta \\ \sin\theta & \cos\theta \end{bmatrix} \begin{bmatrix} \\ \end{bmatrix}$$

The optic axis is now along the ξ-direction, so the E-ray vibrates along the ξ-direction, the O-ray along the η-direction. The original vibration is

$x = A\sin pt, \quad y = 0$

Therefore,

$$\begin{bmatrix} \text{E-vibration} \\ \text{O-vibration} \end{bmatrix} = \begin{bmatrix} C_1 & S_1 \\ -S_1 & C_1 \end{bmatrix} \begin{bmatrix} A\sin pt \\ 0 \end{bmatrix} = \begin{bmatrix} AC_1 \sin pt \\ -AS_1 \sin pt \end{bmatrix}$$

In the phase plate, the O-vibration is retarded by $\pi/2$:

$$\sin(pt - \pi/2) = \sin pt \cos \pi/2 - \cos pt \sin \pi/2 = -\cos pt$$

Therefore, after phase plate,

$$\begin{bmatrix} \text{E-vibration} \\ \text{O-vibration} \end{bmatrix} = \begin{bmatrix} AC_1 \sin pt \\ AS_1 \cos pt \end{bmatrix}$$

To find the Stokes parameters, we need H and K, the components of the final vibration parallel to the x- and y-directions. Using the rotation matrices again,

$$\begin{bmatrix} x \\ y \end{bmatrix} = \begin{bmatrix} C_1 & -S_1 \\ S_1 & C_1 \end{bmatrix} \begin{bmatrix} AC_1 \sin pt \\ AS_1 \cos pt \end{bmatrix} = A \begin{bmatrix} C_1^2 \sin pt - S_1^2 \cos pt \\ S_1 C_1 \sin pt + S_1 C_1 \cos pt \end{bmatrix}$$

$$= A \begin{bmatrix} C_1^2 & -S_1^2 \\ S_1 C_1 & S_1 C_1 \end{bmatrix} \begin{bmatrix} \sin pt \\ \cos pt \end{bmatrix}$$

Now suppose $x = H\sin(pt + \phi)$, $y = K\sin(pt + \beta)$, so that H and K are the components of the resultant vibration in the x- and y-directions and $\Delta = (\beta - \phi)$ is the phase difference between the resultant vibrations in the x- and y-directions.

Then $x = H\sin pt \cos\phi + H\cos pt \sin\phi$

$\qquad y = K\sin pt \cos\beta + K\cos pt \sin\beta$

That is

$$\begin{bmatrix} x \\ y \end{bmatrix} = \begin{bmatrix} H\cos\phi & H\sin\phi \\ K\cos\beta & K\sin\beta \end{bmatrix} \begin{bmatrix} \sin pt \\ \cos pt \end{bmatrix}$$

In the last two equations, we have $\begin{bmatrix} x \\ y \end{bmatrix}$ produced

from $\begin{bmatrix} \sin pt \\ \cos pt \end{bmatrix}$ in two different ways. The matrices

multiplying $\begin{bmatrix} \sin pt \\ \cos pt \end{bmatrix}$ to produce $\begin{bmatrix} x \\ y \end{bmatrix}$ must therefore

be the same in the two equations, that is (setting A = unity, for convenience)

$$\begin{bmatrix} H\cos\phi & H\sin\phi \\ K\cos\beta & K\sin\beta \end{bmatrix} = \begin{bmatrix} C_1^2 & -S_1^2 \\ S_1C_1 & S_1C_1 \end{bmatrix}$$

We now postmultiply each of these equal matrices by its transpose (the matrix obtained by interchanging the rows and the columns of the original matrix - that is the transpose of $\begin{bmatrix} A & B \\ C & D \end{bmatrix}$ is $\begin{bmatrix} A & C \\ B & D \end{bmatrix}$) This gives

$$\begin{bmatrix} H\cos\phi & H\sin\phi \\ K\cos\beta & K\sin\beta \end{bmatrix} \begin{bmatrix} H\cos\phi & K\cos\beta \\ H\sin\phi & K\sin\beta \end{bmatrix}$$

$$= \begin{bmatrix} C_1^2 & -S_1^2 \\ S_1C_1 & S_1C_1 \end{bmatrix} \begin{bmatrix} C_1^2 & S_1C_1 \\ -S_1^2 & S_1C_1 \end{bmatrix}$$

that is

$$\begin{bmatrix} H^2 & HK\cos(\phi-\beta) \\ HK\cos(\phi-\beta) & K^2 \end{bmatrix} = \begin{bmatrix} C_1^4 + S_1^4 & C_1S_1(C_1^2- \\ C_1S_1(C_1^2-S_1^2) & 2S_1^2C_1^2 \end{bmatrix}$$

Therefore, using the fact that corresponding elements in the two matrices are equal, we find

$$I = H^2 + K^2 = C_1^4 + S_1^4 + 2S_1^2C_1^2 = (C_1^2 + S_1^2)^2 = 1$$

$$Q = H^2 - K^2 = C_1^4 + S_1^4 - 2S_1^2 C_1^2 = (C_1^2 - S_1^2)^2 = C_2^2$$

$$U = 2HK\cos\Delta = 2HK\cos(\phi - \beta) = 2C_1 S_1 (C_1^2 - S_1^2) = S_2 C_2$$

giving

$$V^2 = I^2 - Q^2 - U^2 = 1 - C_2^4 - S_2^2 C_2^2 = 1 - C_2^2 (C_2^2 + S_2^2) = S_2^2$$

Here, however, V could be either $+S_2$ or $-S_2$. To find which, we return to the equation

$$\begin{bmatrix} H\cos\phi & H\sin\phi \\ K\cos\beta & K\sin\beta \end{bmatrix} = \begin{bmatrix} C_1^2 & -S_1^2 \\ S_1 C_1 & S_1 C_1 \end{bmatrix}$$

If we equate the determinants of these two matrices, we find that

$$HK(\cos\phi\sin\beta - \sin\phi\cos\beta) = (C_1^3 S_1 + S_1^3 C_1)$$

Hence immediately,

$$HK\sin(\beta - \phi) = S_1 C_1$$

Therefore,

$$V = 2HK\sin\Delta = 2HK\sin(\beta - \phi) = 2S_1 C_1 = S_2$$

The four Stokes parameters that we have just calculated are the result of operating on type (2a) light (normalised to unit amplitude), so that

$$\begin{bmatrix} 1 \\ C_2^2 \\ C_2 S_2 \\ S_2 \end{bmatrix} = \begin{bmatrix} 1 & B & Y & D \\ 0 & F & G & H \\ 0 & L & M & N \\ 0 & R & X & T \end{bmatrix} \begin{bmatrix} 1 \\ 1 \\ 0 \\ 0 \end{bmatrix} = \begin{bmatrix} 1+B \\ F \\ L \\ R \end{bmatrix}$$

that is

$B = 0$, from equation (D.9) $Y = 0$

$F = C_2^2$, from (D.10) $GS_2 = C_2 - C_2^3 = C_2(1 - C_2^2)$

$$= C_2 S_2^2 \qquad G = C_2 S_2$$

$$L = S_2 C_2, \text{ from (D.11)} \quad MS_2 = S_2 - S_2 C_2^2$$

$$M = (1 - C_2^2) = S_2^2$$

$$R = S_2, \text{ from (D.12)} \quad XS_2 = -S_2 C_2 \quad X = -C_2$$

(d) Action of the device on right-handed circularly polarized light, for which, originally, for a beam of intensity 2 units, $x = \cos\omega t$, $y = -\sin\omega t$. As in (c) $H = K = 1$, $\Delta = \pi/2$, $I = V = 2$, $Q = U = 0$

$$\begin{bmatrix} \xi \\ \eta \end{bmatrix} = \begin{bmatrix} C_1 & S_1 \\ -S_1 & C_1 \end{bmatrix} \begin{bmatrix} \cos\omega t \\ -\sin\omega t \end{bmatrix} = \begin{bmatrix} C_1\cos\omega t - S_1\sin\omega t \\ -S_1\cos\omega t - C_1\sin\omega t \end{bmatrix}$$

In the plate, the O-vibration along the η-direction is retarded by $\pi/2$:

$$\sin(\omega t - \pi/2) = -\cos\omega t, \quad \cos(\omega t - \pi/2) = \sin\omega t$$

So, after the plate,

$$\begin{bmatrix} \xi \\ \eta \end{bmatrix} = \begin{bmatrix} C_1\cos\omega t - S_1\sin\omega t \\ -S_1\sin\omega t + C_1\cos\omega t \end{bmatrix} = \begin{bmatrix} -S_1 & C_1 \\ -S_1 & C_1 \end{bmatrix} \begin{bmatrix} \sin\omega t \\ \cos\omega t \end{bmatrix}$$

Therefore, final x and y are given by

$$\begin{bmatrix} x \\ y \end{bmatrix} = \begin{bmatrix} C_1 & -S_1 \\ S_1 & C_1 \end{bmatrix} \begin{bmatrix} -S_1 & C_1 \\ -S_1 & C_1 \end{bmatrix} \begin{bmatrix} \sin\omega t \\ \cos\omega t \end{bmatrix}$$

Thus, after the plate,

$$\begin{bmatrix} H\cos\phi & H\sin\phi \\ K\cos\beta & K\sin\beta \end{bmatrix} \begin{bmatrix} \sin\omega t \\ \cos\omega t \end{bmatrix} = \begin{bmatrix} -C_1 S_1 + S_1^2 & C_1^2 - S_1 C_1 \\ -S_1^2 - C_1 S_1 & S_1 C_1 + C_1^2 \end{bmatrix} \begin{bmatrix} \text{si} \\ \text{co} \end{bmatrix}$$

as in (c). Again, as in (c), multiplying the two-by-two matrix on each side by its transpose, we get

$$\begin{bmatrix} H^2 & HK\cos\Delta \\ HK\cos\Delta & K^2 \end{bmatrix} = \begin{bmatrix} C_1^2 - 2C_1 S_1 + S_1^2 & C_1^2 - S_1^2 \\ C_1^2 - S_1^2 & C_1^2 + 2S_1 C_1 + S_1^2 \end{bmatrix}$$

Therefore,

$$I = H^2 + K^2 = 2C_1^2 + 2S_1^2 = 2$$

$$Q = H^2 - K^2 = -4C_1 S_1 = -2S_2$$

$$U = 2HK\cos\Delta = 2(C_1^2 - S_1^2) = 2C_2$$

$$V^2 = I^2 - Q^2 - U^2 = 4 - 4S_2^2 - 4C_2^2 = 0, \quad V = 0$$

Thus

$$
\begin{bmatrix} 2 \\ -2S_2 \\ 2C_2 \\ 0 \end{bmatrix}
=
\begin{bmatrix} 1 & B & Y & D \\ 0 & F & G & H \\ 0 & L & M & N \\ 0 & R & X & T \end{bmatrix}
\begin{bmatrix} 2 \\ 0 \\ 0 \\ 2 \end{bmatrix}
=
\begin{bmatrix} 2+2D \\ 2H \\ 2N \\ 2T \end{bmatrix}
$$

That is $D = 0$, $H = -S_2$, $N = C_2$, $T = 0$

Therefore, the Mueller matrix of the quarter-wave plate is

$$
\begin{bmatrix}
1 & 0 & 0 & 0 \\
0 & C_2^2 & C_2 S_2 & -S_2 \\
0 & C_2 S_2 & S_2^2 & C_2 \\
0 & S_2 & -C_2 & 0
\end{bmatrix}
$$

D.4 GENERAL PHASE PLATE, PRODUCING PHASE RETARDATION δ IN ORDINARY VIBRATION AND HAVING OPTIC AXIS AT ANGLE θ WITH x-DIRECTION

Let the Mueller matrix be

$$
Z =
\begin{bmatrix}
W & B & Y & D \\
E & F & G & H \\
K & L & M & N \\
P & R & X & T
\end{bmatrix}
$$

and let $C_1 = \cos\theta$, $C_2 = \cos2\theta$, $S_1 = \sin\theta$, $S_2 = \sin2\theta$, $\beta = \cos\Delta$, $\mu = \sin\Delta$.

(a) The device leaves unpolarized light unaffected, that is it turns type (1) light into type (1) light. Therefore,

$$
\begin{bmatrix} 1 \\ 0 \\ 0 \\ 0 \end{bmatrix} = \begin{bmatrix} W & B & Y & D \\ E & F & G & H \\ K & L & M & N \\ P & R & X & T \end{bmatrix} \begin{bmatrix} 1 \\ 0 \\ 0 \\ 0 \end{bmatrix} = \begin{bmatrix} W \\ E \\ K \\ P \end{bmatrix} \quad \text{that is} \quad \begin{array}{l} W = 1 \\ E = 0 \\ K = 0 \\ P = 0 \end{array}
$$

(b) The device leaves unchanged any vibrations along its own optic axis, that is it turns type (2c) light into type (2c) light:

$$
\begin{bmatrix} 1 \\ C_2 \\ S_2 \\ 0 \end{bmatrix} = \begin{bmatrix} 1 & B & Y & D \\ 0 & F & G & H \\ 0 & L & M & N \\ 0 & R & X & T \end{bmatrix} \begin{bmatrix} 1 \\ C_2 \\ S_2 \\ 0 \end{bmatrix} = \begin{bmatrix} 1+BC_2+YS_2 \\ FC_2 + GS_2 \\ LC_2 + MS_2 \\ RC_2 + XS_2 \end{bmatrix}
$$

that is

$$BC_2 + YS_2 = 0, \quad FC_2 + GS_2 = C_2, \quad LC_2 + MS_2 = S_2,$$
$$RC_2 + XS_2 = 0.$$

(c) Action of the device on vibration of unit amplitude along the x-direction that is for which $x = \sin\omega t$, $y = 0$, $I = Q = 1$, $U = V = 0$. Using the argument on rotation of axes, decribed in connection with the quarter-wave plate, if E-vibration is represented by ξ and the O-vibration by η, then $\xi = C_1\sin\omega t$ and $\eta = - S_1\sin\omega t$.

The phase plate retards the O-vibration by δ with respect to the E-vibration. Therefore, after the phase plate,

$$\eta = - S_1\sin(\omega t - \delta) = - S_1\sin\omega t\cos\delta + S_1\cos\omega t\sin\delta$$

$$\xi = C_1\sin\omega t$$

Thus the final components along the x- and y-axes are

$$\begin{bmatrix} x \\ y \end{bmatrix} = \begin{bmatrix} C_1 & -S_1 \\ S_1 & C_1 \end{bmatrix} \begin{bmatrix} C_1\sin\omega t \\ -S_1\beta\sin\omega t + S_1\mu\cos\omega t \end{bmatrix}$$

$$= \begin{bmatrix} C_1^2\sin\omega t + S_1^2\beta\sin\omega t - S_1^2\mu\cos\omega t \\ S_1C_1\sin\omega t - S_1C_1\beta\sin\omega t + S_1C_1\mu\cos\omega t \end{bmatrix}$$

$$= \begin{bmatrix} C_1^2 + S_1^2\beta & -S_1^2\mu \\ S_1C_1(1-\beta) & S_1C_1\mu \end{bmatrix} \begin{bmatrix} \sin\omega t \\ \cos\omega t \end{bmatrix}$$

where $\beta = \cos\delta$ and $\mu = \sin\delta$.

By reference to the work on the quarter-wave plate, if, finally, $x = H\sin(\omega t + \phi)$ and $y = K\sin(\omega t + \alpha)$, so that $\Delta = (\alpha - \phi)$, then, also,

$$\begin{bmatrix} x \\ y \end{bmatrix} = \begin{bmatrix} H\cos\phi & H\sin\phi \\ K\cos\alpha & K\sin\alpha \end{bmatrix} \begin{bmatrix} \sin\omega t \\ \cos\omega t \end{bmatrix}$$

so that

$$\begin{bmatrix} H\cos\phi & H\sin\phi \\ K\cos\alpha & K\sin\alpha \end{bmatrix} = \begin{bmatrix} C_1^2 + S_1^2\beta & -S_1^2\mu \\ S_1C_1(1-\beta) & S_1C_1\mu \end{bmatrix}$$

By multiplying each matrix by its transpose, we get, as before,

$$\begin{bmatrix} H^2 & HK\cos\Delta \\ HK\cos\Delta & K^2 \end{bmatrix} = \begin{bmatrix} C_1^4 + 2S_1^2C_1^2\beta + S_1^4 & S_2C_2(1-\beta)/2 \\ S_2C_2(1-\beta)/2 & 2S_1^2C_1^2(1-\beta) \end{bmatrix}$$

This gives $I = H^2 + K^2 = 1$, $Q = H^2 - K^2 = C_2^2 + S_2^2\beta$, $U = 2HK\cos\Delta = S_2C_2(1-\beta)$, so that (from $I^2 = Q^2 + U^2 + V^2$) $V^2 = S_2^2\mu^2$. It follows that V is either $+ S_2\mu$ or $- S_2\mu$. To find which, we return to the equation

$$\begin{bmatrix} H\cos\phi & H\sin\phi \\ K\cos\alpha & K\sin\alpha \end{bmatrix} = \begin{bmatrix} C_1^2 + S_1^2\beta & -S_1^2\mu \\ S_1C_1(1-\beta) & S_1C_1\mu \end{bmatrix}$$

If we equate the determinants of these two matrices we find that

$$HK(\sin\alpha\cos\phi - \cos\alpha\sin\phi) = \mu S_1 C_1 (C_1^2 + S_1^2\beta) + \mu S_1^3 C_1 ($$

$$= \mu S_1 C_1$$

Hence, immediately,

$$HK\sin(\alpha - \phi) = S_1 C_1 \mu$$

Therefore,

$$V = 2HK\sin\Delta = 2HK\sin(\alpha - \phi) = 2S_1 C_1 \mu = S_2 \mu$$

Thus

$$\begin{bmatrix} 1 \\ C_2^2 + S_2^2\beta \\ S_2 C_2 (1-\beta) \\ S_2\mu \end{bmatrix} = \begin{bmatrix} 1 & B & Y & D \\ 0 & F & G & H \\ 0 & L & M & N \\ 0 & R & X & T \end{bmatrix} \begin{bmatrix} 1 \\ 1 \\ 0 \\ 0 \end{bmatrix} = \begin{bmatrix} 1+B \\ F \\ L \\ R \end{bmatrix}$$

that is $B = 0$, $F = C_2^2 + S_2^2\beta$, $L = S_2 C_2 (1 - \beta)$, $R = S_2\mu$.

Combining these with the four equations in section (b) of this derivation gives $Y = 0$, $G = C_2 S_2 (1 - \beta)$, $M = S_2^2 + C_2^2\beta$, $X = -C_2\mu$.

(d) Action of the plate on circularly polarized light, for which $x = \cos\omega t$, $y = -\sin\omega t$, and Stokes parameters are $(2, 0, 0, 2)$. As in section (c), we transform to $\xi\eta$-axes, along and perpendicular to the optic axis of the plate, retard the O-vibration by δ relative to the E-vibration, then return to the axes along the x- and y-directions. The result is

$$\begin{bmatrix} x \\ y \end{bmatrix} = \begin{bmatrix} C_1 & -S_1 \\ S_1 & C_1 \end{bmatrix} \begin{bmatrix} 1 & 0 \\ \mu & \beta \end{bmatrix} \begin{bmatrix} C_1 & S_1 \\ -S_1 & C_1 \end{bmatrix} \begin{bmatrix} 0 & 1 \\ -1 & 0 \end{bmatrix} \begin{bmatrix} \sin\omega t \\ \cos\omega t \end{bmatrix}$$

$$= \begin{bmatrix} C_1 - \mu S_1 & -S_1\beta \\ S_1 + \mu C_1 & C_1\beta \end{bmatrix} \begin{bmatrix} -S_1 & C_1 \\ -C_1 & -S_1 \end{bmatrix} \begin{bmatrix} \sin\omega t \\ \cos\omega t \end{bmatrix}$$

$$= G \begin{bmatrix} \sin\omega t \\ \cos\omega t \end{bmatrix}, \text{ say}$$

As in section (c), we now want to multiply the matrix connecting $\begin{bmatrix} x \\ y \end{bmatrix}$ and $\begin{bmatrix} \sin\omega t \\ \cos\omega t \end{bmatrix}$ by its transpose. It is here convenient to use a standard theorem of matrix algebra, on the transpose of a product (see section I.8). If the symbol T denotes transposition, it is easy to verify that $(AB)^T = (B)^T(A)^T$, that is that the transpose of the produce of two matrices is equal to the product of the transposes of the matrices separately, but with the order reversed. Thus, $(AB)(AB)^T = (A)(B)(B)^T(A)^T$ $= (A)(BB^T)(A)^T$ (since matrix multiplication is associative.

Applying this result

$$GG^T = \begin{bmatrix} C_1 - \mu S & -S_1\beta \\ S_1 + \mu C_1 & C_1\beta \end{bmatrix} \begin{bmatrix} -S_1 & C_1 \\ -C_1 & -S_1 \end{bmatrix} \begin{bmatrix} -S_1 & -C_1 \\ C_1 & -S_1 \end{bmatrix} \begin{bmatrix} C_1 - \mu S_1 & S_1 + \mu C_1 \\ -S_1\beta & C_1\beta \end{bmatrix}$$

The two middle matrices multiply to give the unit matrix $\begin{bmatrix} 1 & 0 \\ 0 & 1 \end{bmatrix}$, and so can be ignored. The two outer ones, on multiplication and simplification give $\begin{bmatrix} 1 - \mu S_2 & + \mu C_2 \\ + \mu C_2 & 1 + \mu S_2 \end{bmatrix}$. As before, this must equal

$\begin{bmatrix} H^2 & HK\cos\Delta \\ HK\cos\Delta & K^2 \end{bmatrix}$. Thus $I = H^2 + K^2 = 2$,

$Q = H^2 - K^2 = -2\mu S_2$, $U = 2HK\cos\Delta = 2\mu C_2$. Then $I^2 = Q^2 + U^2 + V^2$ leads to $V^2 = 4\beta^2$, but V may be $+ 2\beta$ or $- 2\beta$. To find out which, we again equate the two forms that we have found for the two-by-two matrix which links the column matrices $\begin{bmatrix} x \\ y \end{bmatrix}$

and $\begin{bmatrix} \sin\omega t \\ \cos\omega t \end{bmatrix}$ The equation reads

$$\begin{bmatrix} H\cos\phi & H\sin\phi \\ K\cos\alpha & K\sin\alpha \end{bmatrix} = \begin{bmatrix} C_1-\mu S_1 & -S_1\beta \\ S_1+\mu C_1 & C_1\beta \end{bmatrix}\begin{bmatrix} -S_1 & C_1 \\ -C_1 & -S_1 \end{bmatrix}$$

Equating the determinants of this matrix on the left to the product of the determinants on the right, we obtain

$$HK(\sin\alpha\cos\phi - \cos\alpha\sin\phi) = (C_1^2\beta-\mu\beta S_1 C_1+S_1^2\beta+\mu\beta S_1 C_1)(S_1^2$$

$$= \beta$$

Hence, immediately,

$HK\sin(\alpha - \phi) = \beta$, so that $V = 2HK\sin\Delta = 2HK\sin(\alpha - \phi$

$= 2\beta$.   Thus:

$$\begin{bmatrix} 2 \\ -2\mu S_2 \\ 2\mu C_2 \\ 2\beta \end{bmatrix} = \begin{bmatrix} 1 & B & Y & D \\ 0 & F & G & H \\ 0 & L & M & N \\ 0 & R & X & T \end{bmatrix}\begin{bmatrix} 2 \\ 0 \\ 0 \\ 2 \end{bmatrix} = \begin{bmatrix} 2+2D \\ 0+2H \\ 0+2N \\ 0+2T \end{bmatrix}$$

that is $D = 0$, $H = -\mu S_2$, $N = \mu C_2$, $T = \beta$.

Therefore, the matrix of the general phase plate is

$$\begin{bmatrix} 1 & 0 & 0 & 0 \\ 0 & C_2^2 + S_2^2\beta & C_2 S_2(1-\beta) & -S_2\mu \\ 0 & S_2 C_2(1-\beta) & S_2^2 + \beta C_2^2 & C_2\mu \\ 0 & S_2\mu & -C_2\mu & \beta \end{bmatrix}$$

Notes

(1) This matrix reduces to the matrices representing plain glass, half-wave plate and quarter-wave plate on putting $\delta = 0$, π and $\pi/2$ respectively (see Table 3 in section IV.3).

(2) The matrix $\begin{bmatrix} 0 & 1 \\ -1 & 0 \end{bmatrix}$ was inserted to interchange $\cos\omega t$ and $\sin\omega t$ in the right-hand column.

(3) The matrix $\begin{bmatrix} 1 & 0 \\ \mu & \beta \end{bmatrix}$ has the effect of retarding the O-wave by δ with respect to the E-wave. As in section (c), we write

$$\begin{bmatrix} \xi \\ \eta \end{bmatrix} = \begin{bmatrix} S_1 & C_1 \\ C_1 & -S_1 \end{bmatrix} \begin{bmatrix} \sin\omega t \\ \cos\omega t \end{bmatrix}$$

before and $\begin{bmatrix} S_1 & C_1 \\ \mu S_1 + \beta C_1 & \mu C_1 - \beta S_1 \end{bmatrix} \begin{bmatrix} \sin\omega t \\ \cos\omega t \end{bmatrix}$ after the

phase plate. This matrix is obtained from the previous one on substituting $(\omega t - \delta)$ for ωt in the expression for η. Also,

$$\begin{bmatrix} 1 & 0 \\ \mu & \beta \end{bmatrix} \begin{bmatrix} S_1 & C_1 \\ C_1 & -S_1 \end{bmatrix} = \begin{bmatrix} S_1 & C_1 \\ \mu S_1 + \beta C_1 & \mu C_1 - \beta S_1 \end{bmatrix}$$

Appendix E
Derivation of
Jones Matrices

E.1 THE POLARIZER, SUCH AS A SHEET OF POLAROID

(a) First we will consider an ideal linear polarizer, whose pass-plane is horizontal, that is parallel to the x-direction. This allows only vibrations parallel to the x-direction to pass through the device, so that in the general equations describing the behaviour of an ideal device H_2 is equal to H_1 and K_2 is zero, no matter what the values of H_1 and K_1 may be. Thus, the equations become

$$H_1 = J_{11}H_1 + J_{12}K_1\exp(i\Delta_1)$$

$$0 = J_{21}H_1 + J_{22}K_1\exp(i\Delta_1)$$

for all values of H_1, K_1 and Δ_1.
Set $K_1 = 0$. The equations become $H_1 = J_{11}H_1$, that is $J_{11} = 1$, and $0 = J_{21}H_1$ for all H_1, that is $J_{21} = 0$.
Set $H_1 = 0$. The equations become $0 = J_{12}K_1\exp(i\Delta_1)$ and $0 = J_{22}K_1\exp(i\Delta_1)$ for all K_1 and Δ_1, that is J_{12} and J_{22} are both zero. The Jones matrix for this ideal polarizer is thus $\begin{bmatrix} 1 & 0 \\ 0 & 0 \end{bmatrix}$

In practice, of course, a sheet of polaroid introduces some attenuation, even for the preferred plane of vibration, and its optical thickness will be sufficient to introduce at least several hundred wavelengths of retardation. For some purposes, for example in an interferometer calculation, the above matrix needs to be multiplied by a complex scalar,

which represents the complex amplitude transmittance
of the polarizer. If only one Maxwell column is involved, information about the absolute phase of the
vibration is unlikely to be needed. As a matter of
convenience, however, a given Jones matrix can often
be converted into a simpler or more symmetrical form
by multiplying it by a suitable phase factor. The
above matrix, for example, could equally well be

be represented as $\begin{bmatrix} -1 & 0 \\ 0 & 0 \end{bmatrix}$ or $\begin{bmatrix} i & 0 \\ 0 & 0 \end{bmatrix}$ if this

seemed desirable.
It is just as easy to show that, for a polarizer
with its pass-plane vertical, that is parallel to
the y-axis, the Jones matrix can be represented by

$\begin{bmatrix} 0 & 0 \\ 0 & 1 \end{bmatrix}$ (or $\begin{bmatrix} 0 & 0 \\ 0 & e^{i\phi} \end{bmatrix}$ for example).

(b) We now consider the more general case of a polarizer
with its pass-plane making an angle θ with the x-
direction (see Figure E.1). Suppose we have incident on the polarizer a plane-polarized vibration
with its vibration direction making an angle α with
the x-axis. Suppose the amplitude of this incident
vibration is A; then by our definition $X_1 = A\cos\alpha$
and $Y_1 = A\sin\alpha$. Only the component of this vibration along the direction of the pass-plane of the
polaroid will pass through the device. The amplitude of this component is

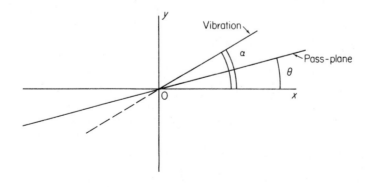

Figure E.1

$$U = A\cos(\alpha - \theta) = A\cos\alpha\cos\theta + A\sin\alpha\sin\theta$$
$$= X_1\cos\theta + Y_1\sin\theta$$

The component of this vibration parallel to the x-axis is

$$X_2 = U\cos\theta = X_1\cos^2\theta + Y_1\sin\theta\cos\theta$$

The component of the emerging vibration parallel to the y-axis is

$$Y_2 = U\sin\theta = X_1\cos\theta\sin\theta + Y_1\sin^2\theta$$

The last two equations can be written in matrix form

$$\begin{bmatrix} X_2 \\ Y_2 \end{bmatrix} = \begin{bmatrix} \cos^2\theta & \sin\theta\cos\theta \\ \sin\theta\cos\theta & \sin^2\theta \end{bmatrix} \begin{bmatrix} X_1 \\ Y_1 \end{bmatrix}$$

so that the matrix of the device is

$$\begin{bmatrix} \cos^2\theta & \sin\theta\cos\theta \\ \sin\theta\cos\theta & \sin^2\theta \end{bmatrix}$$

E.2 JONES MATRIX OF THE GENERAL LINEAR RETARDER

We consider a crystal plate with its optic axis at an angle α to the x-axis, and we suppose that this plate *retards* the phase of the ordinary vibration (perpendicular to the optic axis) by δ relative to the vibration along the optic axis, that is to the E-vibration. The device makes no difference to the Maxwell column of a plane-polarized vibration parallel to the optic axis. Suppose the amplitude of this vibration is A, then, as in the section on the polaroid, we know that its components are $X_1 = A\cos\alpha$ and $Y_1 = A\sin\alpha$. Since this vibration is unaffected by the device, the emerging components are the same as the entering ones, that is $X_2 = A\cos\alpha$ and $Y_2 = A\sin\alpha$.

Substituting in the general equation for the behaviour of the device, we find the equations

$$A\cos\alpha = J_{11}A\cos\alpha + J_{12}A\sin\alpha$$

$$A\sin\alpha = J_{21}A\cos\alpha + J_{22}A\sin\alpha$$

which lead immediately to

$$\tan\alpha = \frac{\sin\alpha}{\cos\alpha} = \frac{1 - J_{11}}{J_{12}} = \frac{J_{21}}{1 - J_{22}}$$

(given that $\cos\alpha$ does not vanish). We now consider the action of the device on a beam of plane-polarized light with its vibration plane parallel to the x-axis and of amplitude A. For this beam the Maxwell column is

$$E_1 = \begin{bmatrix} A \\ 0 \end{bmatrix}$$

We now need to consider the components of this vibration along and parallel to the optic axis. To find these components we use the rotation matrices which we introduced while developing the Mueller matrix of a quarter-wave plate. The result is

$$\begin{bmatrix} \text{Component along optic axis} \\ \text{Component perpendicular to optic axis} \end{bmatrix} = \begin{bmatrix} U_1 \\ V_1 \end{bmatrix} \quad \text{(say)}$$

$$= \begin{bmatrix} \cos\alpha & \sin\alpha \\ -\sin\alpha & \cos\alpha \end{bmatrix} \begin{bmatrix} A \\ 0 \end{bmatrix} = \begin{bmatrix} A\cos\alpha \\ -A\sin\alpha \end{bmatrix}$$

The component perpendicular to the optic axis is now *retarded* by a phase angle δ, that is to say V_1 is multiplied by $\exp(-i\delta)$. The matrix of components along and perpendicular to the optic axis thus becomes

$$\begin{bmatrix} U_2 \\ V_2 \end{bmatrix} = \begin{bmatrix} A\cos\alpha \\ -A\sin\alpha\exp(-i\delta) \end{bmatrix}$$

We now need to return to the former axes. To do this, we again use the rotation matrix, but this time in reverse to get from our UV-axes to our former XY-axes. The final form of the Maxwell column is thus

$$E_2 = \begin{bmatrix} \cos\alpha & -\sin\alpha \\ \sin\alpha & \cos\alpha \end{bmatrix} \begin{bmatrix} A\cos\alpha \\ -A\sin\alpha\exp(-i\delta) \end{bmatrix}$$

$$= \begin{bmatrix} A\cos^2\alpha + A\sin^2\alpha\exp(-i\delta) \\ A\cos\alpha\sin\alpha - A\sin\alpha\cos\alpha\exp(-i\delta) \end{bmatrix}$$

$$= \begin{bmatrix} (\cos^2\alpha + \sin^2\alpha\exp(-i\delta))A \\ \cos\alpha\sin\alpha(1 - \exp(-i\delta))A \end{bmatrix}$$

Now, here,

$$E_2 = \begin{bmatrix} J_{11} & J_{12} \\ J_{21} & J_{22} \end{bmatrix} \begin{bmatrix} A \\ O \end{bmatrix} = \begin{bmatrix} J_{11}A \\ J_{21}A \end{bmatrix}$$

Therefore, comparing expressions for E_2,

$J_{11} = \cos^2\alpha + \sin^2\alpha\exp(-i\delta)$ and $J_{21} = \cos\alpha\sin\alpha(1 - \exp($

Use of the two equations developed above for $\tan\alpha$ now leads, by a very easy manipulation, to

$J_{12} = \sin\alpha\cos\alpha(1 - \exp(-i\delta))$ and $J_{22} = \sin^2\alpha + \cos^2\alpha\text{ex}$

so that the Jones matrix of the general phase plate in general orientation is

$$\begin{bmatrix} \cos^2\alpha + \sin^2\alpha\exp(-i\delta) & \cos\alpha\sin\alpha(1 - (\exp-i\delta)) \\ \cos\alpha\sin\alpha(1 - \exp(-i\delta)) & \sin^2\alpha + \cos^2\alpha\exp(-i\delta) \end{bmatrix}$$

It is left as an exercise to the student to prove that the Jones matrix for a simple rotator, that is a device which twists the plane of vibration of a beam of plane-polarized light through an angle θ counterclockwise, like certain organic liquids, is

$$R(-\theta) = \begin{bmatrix} \cos\theta & -\sin\theta \\ \sin\theta & \cos\theta \end{bmatrix}$$

In general, if J represents the Jones matrix already calculated for a particular device, the new matrix for the same device rotated through an angle θ will be given by the triple matrix product

$R(-\theta) \times J \times R(\theta)$

As an example of this rule, the student should verify that the matrix which has already been given above for a general linear retarder at angle θ is equal to the product

$$\begin{bmatrix} \cos\theta & -\sin\theta \\ \sin\theta & \cos\theta \end{bmatrix} \begin{bmatrix} 1 & 0 \\ 0 & e^{-i\delta} \end{bmatrix} \begin{bmatrix} \cos\theta & \sin\theta \\ -\sin\theta & \cos\theta \end{bmatrix}$$

See Table 4 in section IV.5 for several particular cases of this type of matrix. It should be noted that it is the azimuth of the fast axis which is specified in each case, so that, for a wave plate cut from a negative uniaxial crystal, the δ-values refer to the *retardation* of the ordinary vibration.

Appendix F
Connection between Jones and Mueller Calculi

F.1 INTRODUCTION

Since the Mueller and the Jones matrices both describe the behaviour of optical devices, and for fully
polarized light calculations by the two methods always
lead to the same final results, there must be a close
relationship between the two systems. Nathaniel H.
Parkes showed, by using the Wiener coherency matrices,
that formulae can be developed by which the Mueller
matrix elements of the system can be found in terms of
the Jones matrix elements. It is the purpose of this
section to show that this can be done within the formalism of the Jones and Mueller systems, and also to
provide formulae by which the process can be reversed,
and the real and imaginary parts of the Jones matrix
elements calculated from a knowledge of the Mueller
matrix (provided that the latter represents a physicall
possible polarizing device).

F.2 DEVELOPMENT OF THE CONNECTION BETWEEN THE STOKES
    AND THE MAXWELL COLUMNS

We shall now show that all the Stokes parameters can
be expressed in terms of the Maxwell column, its transposed complex conjugate and some constant two-by-two
matrices, to be introduced in this section.

For a general beam of light, the Maxwell column is

$$E = \begin{bmatrix} H \\ K\exp(i\Delta) \end{bmatrix}$$ Its transpose is the row matrix

$\begin{bmatrix} H & K\exp(i\Delta) \end{bmatrix}$ The complex conjugate of this transpose is $\begin{bmatrix} H & K\exp(-i\Delta) \end{bmatrix}$. (This may be called the
Hermitian conjugate of E. As in section IV.7 it will
be denoted by $\mathbf{E}$.)

Then (by matrix multiplication),

(a) $\mathbf{E} \times E = \begin{bmatrix} H & K\exp(-i\Delta) \end{bmatrix} \times \begin{bmatrix} H \\ K\exp(i\Delta) \end{bmatrix} = H^2 + K^2$

(because $\exp(i\Delta) \times \exp(-i\Delta) = 1$). This is I, the first of the Stokes parameters of the beam. For convenience later, we shall insert, between the $\mathbf{E}$ and the E, the unit two-by-two matrix $\begin{bmatrix} 1 & 0 \\ 0 & 1 \end{bmatrix}$ here called A_1, which obviously cannot affect the result. We can thus write

$$\mathbf{E} \times A_1 \times E = I \qquad \text{(F.1)}$$

where
$$A_1 = \begin{bmatrix} 1 & 0 \\ 0 & 1 \end{bmatrix}$$

(b) $\mathbf{E} \times \begin{bmatrix} 1 & 0 \\ 0 & -1 \end{bmatrix} \times E = \mathbf{E} \times \begin{bmatrix} 1 & 0 \\ 0 & -1 \end{bmatrix} \times \begin{bmatrix} H \\ K\exp(i\Delta) \end{bmatrix}$

$$= \begin{bmatrix} H & K\exp(-i\Delta) \end{bmatrix} \times \begin{bmatrix} H \\ -K\exp(i\Delta) \end{bmatrix}$$

$$= H^2 - K^2 = Q$$

We write this in the form:

$$\mathbf{E} \times A_2 \times E = Q \qquad \text{(F.2)}$$

where
$$A_2 = \begin{bmatrix} 1 & 0 \\ 0 & -1 \end{bmatrix}.$$

(c) $\mathbf{E} \times \begin{bmatrix} 0 & 1 \\ 1 & 0 \end{bmatrix} \times E = \mathbf{E} \times \begin{bmatrix} 0 & 1 \\ 1 & 0 \end{bmatrix} \times \begin{bmatrix} H \\ K\exp(i\Delta) \end{bmatrix}$

$$= \begin{bmatrix} H & K\exp(-i\Delta) \end{bmatrix} \times \begin{bmatrix} K\exp(i\Delta) \\ H \end{bmatrix}$$

$$= HK\exp(i\Delta) + HK\exp(-i\Delta)$$
$$= HK(\cos\Delta + i\sin\Delta + \cos\Delta - i\sin\Delta)$$
$$= 2HK\cos\Delta = U$$

We write this in the form

$$\mathbf{E} \times A_3 \times E = U \tag{F.3}$$

where
$$A_3 = \begin{bmatrix} 0 & 1 \\ 1 & 0 \end{bmatrix}$$

(d) $\mathbf{E} \times \begin{bmatrix} 0 & -i \\ i & 0 \end{bmatrix} \times E = \mathbf{E} \times \begin{bmatrix} 0 & -i \\ i & 0 \end{bmatrix} \times \begin{bmatrix} H \\ K\exp(i\Delta) \end{bmatrix}$

$$= \begin{bmatrix} H & K\exp(-i\Delta) \end{bmatrix} \times \begin{bmatrix} -iK\exp(i\Delta) \\ iH \end{bmatrix}$$

$$= -iHK\exp(i\Delta) + iHK\exp(-i\Delta)$$

$$= HK\{-i(\cos\Delta + i\sin\Delta) + i(\cos\Delta - i$$

$$= HK\{-i\cos\Delta - i^2\sin\Delta + i\cos\Delta - i^2s$$

$$= 2HK\sin\Delta \text{ (because } i^2 = -1) = V$$

We write this in the form

$$\mathbf{E} \times A_4 \times E = V \tag{F.4}$$

where
$$A_4 = \begin{bmatrix} 0 & -i \\ i & 0 \end{bmatrix}$$

(The above equations (F.1) to (F.4) enable us to represent each of the four Stokes parameters in terms of a 'matrix sandwich'. The three matrices A_2, A_3 and A_4 are closely related to Pauli spin matrices.)

F.3 DEVELOPMENT OF THE SANDWICH MATRICES IN TERMS OF THE MUELLER AND OF THE JONES MATRIX ELEMENTS

F.3.1 Using Mueller elements

The first equation defining the Mueller elements is

$$I_1 = M_{11}I_0 + M_{12}Q_0 + M_{13}U_0 + M_{14}V_0$$

Inserting the expressions developed in the last section for the Stokes parameter in terms of the Maxwell matrix elements, this becomes

$$E_1 \times A_1 \times E_1 = (M_{11} \times \mathbf{E}_0 \times A_1 \times E_0) + (M_{12} \times E_0 \times A_2 \times E_0)$$

$$+ (M_{13} \times \mathbf{E}_0 \times A_3 \times E_0) + (M_{14} \times E_0 \times A_4 \times E_0)$$

$$= \mathbf{E}_0 \times \left[M_{11}A_1 + M_{12}A_2 + M_{13}A_3 + M_{14}A_4 \right] \times E_0$$

Substituting the values of the A-matrices and simplifying, this becomes

$$E_1 \times A_1 \times E_1 = E_0 \times \begin{bmatrix} M_{11} + M_{12} & M_{13} - iM_{14} \\ M_{13} + iM_{14} & M_{11} - M_{12} \end{bmatrix} \times E_0 \qquad (F.5)$$

The other three defining equations for the Mueller elements will obviously develop into equations identical with this except for the subscript on the As on the left side, and the first subscripts on the Ms on the right side. Thus:

$$E_1 \times A_2 \times E_1 = E_0 \times \begin{bmatrix} M_{21} + M_{22} & M_{23} - iM_{24} \\ M_{23} + iM_{24} & M_{21} - M_{22} \end{bmatrix} \times E_0 \qquad (F.6)$$

$$E_1 \times A_3 \times E_1 = E_0 \times \begin{bmatrix} M_{31} + M_{32} & M_{33} - iM_{34} \\ M_{33} + iM_{34} & M_{31} - M_{32} \end{bmatrix} \times E_0 \qquad (F.7)$$

$$E_1 \times A_4 \times E_1 = E_0 \times \begin{bmatrix} M_{41} + M_{42} & M_{43} - iM_{44} \\ M_{43} + iM_{44} & M_{41} - M_{42} \end{bmatrix} \times E_0 \qquad (F.8)$$

F.3.2 Using Jones elements

The defining equation for the Jones matrix is $E_1 = JE_0$. Taking Hermitians on both sides, $\mathbf{E}_1$ = (Hermitian of JE_0). Now the transpose of a product of two factors is the product of the transposes of the two factors *taken in reverse order*, that is (transpose of AB) = (transpose of B) × (transpose of A). Since the Hermitian is merely the complex conjugate of the transpose, a similar theorem is clearly true for Hermitians. Thus (Hermitian of JE_0) = (Hermitian of E_0) × (Hermitian of J), that is $\mathbf{E}_1 = \mathbf{E}_0 \times \mathbf{J}$.

Thus (inserting A_1, the identity matrix, for consistency with later equations),

$$\mathbf{E}_1 \times A_1 \times E_1 = (\mathbf{E}_0 \times \mathbf{J}) \times A_1 \times (J \times E_0)$$

$$= \mathbf{E}_0 \times (\mathbf{J} \times A_1 \times J) \times E_0 \tag{F.9}$$

(using the associative property of matrix multiplication). Similarly, we can obtain

$$\mathbf{E}_1 \times A_2 \times E_1 = \mathbf{E}_0 \times (\mathbf{J} \times A_2 \times J) \times E_0 \tag{F.10}$$

$$\mathbf{E}_1 \times A_3 \times E_1 = \mathbf{E}_0 \times (\mathbf{J} \times A_3 \times J) \times E_0 \tag{F.11}$$

$$\mathbf{E}_1 \times A_4 \times E_1 = \mathbf{E}_0 \times (\mathbf{J} \times A_4 \times J) \times E_0 \tag{F.12}$$

F.4 COMPARISON OF THE PAIRS OF SANDWICH MATRICES IN THE TWO SYSTEMS

The left sides of equations (F.5) and (F.9) are identical, so the right sides must be equal. On each of the right sides appears $\mathbf{E}_0 \times$ (another matrix) $\times E_0$. Since the two right sides are equal, the matrices sandwiched between $\mathbf{E}_0$ and E_0 must be equal in the two equations. The same argument applies to the sandwiched matrices in equations (F.6) and (F.10), those in equations (F.7) and (F.11), and those in equations (F.8) and (F.12).

The sandwiched matrix in equation (F.9) is $(\mathbf{J} \times A_1 \times J)$. $\mathbf{J}$ is the matrix obtained by transposing J, then replacing each of the matrix elements in the transposed J by its complex conjugate. Since

$$J = \begin{bmatrix} J_{11} & J_{12} \\ J_{21} & J_{22} \end{bmatrix} \text{ its transpose is } \begin{bmatrix} J_{11} & J_{21} \\ J_{21} & J_{22} \end{bmatrix}, \text{ so that}$$

$$\mathbf{J} = \begin{bmatrix} G_{11} & G_{21} \\ G_{12} & G_{22} \end{bmatrix}, \text{ where the } Gs \text{ are defined as the com-}$$

plex conjugates of the corresponding Js.

Thus, the sandwiched matrix in equation (F.9) becomes

$$\begin{bmatrix} G_{11} & G_{21} \\ G_{12} & G_{22} \end{bmatrix} \begin{bmatrix} 1 & 0 \\ 0 & 1 \end{bmatrix} \begin{bmatrix} J_{11} & J_{12} \\ J_{21} & J_{22} \end{bmatrix}$$

$$= \begin{bmatrix} G_{11}J_{11} + G_{21}J_{21} & G_{11}J_{12} + G_{21}J_{22} \\ G_{12}J_{11} + G_{22}J_{21} & G_{12}J_{12} + G_{22}J_{22} \end{bmatrix}$$

The elements in the final matrix can now be equated, each with its corresponding element in the sandwiched matrix in equation (F.5). Thus:

$$M_{11} + M_{12} = G_{11}J_{11} + G_{21}J_{21} \qquad (F.13)$$

$$M_{13} - iM_{14} = G_{11}J_{12} + G_{21}J_{22} \qquad (F.14)$$

$$M_{13} + iM_{14} = G_{12}J_{11} + G_{22}J_{21} \qquad (F.15)$$

$$M_{11} - M_{12} = G_{12}J_{12} + G_{22}J_{22} \qquad (F.16)$$

Treating the other pairs of sandwiched matrices in the same way, we get, from equations (F.6) and (F.10),

$$M_{21} + M_{22} = G_{11}J_{11} - G_{21}J_{21} \qquad (F.17)$$

$$M_{23} - iM_{24} = G_{11}J_{12} - G_{21}J_{22} \qquad (F.18)$$

$$M_{23} + iM_{24} = G_{12}J_{11} - G_{22}J_{21} \qquad (F.19)$$

$$M_{21} - M_{22} = G_{12}J_{12} - G_{22}J_{22} \qquad (F.20)$$

from equations (F.7) and (F.11),

$$M_{31} + M_{32} = G_{11}J_{21} + G_{21}J_{11} \qquad (F.21)$$

$$M_{33} - iM_{34} = G_{11}J_{22} + G_{21}J_{12} \qquad (F.22)$$

$$M_{33} + iM_{34} = G_{12}J_{21} + G_{22}J_{11} \qquad (F.23)$$

$$M_{31} - M_{32} = G_{12}J_{22} + G_{22}J_{12} \qquad (F.24)$$

from equations (F.8) and (F.12),

$$M_{41} + M_{42} = i(G_{21}J_{11} - G_{11}J_{21}) \qquad (F.25)$$

$$M_{43} - iM_{44} = i(G_{21}J_{12} - G_{11}J_{22}) \qquad (F.26)$$

$$M_{43} + iM_{44} = i(G_{22}J_{11} - G_{12}J_{21}) \qquad (F.27)$$

$$M_{41} - M_{42} = i(G_{22}J_{12} - G_{12}J_{22}) \qquad (F.28)$$

F.5 DERIVATION OF THE MUELLER MATRIX ELEMENTS IN TERMS
OF THE JONES MATRIX ELEMENTS

The derivation of the Mueller from the Jones matrix
elements is now simple: we add together or subtract
suitable pairs of equations from numbers (F.13) to
(F.28) and halve the results. Thus:

(13+16) $\quad M_{11} = (G_{11}J_{11} + G_{21}J_{21} + G_{12}J_{12} + G_{22}J_{22})/2$ (F.2

(13−16) $\quad M_{12} = (G_{11}J_{11} + G_{21}J_{21} - G_{12}J_{12} - G_{22}J_{22})/2$ (F.3

(14+15) $\quad M_{13} = (G_{11}J_{12} + G_{21}J_{22} + G_{12}J_{11} + G_{22}J_{21})/2$ (F.3

(15−14) $\quad M_{14} = i(G_{11}J_{12} + G_{21}J_{22} - G_{12}J_{11} - G_{22}J_{21})/2$ (F.3

(17+20) $\quad M_{21} = (G_{11}J_{11} + G_{12}J_{12} - G_{21}J_{21} - G_{22}J_{22})/2$ (F.3

(17−20) $\quad M_{22} = (G_{11}J_{11} + G_{22}J_{22} - G_{21}J_{21} - G_{12}J_{12})/2$ (F.3

(18+19) $\quad M_{23} = (G_{12}J_{11} + G_{11}J_{12} - G_{22}J_{21} - G_{21}J_{22})/2$ (F.3

(19−18) $\quad M_{24} = i(G_{11}J_{12} + G_{22}J_{21} - G_{21}J_{22} - G_{12}J_{11})/2$ (F.3

(21+24) $\quad M_{31} = (G_{11}J_{21} + G_{22}J_{11} + G_{12}J_{22} + G_{22}J_{12})/2$ (F.3

(21−24) $\quad M_{32} = (G_{11}J_{21} + G_{21}J_{11} - G_{12}J_{22} - G_{22}J_{12})/2$ (F.3

(22+23) $\quad M_{33} = (G_{11}J_{22} + G_{21}J_{12} + G_{12}J_{21} + G_{22}J_{11})/2$ (F.3

(23−22) $\quad M_{34} = i(G_{11}J_{22} + G_{21}J_{12} - G_{12}J_{21} - G_{22}J_{11})/2$ (F.4

(25+28) $\quad M_{41} = i(G_{21}J_{11} + G_{22}J_{12} - G_{11}J_{21} - G_{12}J_{22})/2$ (F.4

(25−28) $\quad M_{42} = i(G_{21}J_{11} + G_{12}J_{22} - G_{11}J_{21} - G_{22}J_{12})/2$ (F.4

(26+27) $\quad M_{43} = i(G_{21}J_{12} + G_{22}J_{11} - G_{11}J_{22} - G_{12}J_{21})/2$ (F.4

(27−26) $\quad M_{44} = (G_{22}J_{11} + G_{11}J_{22} - G_{12}J_{21} - G_{21}J_{12})/2$ (F.4

These expressions are identical with those given by
Parkes.

F.6 DERIVATION OF THE JONES MATRIX ELEMENTS IN TERMS
   OF THE MUELLER MATRIX ELEMENTS

The Jones elements are, in general, complex, and it is necessary to determine both the real and the imaginary parts. It proves convenient in the derivation to use the polar form of the matrix elements, expressing a complex number such as $X + iY$ in the form $R\exp(i\theta)$, where $R = (X^2 + Y^2)^{\frac{1}{2}}$ and $\tan\theta = Y/X$. The complex conjugate $X - iY$ is then $R\exp(-i\theta)$. Thus, we express J_{11} as $R_{11}\exp(i\theta_{11})$, etc. Thus $G_{11}J_{11} = (R_{11}\exp(-i\theta_{11}))(R_{11}\exp(i\theta_{11})) = R_{11}^2$, etc.

Then, adding equations (F.13) and (F.17),

$$M_{11} + M_{21} + M_{12} + M_{22} = 2G_{11}J_{11} = 2R_{11}^2$$

Therefore,

$$R_{11} = \{(M_{11} + M_{21} + M_{12} + M_{22})/2\}^{\frac{1}{2}} \tag{F.45}$$

Similarly, subtracting equation (F.17) from equation (F.13), we get

$$R_{21} = \{(M_{11} + M_{12} - M_{21} - M_{22})/2\}^{\frac{1}{2}} \tag{F.46}$$

Adding equations (F.16) and (F.20),

$$R_{12} = \{(M_{11} - M_{12} + M_{21} - M_{22})/2\}^{\frac{1}{2}} \tag{F.47}$$

Subtracting equation (F.20) from equation (F.16),

$$R_{22} = \{(M_{11} - M_{12} - M_{21} + M_{22})/2\}^{\frac{1}{2}} \tag{F.48}$$

This gives the magnitudes of the Jones matrix elements. For the angles, it is sufficient to find the difference between one of them, chosen arbitrarily, and the other three, because adding a constant to all four of the Jones angles is, in effect, only the same as adding the same constant to ωt on the right sides of equations (F.1) and (F.2), that is in the place of ϕ which, as already remarked, produces no effect on any intensities.

Adding equations (F.31) and (F.35),

$$M_{13} + M_{23} = G_{11}J_{12} + G_{12}J_{11}$$
$$= R_{11}R_{12}\left[(\exp-i\theta_{11})(\exp i\theta_{12})\right]$$
$$+ R_{12}R_{11}\left[(\exp-i\theta_{12})(\exp i\theta_{11})\right]$$

$$= R_{11}R_{12}\left[\exp i\,(\theta_{11}-\theta_{12}) + \exp{-i}\,(\theta_{11}-\theta_{12})\right]$$

$$= R_{11}R_{12} \times 2\cos(\theta_{11}-\theta_{12})$$

Therefore,

$$\cos(\theta_{11}-\theta_{12}) = \frac{(M_{13} + M_{23})}{2R_{11}R_{12}}$$

$$= \frac{(M_{13} + M_{23})}{\{(M_{11} + M_{21})^2 - (M_{12} + M_{22})^2\}^{\frac{1}{2}}} \qquad (F.49$$

(on substituting for R_{11} and R_{12} from equations (F.45) and (F.47) respectively). Adding equations (F.32) and (F.34),

$$M_{14} + M_{24} = i(G_{11}J_{12} - G_{12}J_{11})$$

$$= iR_{11}R_{12}\left[(\exp{-i}\theta_{11})(\exp i\theta_{12})\right.$$

$$\left. - (\exp{-i}\theta_{12})(\exp i\theta_{11})\right]$$

$$= iR_{11}R_{12}\left[\exp i\,(\theta_{12}-\theta_{11}) - \exp{-i}\,(\theta_{12}-\theta_{11})\right]$$

$$= iR_{11}R_{12} \times 2i\sin(\theta_{12}-\theta_{11})$$

$$= 2R_{11}R_{12}\sin(\theta_{11}-\theta_{12})$$

Therefore, as before,

$$\sin(\theta_{11}-\theta_{12}) = \frac{M_{14} + M_{24}}{\{(M_{11} + M_{21})^2 - (M_{12} + M_{22})^2\}^{\frac{1}{2}}} \qquad (F.5$$

A knowledge of both $\sin(\theta_{11}-\theta_{12})$ and $\cos(\theta_{11}-\theta_{12})$ is enough to determine the angle completely in the range from zero to 2π.

The sines and cosines of the differences between the other two angles in the Jones matrix elements and θ_{11} can be found in the same way, by suitable combination of the equations (F.39) to (F.44). The results are,

adding equations (F.37) and (F.38),

$$\cos(\theta_{21}-\theta_{11}) = \frac{M_{31} + M_{32}}{\{(M_{11} + M_{12})^2 - (M_{21} + M_{22})^2\}^{\frac{1}{2}}} \qquad (F.5$$

adding equations (F.41) and (F.42),

$$\sin(\theta_{21}-\theta_{11}) = \frac{M_{41} + M_{42}}{\{(M_{11} + M_{12})^2 - (M_{21} + M_{22})^2\}^{\frac{1}{2}}} \qquad (F.52)$$

adding equations (F.39) and (F.44),

$$\cos(\theta_{11}-\theta_{22}) = \frac{M_{33} + M_{44}}{\{(M_{11} + M_{22})^2 - (M_{21} + M_{12})^2\}^{\frac{1}{2}}} \qquad (F.53)$$

subtracting equation (F.40) from equation (F.43),

$$\sin(\theta_{22}-\theta_{11}) = \frac{M_{43} - M_{34}}{\{(M_{11} + M_{22})^2 - (M_{21} + M_{12})^2\}^{\frac{1}{2}}} \qquad (F.54)$$

In all of the ten equations (F.45) to (F.54) any square root term representing an R_{ij} value should be taken to have a positive sign.

In using the conversion formulae some caution will be needed, since the Mueller matrix, with its *sixteen* real elements, cannot always be expressed in terms of a Jones matrix, which contains only *four* real and *four* imaginary elements. It can be shown that to every Jones matrix which can be written down there corresponds a physically possible polarizing device, so that the corresponding Mueller matrix also exists. If, on the other hand, sixteen numbers are arbitrarily written down to form a Mueller matrix, then it may not be possible to solve equations (F.45) to (F.54) without obtaining *imaginary* answers for what should be *real* R-values and θ-differences. Even if the terms within the square root signs are all positive, the equations may predict impossible pairs of sine and cosine values, whose squares do not sum to unity. This happens, for example, if one considers the Mueller matrix for an 'ideal depolarizer', for which $M_{11} = 1$ and all other elements vanish. Except in terms of time-averaging, this is a physically impossible device, for which no Jones matrix exists.

Bibliography and conclusion

Since this is an introductory text, our final task is to make some suggestions for further study of matrix methods and their applications in optics. The limited bibliography given below contains 26 selected references where those topics which have been discussed in this book can be pursued at greater length. Also included at the end of the list are 10 references pointing towards other aspects of matrices which we have been unable to cover but which may repay exploration.

SELECTED REFERENCES

Matrix methods

1. A. Coulson, *Introduction to Matrices*, Longmans Green, London, 1965.
2. A. C. Aitken, *Determinants and Matrices*, Oliver and Boyd, Ltd., Edinburgh, 1939
3. R. Bellman, *Introduction to Matrix Analysis*, McGraw-Hill, New York, 1960.

Matrices in paraxial optics

4. R.A. Sampson, A new treatment of optical aberrations, *Phil.Trans.R.Soc.*, **212**, pp.149-185, 1913.
5. T. Smith, On tracing rays through an optical system, *Proc.phys.Soc.Lond.*, **27**, 502, 1915: *Proc.phys.Soc.Lond.*, **57**, 286, 1945.
6. E. L. O'Neill, *Introduction to Statistical Optics*, Addison-Wesley, Reading, Mass., 1963.
7. W. Brouwer, *Matrix Methods in Optical Instrument Design*, W.A. Benjamin, Inc., New York, 1964.

8. K. Halbach, Matrix representation of Gaussian optics, *Am.J.Phys.*, **32**, 90, 1964.

9. P. I. Richards, Conventions in matrix algebra, *Am.J.Phys.*, **32**, 890, 1964.

10. D. C. Sinclair, *Image-forming Optics, Notes for 1971 Summer School on Contemporary Optics,* The Institute of Optics, University of Rochester, New York, 1971.

Optical resonators and Gaussian beams

11. H. Kogelnik, On the propagation of Gaussian beams of light through lenslike media including those with a loss or gain variation, *Applied Optics,* **4**, 1562, 1965.

12. H. Kogelnik and T. Li, Laser beams and resonators, (an excellent review which appears in *two* journals) *Proc.I.E.E.E.*, **54**, 1312, 1966: *Applied Optics,* **5**, 1550, 1966.

13. A. E. Siegman, Unstable optical resonators for laser applications, *Proc.I.E.E.E.*, **53**, 1312, 1965.

14. A. E. Siegman, Stabilising output with unstable resonators, *Laser Focus,* **7**, 42, May 1971.

15. S. Yamamoto and T. Makimoto, On the ray-transfer matrix of a tapered lenslike medium, *Proc.I.E.E.E.*, **59**, 1254, 1971.

16. J. A. Arnaud, Modes in helical gas lenses, *J.opt. Soc.Am.*, **61**, 751, 1971.

17. A. N. Chester, Mode selectivity and mirror mis-alignment effects in unstable laser resonators, *Applied Optics,* , 2584, 1972.

18. G. J. Ernst and W. J. Witteman, Mode structure of active resonators, *I.E.E.E. J.Quant.Electr.*, **QE-9**, 911, 1973.

Matrices in polarization optics

19. G. G. Stokes, *Trans.Camb.phil.Soc.*, **9**, 399, 1852.

20. W. A. Shurcliff, *Polarized Light: Production and Use,* Harvard Univ. Press, 1962.

21. R. C. Jones, *J.opt.Soc.Am.*, **46**, 126, 1956.

22. E. L. O'Neill, (*see* reference 6) pp.133-156.

23. N. G. Parke, *J.Math. and Phys.*, **28**, 131, 1949.

24. M. Born and E. Wolf, *Principles of Optics,* 4th Edn., Pergamon Press, 1970, pp.544-555.

25. W. H. McMaster, Matrix representation of polariza-
    tion, *Rev.mod.Phys.*, **33**, 8, 1961.
26. J. W. Simmons and M. J. Guttmann, *States Waves
    and Photons: A Modern Introduction to Light*,
    Addison-Wesley, Reading, Mass., 1970.

ADDITIONAL REFERENCES COVERING RELATED TOPICS

27. R. Braae, *Matrix Algebra for Electrical Engineers*,
    Pitman, London, 1963.
28. W. E. Lewis and D. G. Pryce, *Application of Matrix
    Theory to Electrical Engineering*, Spon, London,
    1965.
29. O. N. Stavroudis, *The Optics of Rays, Wavefronts,
    and Caustics*, Academic Press, New York and London,
    1972.
30. M. Herzberger, *Modern Geometrical Optics*, Inter-
    science Publishers Inc., New York, 1958.
31. J. H. Harrold, Matrix algebra for ideal lens
    problems, *J.opt.Soc.Am.*, **44**, 254, 1954.
32. R. E. Hopkins, *Mirror and Prism Systems, Chapter
    VIII in Vol.III of Applied Optics and Optical
    Engineering*, R. Kingslake ed., Academic Press,
    London and New York, 1965.
33. M. Born and E. Wolf (*see* reference 24) pp.51-70
    on characteristic matrices for layered media.
34. F. Abelès, Transmission of light by a system of
    alternate layers, *C.r.Acad.Sci.Paris*, **226**, 1808,
    1948.
35. W. T. Welford (formerly W. Weinstein), Thin film
    optics, *Vacuum*, **IV**, 3, 1954.
36. H. Gamo, *Matrix Treatment of Partial Coherence,
    Chapter III in Progress in Optics Vol.III*, E. Wolf
    ed., North Holland, Amsterdam, 1964.

For those who are concerned in any way with prism
design or with thin film optics, the matrix methods
discussed in references 32-35 should prove particularly
useful.

For the student who enjoys tackling theory which is
both elegant and relevant to current problems, we
should recommend the discussions of partial coherence
and of coherency matrices which appear in references
22, 24-26 and 36.

Finally, for those who wish merely to explore a little further, we would suggest that their next steps should be to diagonalize some of the polarization matrices in chapter IV and then interpret the resultant eigen-vectors. Particularly in laser optics, they will find several experimental situations where these eigen-vectors are of considerable importance.

Index

348